D0081603

Thermodynamics of Materials

Volume I

MIT Series in Materials Science and Engineering

In response to the growing economic and technological importance of polymers, ceramics, advanced metals, composites, and electronic materials, many departments concerned with materials are changing and expanding their curricula. The advent of new courses calls for the development of new textbooks that teach the principles of materials science and engineering as they apply to all classes of materials.

The MIT Series in Materials Science and Engineering is designed to fill the needs of this changing curriculum.

Based on the curriculum of the Department of Materials Science and Engineering at the Massachusetts Institute of Technology, the series will include textbooks for the undergraduate core sequence of courses on Thermodynamics, Physical Chemistry, Chemical Physics, Structures, Mechanics, and Transport Phenomena as they apply to the study of materials. More advanced texts based on this core will cover the principles and technologies of different material classes, such as ceramics, metals, polymers, and electronic materials.

The series will define the modern curriculum in materials science and engineering as the discipline changes with the demands of the future.

The MIT Series Committee

Samuel L. Allen
Yet-Ming Chiang
Merton C. Flemings
David V. Ragone
Julian Szekely
Edwin L. Thomas

Thermodynamics of Materials

Volume I

David V. Ragone
Massachusetts Institute of Technology

John Wiley & Sons, Inc.
New York • Chichester • Brisbane • Toronto • Singapore

Acquisitions Editor	Cliff Robichaud
Marketing Manager	Susan Elbe
Senior Production Editor	Charlotte Hyland
Designer	Kevin Murphy
Manufacturing Manager	Susan Stetzer
Illustration	Sandra L. Rigby

This book was set in 10/12 Times Roman by University Graphics, Inc. and printed and bound by R. R. Donnelley. The cover was printed by Phoenix Color.

Recognizing the importance of preserving what has been written, it is a policy of John Wiley & Sons, Inc. to have books of enduring value published in the United States printed on acid-free paper, and we exert our best efforts to that end.

Library of Congress Cataloging in Publication Data:
Ragone, David V., 1930–
Thermodynamics of materials / David V. Ragone.
p. cm.
Includes bibliographical references.
ISBN 0-471-30885-4 (v. 1 : cloth).—ISBN 0-471-30886-2
(v. 2 : cloth)
1. Materials—Thermal properties. 2. Thermodynamics.
TA418.52.R34 1995
536′.7—dc20 94-25647
CIP

Printed in the United States of America

10 9 8 7 6 5 4 3 2 1

To Kit

Preface—Vol. I

Thermodynamics, the study of energy and its transformations, is applicable to all fields of science and engineering. The text for a basic course in thermodynamics need not be directed at students with focused professional aims. But general treatments of thermodynamics tend to be too abstract for engineering and science students who have already committed to a particular area of study. This is especially true of students in materials science and engineering who are interested in how thermodynamics applies to the different classes of materials that are studied as part of a modern curriculum in materials science and engineering.

This volume is a textbook for the first part of a two-course sequence at the undergraduate level on the physical chemistry of materials. The subject matter of the first course, and part of the second, is thermodynamics. The second course, the subject of Volume II, covers the thermodynamics of defects and surfaces, the statistical thermodynamics of molecular motion, and some aspects of solid-state kinetics, such as diffusion, nucleation, reaction kinetics, and nonequilibrium thermodynamics. Together these two courses provide background for a series of other courses that follow, including those devoted to general principles such as physics of materials, and structure of materials, and those devoted to specific classes of materials such as electronic materials, ceramics, polymers, and metals.

In addition to a presentation based on macroscopic thermodynamics, this book includes an introduction to microscopic thermodynamics, or statistical thermodynamics. Macroscopic thermodynamics is an empirical science based on the macroscopic properties of matter, and does not rely on any hypotheses concerning the structure of matter. It predicts many of the relationships among these macroscopic properties, but does not attempt to predict their absolute values. Statistical thermodynamics aims to provide an understanding of the macroscopic thermodynamic properties of materials, and to predict absolute values based on atomic and molecular structure and motion. It also provides background for the understanding of differences among various classes of materials. The chapter on statistical thermodynamics, for example, deals with the differences in thermodynamic properties of metallic and polymer solutions.

A few words on the sequence of study are in order. The chapter on statistical thermodynamics is placed last, as Chapter 10. It need not be covered last. In fact, the basic sections of Chapter 10 can usefully be studied immediately following Chapter 3, the study of property relations. The same flexibility is true of Chapter 8 on the phase rule. It can, if desired, be presented as part of the background to Chapter 5, the study of chemical reactions. It also fits quite well where it is, as an introduction to phase diagrams, Chapter 9.

The text assumes that students using it will have completed a college-level course in chemistry and a year-long course in physics in addition to a year of calculus.

Those who work in the field of materials, broadly defined, must be multilingual with respect to science and engineering. By this I mean that they must be able to understand and apply the knowledge generated by physicists, chemists, and other scientists, and they must also be able to communicate with, and serve the needs of, engineers from diverse fields. This book intends to provide background for both.

This broad charter influences the text in several ways. The first and second laws of thermodynamics, for example, are presented using an analysis of open systems, as engineers do. Open systems are those in which material may enter, leave, or do both. By the elimination of the flow terms, of course, open system analyses can easily be reduced to closed systems analyses, which are used primarily by scientists. Another influence of the broad charter will be seen in the units used in the statement of problems at the end of each chapter. The text uses the International System of Units (SI), the modern metric system, using joules as a measure of energy, for example. However, the data given in problem statements will use whatever units were attached to the data when they were found in the literature. This approach was adopted as a way of familiarizing students with the variety of units they will encounter in the literature and in the practice of their professions.

I acknowledge, with thanks, the efforts of my colleague, Prof. Gerbrand Ceder, who taught from a draft of this book for three years, and to the student assistants who carefully reviewed the drafts to search out and correct the seemingly endless errors: John Zaroulis, Patrick Tedisch, Gerardo Garbulski, Adrian Kohan, Julie Ngau, and Andrew Kim. Thanks are also due for the helpful and thoughtful suggestions of the reviewers of the manuscript, Profs. John Angus, Robert Auerback, William Bitler, Charles Brooks, Richard Heckel, and George St. Pierre. And finally, I give a special expression of thanks to my wife, Kit, for her patience and constant encouragement during the preparation of this text.

David V. Ragone
Cambridge, Massachusetts
March 1994

Contents

Chapter 10 Statistical Thermodynamics 245

Appendix Background 293

Chapter 1

The First Law

The science of thermodynamics is concerned with heat and work, and transformations between the two. It is based on two laws of nature, the first and second laws of thermodynamics. By logical reasoning and skillful manipulation of these laws, it is possible to correlate many of the properties of materials and to gain insight into the many chemical and physical changes that materials undergo.

The first law of thermodynamics is simply the principle of conservation of energy; that is, energy can be neither created nor destroyed. To derive the practical benefits of knowing this principle, we must construct an accounting system for energy, sometimes called an energy balance. This system must handle flows of energy, such as heat and work, as well as the various forms of energy that matter possesses. To operate this accounting system, we will need to understand a series of basic notions and definitions.

1.1 SYSTEM AND SURROUNDINGS

As the word is used in thermodynamics, a *system* is any portion of space or matter set aside for study. A system can be open or closed. An open system is one in which matter is allowed to enter, leave the system, or do both. In a closed system no matter enters or leaves.

Whether a system is open or closed does not depend on any particular physical configuration of the apparatus being studied. It depends on how the system chosen for study is defined. Consider, for example, a fluid flowing through a pipe. A specific portion of the pipe may be considered to be the system. In this case we have defined an open system. Fluid enters and leaves the system. Alternatively, one can define the system as a particular piece of fluid traveling through the pipe. In this case, we have defined a closed system. Although the system moves, matter neither enters nor leaves the system as we have defined it. Note that if the fluid is a compressible fluid, such as a gas, its boundaries, in addition to moving through space, may expand or contract. Its physical volume may change. However, as long as no matter enters or leaves the system as defined, it is a closed system.

Having thus defined a *system,* we say that *the surroundings* is everything else: that is, the universe outside the system.

1.2 ENERGY TRANSFER

For the purposes of thermodynamics, energy transferred between system and surroundings is divided into two categories, heat and work. The reason for this division may seem arbitrary at this point. The importance of the distinction will become apparent when we discuss the second law of thermodynamics, which deals with restrictions on the conversion of heat into work.

Heat (Q) is energy transferred between the system and surroundings *because of a temperature difference.* Take, for example, a cold piece of iron placed in a high temperature furnace. Energy flows into the iron to heat it. If one defines the iron as the system, then the energy that flows into the system is called heat. The algebraic sign of the heat term is positive when heat flows from the surroundings into the system. The increase in temperature of the iron (the system) is caused by an increase in the *thermal energy* of the system. The system, by our definition, may not possess heat. Heat is a flow quantity. *Heat is energy in transit.*

Work (W) is defined as all other forms of energy transferred between the system and its surroundings. Thermodynamic work can take many forms (mechanical, electrical, magnetic, etc.). To examine one form of mechanical work, consider the compression of a gas in a cylinder of an automobile engine. If we consider the gas to be the system, then work is done on the system by the face of the piston. The magnitude of the work done is the force $F,$ multiplied by the distance Δl through which the piston moved (Figure 1.1).

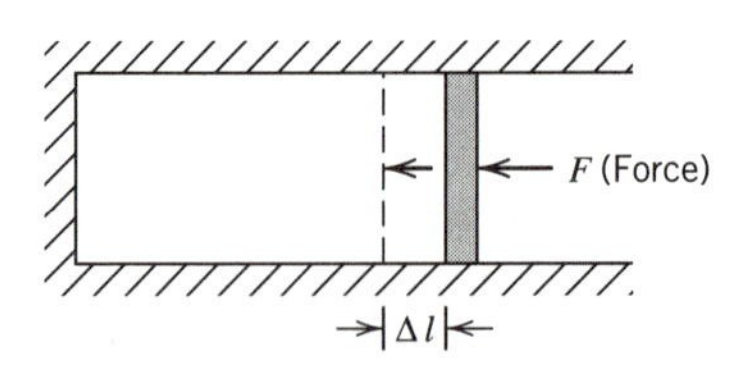

Figure 1.1 Mechanical work.

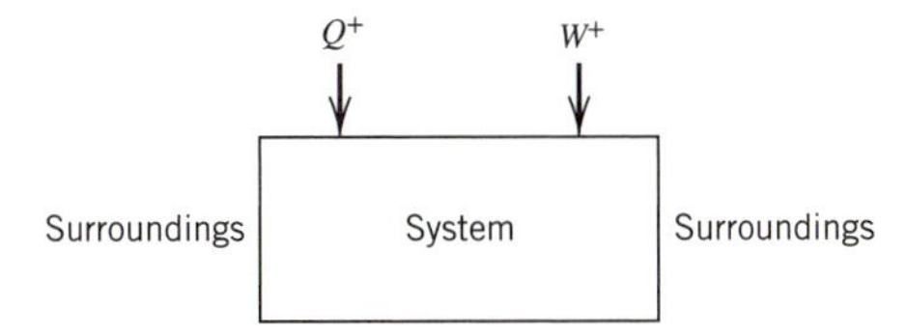

Figure 1.2 The sign convention for heat and work.

If the cross-sectional area of the piston is taken as A, this work term can be converted into:

$$W = F\Delta l = \frac{F}{A} A\Delta l = P\Delta V \tag{1.1}$$

The work done is, thus, the resisting pressure P multiplied by ΔV, the change in volume of the system. The work term is expressed in Joules (J) when the pressure is expressed in newtons per square meter [N/m^2, or pascals (Pa)], and the volume change is expressed in cubic meters.

The direction of force and the distance moved establishes the algebraic sign of the work term (Figure 1.2). The magnitude of the work term is considered *positive* when work is done *on* a system by the surroundings. Work is *negative* when the system does work *on* the surroundings.[1] The sign convention is the same for both heat and work; that is, these terms are considered to be positive when they add energy to the system.

In thermodynamics, a distinction is made between reversible and irreversible processes. In general, a process is called reversible if the initial state of the system can be restored with no observable effects in the system or the surroundings. As an example of irreversible work, consider a gas expanding into a vacuum chamber. If the gas is taken as the system (imagine it to be enclosed in a very flexible balloon), then the resisting force (and pressure) is zero. To restore the system and surroundings to their original conditions, the gas must be compressed by a steadily increasing force. Work will be done on the system. This will result in an energy increase in the system. Clearly there will be an observable effect on the system and on the surroundings; therefore the process is irreversible. If, however, the forces at the boundary of the system are balanced such that the resisting force at the boundary differs from the force inside the system by only an infinitesimal amount, the work is reversible. The direction of the movement of the boundary could be reversed by just an infinitesimal change in the forces. In the case of P-V work, if the pressure resisting an expansion differs only infinitesimally from the pressure of the gas in the system, the work is considered to be reversible. If there is a finite pressure difference between the system and the surroundings in which the movement is taking place, the work is irreversible.

---

[1]According to the opposite sign convention for work, used in some texts on thermodynamics, work done *on* the system is considered negative. The effect of this on the statement of the energy balance is discussed later as footnote 4.

The examples of reversible and irreversible processes just cited refer to mechanical work. The same basic idea applies to other forms of work as well. If a battery whose open circuit voltage is 1.5 V is charged using a source of electrical energy at 2.0 V, we say that the process is irreversible. When the battery tries to return to its original uncharged state, it will discharge the electrical energy at 1.5 V or less. As we will see later, the irreversible work done during the charging process will end up as heat, which must be dissipated from the battery.

1.3 ENERGY OF A SYSTEM

For our purposes, the energy of a system can be divided into three categories: internal energy, potential energy, and kinetic energy. To take them in reverse order, kinetic energy refers to the energy possessed by the system because of its *overall* motion, either translational or rotational. The word "overall" is italicized because the kinetic energy to which we refer is the kinetic energy of the entire system, not the kinetic energy of the molecules in the system. If the system is a gas, the kinetic energy is the energy due to the macroscopic flow of the gas, not the motion of individual molecules. A familiar form of kinetic energy is the translational energy $\frac{1}{2}mv^2$ possessed by a body of mass m moving at a velocity v.

The potential energy of a system is the sum of the gravitational, centrifugal, electrical, and magnetic potential energies. To illustrate using gravitational potential energy, a one-kilogram mass, 10 m above the ground, clearly has a greater potential energy than the same kilogram mass on the ground. That potential energy can be converted into other forms of energy, such as kinetic energy, if the mass is allowed to fall freely. Kinetic and potential energy depend on the environment in which the system exists. In particular, the potential energy of a system depends on the choice of a zero level. In our example, if the ground level is considered to be at zero potential energy, then the potential energy of the mass at 10 m above the ground will have a positive potential energy equal to the mass (1 kg) multiplied by the gravitational constant (g = 9.807 m/s^2) and the height above the ground (10 m). Its potential energy will be 98.07 (kg·m^2)/s^2, or 98.07 newton-meters (N·m), that is, 98.07 J. The datum plane for potential energy can be chosen arbitrarily. If it had been chosen at 10 m above the ground level, the potential energy of the mass would have been zero. Of course, the *difference* in potential energy between the mass at 10 m and the mass at ground level is the same independent of the datum plane.

The internal energy of a system depends on the inherent qualities, or properties, of the materials in the system, such as composition and physical form, as well as the environmental variables (temperature, pressure, electric field, magnetic field, etc.). Internal energy can have many forms, including mechanical, chemical, electrical, magnetic, surface, and thermal.[2] To consider some examples, a spring that is

---

[2]In addition to the forms of internal energy already mentioned, the mass of the system itself can be considered a form of energy according to Einstein's relationship, $E = mc^2$, where c is the speed of light. Such a term must be introduced when the mass of a system changes, as in nuclear reactor fuel.

compressed has a higher internal energy (mechanical energy) than a spring that is not compressed, because the compressed spring can do some work on changing (expanding) to the uncompressed state. As an example of chemical energy, consider two identical vessels, each containing hydrogen and oxygen. In the first, the gases are contained in the elemental form, pure hydrogen and pure oxygen in a ratio of 2:1. In the second, the identical number of atoms is contained, but in the form of water. One can appreciate that the internal energy of the first is different from the second. A spark set off in the first container will result in a violent release of energy. The same will not be true in the second. Clearly, the internal energy present differs in these two situations. Any energy balance will have to take this difference into account.

On the question of thermal energy, it is intuitive that the internal energy of a system increases as its temperature is increased: We know that we have to add energy to a bar of iron to raise its temperature. The form of internal energy of a material that is related to its temperature is called *thermal energy*. To be precise, we should not call it heat. Heat is the energy *in transfer* between a system and the surroundings. Thermal energy is *possessed* by the system.

1.4 ENERGY AS A STATE FUNCTION

The entire structure of the science of thermodynamics is built on the concept of equilibrium states and the postulate that the change in the value of thermodynamic quantities, such as internal energy, between two equilibrium states of a system does not depend on the thermodynamic path the system took to get between the two states. The change is defined by the final and beginning equilibrium states of that system.[3]

This means that the internal energy *change* of a system is determined by a knowledge of all the parameters that specify the system in its final state and in its initial state. The parameters are pressure, temperature, magnetic field, surface area, mass, and so on. If a system changes from a state labeled 1 to a state labeled 2, the change in internal energy (ΔU) will be $U_2 - U_1$, the internal energy in the final state less the internal energy in the initial state. The difference does not depend on how the system got from state 1 to state 2. The internal energy thus is referred to as a state function, or a point function, that is, a function of the *state* of the system only, and not its history.

$$\Delta U = U_2 - U_1 \tag{1.2}$$

1.5 WORK

In contrast to internal energy, the work done on or by the system as it progresses from state 1 to state 2 is a function of the path taken.

---

[3]At this point in our study of thermodynamics, it is sufficient to state that a system is in equilibrium if its temperature, pressure, density, and other physical properties are uniform throughout. For a more general definition of equilibrium, see Chapters 4 and 5.

To illustrate, let us define a system consisting of a gas that will proceed from state 1 to state 2 as indicated in Figure 1.3*a*. The pressure on the system is plotted on the vertical axis, the volume of the system on the horizontal axis. The work done by the system, which is the resisting pressure multiplied by the change in volume, is negative because the system does work on the surroundings. For the purposes of this example, let us assume that the process took place reversibly. The pressure inside the gas is equal to the resisting pressure (i.e., differs from the resisting pressure by only an infinitesimal amount). Path 1 proceeds by an isobaric (constant pressure) expansion from point 1 to intermediate point *A*, followed by an isochoric (constant volume) increase in pressure from *A* to point 2. No work was done during the second part of the process because there was no motion. The total work done is:

$$W = -P_1(V_2 - V_1) \quad \textbf{(1.3)}$$

The system could, of course, have proceeded from point 1 to point 2 by a variety of paths. Path 2, shown in Figure 1.3*b*, is one such path. The system moves from point 1 to *B* on an isochoric path (no work done), and then from *B* to 2 on an isobaric path. The work done by the system in this process (1 to *B* to 2) is:

$$W = -P_2(V_2 - V_1) \quad \textbf{(1.4)}$$

By comparing Eqs. 1.3 and 1.4, one can see that the work done via the two different paths is indeed different. The work done on path 1–*B*–2 is larger than 1–*A*–2 because P_2 is greater than P_1.

The different pressure–volume paths of the gas in the system just described were controlled by changing the temperature of the gas (i.e., by adding or subtracting heat). Note that the system could be operated in a cyclical fashion, that is, by pro-

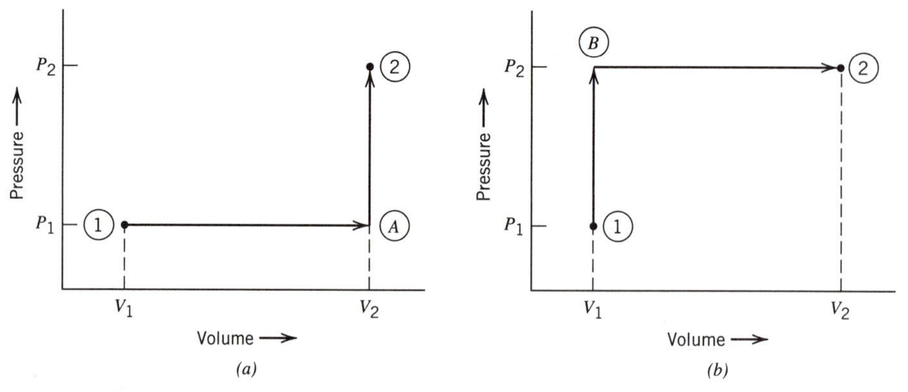

Figure 1.3 Work is a function of path: less work is done in (*a*) than in (*b*).

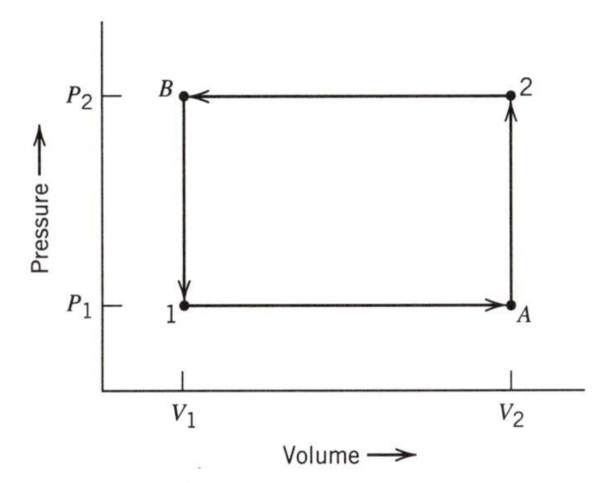

Figure 1.4 Work in a closed cycle.

ceeding from point 1 through A to point 2, and then back through B to point 1. In this cycle, the net work done (Figure 1.4) is the integral of $P\,dV$ around the cycle:

$$W = \oint P\,dV = +(P_2 - P_1)(V_2 - V_1) \tag{1.5}$$

Note that the work done in this sense was positive; that is, the surroundings did work on the system. Had the cycle been operated in the opposite direction (i.e., from point 1 to B to point 2 to A to point 1), the work would have been negative. As the system goes through a cycle of this kind, net work is either added or removed from the system. The difference, of course, is provided by the net amount of heat added or subtracted. This conversion of heat into work or vice versa is the subject of heat engines, which is discussed in Chapter 2.

1.6 THE CLOSED SYSTEM

In a closed system, no matter enters or leaves the system as defined. The boundaries of the system may expand or contract as work is done by or on the system, and thermal energy may flow into or out of the system as heat. The change in the internal energy of the system is a result of these transfers of energy.

The first law of thermodynamics for a closed system that undergoes no changes in potential and kinetic energy is shown in differential form[4] as follows:

$$\delta Q + \delta W = dU \tag{1.6}$$

Note that the term for internal energy (U) is shown by the differential dU. The notation δ is used in front of Q and W to remind us that Q and W are not state

[4]If the sign convention for work is opposite to the one used in this text (see note 1), then the First Law equation for a closed system will be $\delta Q - \delta W = dU$.

functions; rather, they are functions of the path taken. The internal energy U is a point or state function, and its differential is indicated by d.

We could have added the kinetic and potential energy terms to Eq. 1.6 to make it more complete as follows:

$$\delta Q + \delta W = dU + d(\text{PE}) + d(\text{KE}) \tag{1.7}$$

To revisit the example in Section 1.3, consider the process of raising a one-kilogram mass from ground level to a height of 10 m. In the First Law balance (Eq. 1.7), the term δQ is zero because there was no heat transfer involved in the process. There was also no change in internal energy ($dU = 0$) because the properties of the mass were the same at the beginning as at the end of the process. The work term δW is equal to the resisting force mg multiplied by the distance traveled dl:

$$\delta W = +mg\, dl$$

$$W = +mg\, \Delta l = mg(10)$$

The change in potential energy is $+mg(10)$ as in Section 1.3. Thus the energy balance is complete.

$$\delta W = d(\text{PE}) \qquad \text{or} \qquad W = \Delta(\text{PE})$$

$$mg(10) = mg(10)$$

Recognize that this example is a simple one and does not demonstrate the usefulness of the First Law. Ordinarily, one of the terms in the energy balance is unknown, and the First Law is used to determine its value.

1.7 NOTATION

In our study of thermodynamics we will have to adopt a form of notation that differentiates between the properties of the total system and properties per unit mass, called *specific* properties. In this text, the following notation has been adopted:

$$\underline{U} \equiv \frac{U}{m} \tag{1.8}$$

The internal energy of the system divided by the mass is called the specific internal energy of the system. Internal energy is measured in joules. Mass is measured in kilograms. Specific internal energy has units of joules per kilogram (J/kg). The same system of notation applies for volume.

$$\underline{V} \equiv \frac{V}{m} \tag{1.9}$$

The term $\underline{V}$ is the specific volume in cubic meters per kilogram. Note that the *density* of a material is the reciprocal of the specific volume. Density in the SI system of units is given in kilograms per cubic meter. Sometimes, in chemistry texts, the term ''specific gravity'' is used as a measure of density. The specific gravity of a material is the ratio of its density to the density of water, which is assumed to be 1 g/cm^3. The numerical value of specific gravity is the same as that for density, measured in grams per cubic centimeter. However, the specific gravity is, strictly speaking, dimensionless because it is a ratio.

As an example let us calculate the specific volume of water, given that its density is about 1 g/cm^3 at 298 K. The specific volume is the reciprocal of the density ρ, taking appropriate unit conversions into account.

$$\underline{V} = \frac{1}{\rho}$$

$$\underline{V} = \frac{1}{1.0\ (\text{g/cm}^3)} \times 1000 \left(\frac{\text{g}}{\text{kg}}\right) \times 10^{-6} \left(\frac{\text{m}^3}{\text{cm}^3}\right) = 10^{-3}\ \frac{\text{m}^3}{\text{kg}}$$

When the specific volume *per mole of* material is required in thermodynamic calculations, we call this the *molar volume*. The molar volume of elemental copper (molecular weight = 63.54 g/mol and density = 8.96 g/cm^3) is:

$$\underline{V} = 63.54 \left(\frac{\text{g}}{\text{mol}}\right) \times \frac{1}{8.96\ (\text{g/cm}^3)} \times 10^{-6} \left(\frac{\text{m}^3}{\text{cm}^3}\right)$$

$$\underline{V} = 7.09 \times 10^{-6}\ \frac{\text{m}^3}{\text{mol}} \quad \text{or} \quad 7.09\ \frac{\text{cm}^3}{\text{mol}}$$

Note that the same symbol, $\underline{V}$, is used to denote *specific* volume (m^3/kg) and *molar* volume (m^3/mol). The meaning of the symbol is usually obvious from the context of the use. In dealing with chemical reactions or with the properties of solutions, $\underline{V}$ usually represents molar volume.

1.8 INTENSIVE AND EXTENSIVE PROPERTIES

The properties of a system, or a portion of it, are either extensive or intensive, depending on whether they depend on the mass of the system. *Extensive properties* depend on the extent or mass of the system. *Intensive properties* do not depend on the extent or mass of the system. Pressure and temperature are intensive properties. Volume is an extensive property. Both *specific* volume and density are intensive properties.

There is a simple test to determine whether a property is intensive or extensive. Imagine that another identical system is created alongside the system under consideration. If the property or quantity in question has doubled, then that property is extensive. Otherwise it is intensive. Doubling a system does not double the temperature. Hence, temperature is an intensive property. Doubling a system doubles the volume. Volume is an extensive property.

1.9 THE OPEN SYSTEM

In an open system, matter may enter or leave the system as defined. When writing the First Law energy balance for such a system, we must, of course, take into account the internal energy of the materials entering and leaving that system. In addition, we must account for the work done *on* the system when the entering material is pushed into it, and *by* the system when it pushes material out of the system. These two work terms are called flow work (Figure 1.5).

The flow work done in pushing the entering material (designated as material "i") into the system is:

$$\text{flow work}_i = \int_0^V P d V_i$$

$$V_i = \underline{V}_i m_i \qquad \text{or} \qquad dV = \underline{V}_i dm_i$$

$$\text{flow work}_i = \int_0^m P\underline{V}_i dm_i = P\underline{V}_i m_i$$

In differential form:

$$\delta\,(\text{flow work})_i = +P\underline{V}_i \delta m_i$$

A similar term is needed for the flow work involved in pushing the exiting material from the system. Omitting the kinetic and potential energy terms for simplicity, the First Law is:

$$\underline{U}_i\,\delta m_i - (\underline{U}_o\,\delta m_o + \delta Q + \delta W - \delta\,(\text{flow work}) = dU$$

$$(\underline{U}_i + P_i\underline{V}_i)\delta m_i - (\underline{U}_o + P_o\underline{V}_o)\delta m_o + \delta Q + \delta W = dU$$

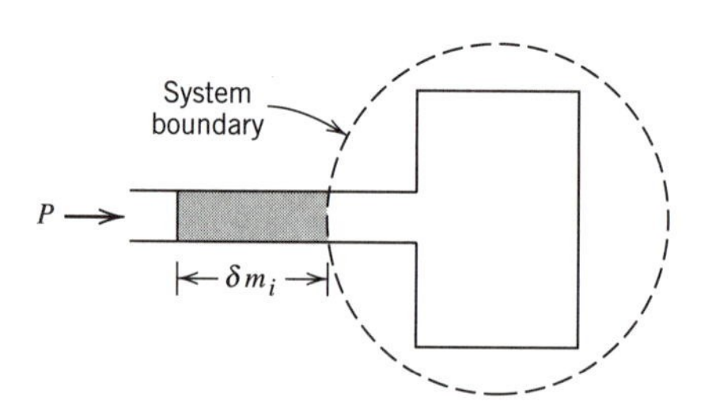

Figure 1.5 Flow work in an open system.

where the terms $\underline{U}_i\ \delta m_i$ and $\underline{U}_o\ \delta m_o$ represent the internal energy carried into and out of the system by the matter entering and leaving the system, and $P_i\underline{V}_i\ \delta m_i$ and $P_o\underline{V}_o\ \delta m_o$ are the flow work terms.

Note that the work term could have been included in the δW term. However, the flow work is always associated with the material entering or leaving the system and is easily and conveniently expressed in terms of the properties of the material in transit. Thus it is more practical to associate the term with the properties of the material flowing into or out of the system:

$$\sum_{m_i} (\underline{U}_i + P_i\underline{V}_i)\delta m_i - \sum_{m_o} (\underline{U}_o + P_o\underline{V}_o)\delta m_o + \delta Q + \delta W = dU \qquad \textbf{(1.10)}$$

where the summation term Σ in conjunction with the flow terms represents the sum over all the mass streams entering (or leaving) the system.

1.10 ENTHALPY

The term $U + PV$ is defined as *enthalpy*. It is designated by the letter H.

$$H \equiv U + PV \qquad \textbf{(1.11a)}$$

Specific enthalpy is enthalpy per unit mass:

$$\underline{H} = \underline{U} + P\underline{V} \qquad \textbf{(1.11b)}$$

Taking potential and kinetic energy terms into account, the First Law statement becomes:

$$\sum_{m_i} (\underline{H}_i + \underline{KE}_i + \underline{PE}_i)\delta m_i - \sum_{m_o} (\underline{H}_o + \underline{KE}_o + \underline{PE}_o)\delta m_o + \delta Q + \delta W = d(U + \text{PE} + \text{KE}) \qquad \textbf{(1.12)}$$

As a limiting case, it is apparent that the First Law written for an open system (Eq. 1.12) is reduced to the closed system statement (Eq. 1.7) when the mass entering, δm_i, and the mass leaving, δm_o, are zero.

1.11 STEADY STATE

Another limiting case that is quite useful especially when considering a materials processing operation is the *steady state*. Steady state is defined as one in which the system does not change with time. In more formal terms we would say that every quantity or property of the system is time invariant. Material may enter and leave the system, but the system itself remains unchanged. This form of analysis is useful in the case of processing apparatus, such as pumps, turbines, chemical reaction vessels, smelters, or blast furnaces. In these devices, once steady operations have been achieved, material enters and material leaves, but the system itself remains basically unchanged with time. In this case the First Law becomes:

$$\sum_{m_i} (\underline{H}_i + \underline{KE}_i + \underline{PE}_i)\delta m_i - \sum_{m_o} (\underline{H}_o + \underline{KE}_o + \underline{PE}_o)\delta m_o + \delta Q + \delta W = 0 \quad \mathbf{(1.13)}$$

For a finite process, this can be expressed as follows:

$$Q + W = \sum_{m_o} (\underline{H}_o + \underline{KE}_o + \underline{PE}_o)m_o - \sum_{m_i} (\underline{H}_i + \underline{KE}_i + \underline{PE}_i)m_i \quad \mathbf{(1.14)}$$

Or, when no changes in kinetic or potential energy are involved:

$$Q + W = \sum_{m_o} \underline{H}_o m_o - \sum_{m_i} \underline{H}_i m_i \quad \mathbf{(1.15)}$$

1.12 HEAT CAPACITY AT CONSTANT VOLUME

The amount of thermal energy required to change the temperature of a material is the ''heat capacity'' of the material. Note that this terminology is inconsistent with the definition of heat as a transfer quantity. The term measuring the energy required to change the temperature of material in a system would more properly be called *energy* capacity of the material. Historically, however, the term *heat capacity* has been used and will be used in this text. Another exception will be made in the notation used for heat capacity. An underscore has been adopted as the notation for specific quantities, that is, quantities per unit mass. However, heat capacities are always stated in terms of unit mass, that is, in joules *per kilogram*-kelvin, joules *per mole*-kelvin, or calories *per gram*-degree centigrade. Hence, the heat capacity term is not underscored. In a strict sense, the heat capacity per mole should be called the *specific* heat capacity. However, the word ''specific'' usually is omitted, and it is assumed that the heat capacity is given per unit of mass.[5]

For a closed system at constant volume ($dV = 0$), no mechanical work is done, and we can write the First Law as

$$\delta Q = dU$$

$$\delta Q = mC\,dT = dU$$

$$C\,dT = \frac{dU}{m} = d\underline{U}$$

where C is the heat capacity.

---

[5]The term *specific heat* is also used to denote heat capacity, in some publications on low temperature physics.

This heat capacity at constant volume is given the notation C_V and is defined as follows:

$$C_V \equiv \left(\frac{\partial \underline{U}}{\partial T}\right)_V \qquad \textbf{(1.16a)}$$

Note that we have implicitly assumed that no other forms of work, such as electrical or magnetic, are involved. The constant volume change of internal energy with temperature of a mass m is:

$$dU = mC_V dT \qquad \textbf{(1.16b)}$$

It is important to recognize that the heat capacity C_V is a function of temperature *and* of the specific volume of the material to which it applies. In special cases, such as ideal gases (discussed in Section 1.14) the heat capacity is independent of specific volume, but in general it is not.

1.13 HEAT CAPACITY AT CONSTANT PRESSURE

A material heated at constant pressure usually expands.[6] The thermal energy added to the material is accounted for by the increase in the internal energy of the material plus the work done by the material as it expands against the constant pressure imposed on it. From the First law (Eq. 1.6):

$$\delta Q + \delta W = dU$$

Considering only mechanical work:

$$\delta Q - P\,dV = dU$$

$$\delta Q = dU + P\,dV$$

At constant pressure ($dP = 0$), the term $P\,dV = d(PV)$, and we can write:

$$\delta Q = d(U + PV) = dH$$

$$dH = mC_P dT \quad \text{or} \quad d\underline{H} = C_P dT \qquad \textbf{(1.17)}$$

$$C_P \equiv \left(\frac{\partial \underline{H}}{\partial T}\right)_P$$

[6]There are some metallic materials (e.g., Invar) that undergo changes in magnetization with temperature, hence exhibit very low thermal expansion. A particular material composed of iron, nickel, chromium, and titanium actually exhibits negative thermal expansion in a certain temperature range.

Table 1.1 Values for Use in Finding C_p for Selected Materials
[$C_p = a + bT + cT^{-2}$ J/mol·K]

Material	a	$b \times 10^3$	$c \times 10^{-5}$	Range (K)
Ag (s)	21.30	8.54	1.51	298–1234
Ag (l)	30.5			1234–1600
Ag_2O	41.92	100.12	[$-55.6 \times 10^{-6}T^2$]	298–500
Al (s)	20.67	12.38		298–932
Al (l)	29.3			932–1273
Al_2O_3	114.8	12.80	35.4	298–1700
Au (s)	23.68	5.19		298–1336
Au (l)	29.3			1336–1600
C (graphite)	17.15	4.27	−8.79	298–2300
C (diamond)	9.12	13.22	−6.19	298–1200
CO	28.41	4.10	−0.46	298–2500
CO_2	44.14	9.04	−8.54	298–2500
CH_4	23.64	47.86	−1.92	298–1500
Ca_α	22.22	13.93		273–713
Ca_β	6.28	32.38	10.46	713–1123
Ca (l)	31.0			1123–1220
CaO	49.62	4.52	−6.95	298–1177
$CaCo_3$ (α,β)	104.52	21.92	−25.94	298–1200
Cr (s)	24.43	9.87	−3.68	298–2176
Cr (l)	39.3			2167–3000
Cr_2O_3	119.37	9.20	−15.65	350–1800
Cu (s)	22.64	6.28		298–1356
Cu (l)	31.4			1356–1600
Cu_2O	62.34	23.85		298–1200
CuO	38.79	20.08		298–1250
Fe_α	17.49	24.77		273–1033
Fe_β	37.7			1033–1181
Fe_γ	7.70	19.50		1181–1674
Fe_δ	43.93			1674–1812
Fe (l)	41.8			1812–1873
$Fe_{.95}O$ (s)	48.79	8.37	−2.80	298–m.p.*
Fe_2O_3 (α)	98.18	77.82	−14.85	298–950
Fe_3O_4 (α)	91.55	201.67		298–950
H_2	27.28	3.26	0.50	298–3000
H_2O (l)	75.44			273–373
H_2O (g)	30.0	10.71	0.33	298–2500
Li (s)	12.76	35.98		273–454
Li (l)	29.3			500–1000
LiCl	46	14.2		298–m.p.
Mg (s)	22.30	10.25	−0.43	293–923
Mg (l)	33.9			923–1130

Table 1.1 (continued)

Material	a	$b \times 10^3$	$c \times 10^{-5}$	Range (K)
MgO	42.59	7.28	−6.19	298–2100
$MgCl_2$ (s)	79.08	5.86	−8.62	298–m.p.
$MgCO_3$	77.91	57.78	−17.41	298–750
Ni (s) (α)	22.52	−10.42	[+ $10.44 \times 10^{-6}T^2$]	300–615
Ni (l)	38.49			1728–1900
NiO ($NiO_{1.005}$)	54.02			523–1100
Pb (s)	23.56	9.75		298–601
Pb (l)	32.43	−3.10		601–1200
PbO (red)	44.53	16.74		298–900
Si (s)	23.22	3.68	−3.81	298–1200
Si (l)	31.0			1683–1900
SiO_2 (α) (quartz)	46.94	34.31	−11.30	298–848

*m.p. = melting point

Source: Values from O. Kubaschewski and E. L. L. Evans, *Metallurgical Thermochemistry,* Wiley, New York, 1956.

Heat capacities are usually tabulated as constant pressure heat capacities because most heating or cooling of materials takes place at constant pressure ambient conditions. Some representative heat capacities for materials are given in Table 1.1. It is important to remember that C_P is, in general, a function of both temperature and pressure. The values tabulated in Table 1.1 are at one atmosphere (1 atm) pressure. In the case of ideal gases, the heat capacity is not a function of pressure, and for most solids and liquids, these values are substantially independent of pressure.

As an example of the use of heat capacities, let us calculate the energy required and the cost of heating a slab of aluminum of mass one metric ton (1000 kg) from 300 K to 800 K, a temperature that might be used to reduce the thickness of the aluminum through rolling. The aluminum will be heated by passing it through a furnace that uses electricity as its source of energy. The cost of electrical energy is assumed to be 5 cents per kilowatt-hour. For the purpose of this problem, assume that there are no extraneous heat losses from the furnace; that is, all the electrical energy entering the furnace is used to heat the aluminum.

Select the furnace as the system. Note that it is an open system. We will assume that it is at steady state. The basis of the calculation will be one metric ton of aluminum passing through the furnace. The appropriate statement of the First Law for this problem is:

$$\underline{H}_i\, m_i - \underline{H}_o\, m_o + Q + W = \Delta U = 0$$

Note that the heat flow term is zero because the energy did not flow into the system because of a temperature difference. Rather, energy entered the system as work because of an electrical potential difference.

To evaluate the enthalpy term in the equation, we make use of the heat capacity of aluminum:

$$W = m_{\text{Al}}\,(\underline{H}_{\text{o}} - \underline{H}_{\text{i}})$$

$$\underline{H}_{\text{o}} - \underline{H}_{\text{i}} = \int_{300}^{800} C_p dT = \int_{300}^{800} (20.67 + 12.38 \times 10^{-3}\,\text{T})dT$$

$$\underline{H}_{\text{o}} - \underline{H}_{\text{i}} = 20.67\,(800 - 300) + \frac{12.38 \times 10^{-3}}{2}(800^2 - 300^2)$$

$$\underline{H}_{\text{o}} - \underline{H}_{\text{i}} = 13{,}730\ \text{J/mol} \qquad \text{or} \qquad \underline{H}_{\text{o}} - \underline{H}_{\text{i}} = 13{,}730 \times \frac{1000}{27} = 508{,}560\ \text{J/kg}$$

$$W = 508{,}560\,(m) = 5.086 \times 10^8\ \text{J} = 141.4\ \text{kW·h}$$

$$\text{cost} = (\$0.05)(141.4) = \$7.07$$

Another way to approach the same problem is to select an aluminum slab as the system (a closed system). The basis of the calculation is a slab weighing one metric ton. The First Law equation, in integrated form, is:

$$Q + W = U_2 - U_1$$

$$Q = U_2 - U_1 - W$$

The work term is not zero because the aluminum expands upon heating. The work is the resisting pressure (atmospheric pressure) multiplied by the change in volume, $V_2 - V_1$, and it is negative because the system does work on the surroundings. Thus the equation becomes:

$$Q = U_2 - U_1 + P(V_2 - V_1) = U_2 + PV_2 - (U_1 + PV_1)$$

$$Q = H_2 - H_1 = (\underline{H}_2 - \underline{H}_1)m$$

This is essentially the same as the result obtained using the open system. The energy supplied to the slab is treated as heat in this case. Because there are no extraneous heat losses from the furnace, the energy transferred as heat is identical to that supplied as electrical energy to the furnace. The approaches, thus, yield the same answer, as they should.

1.14 ADIABATIC FLOW THROUGH A VALVE: JOULE–THOMSON EXPANSION

An *adiabatic* process is one in which no heat is added or removed from the system. A special case that often arises in the analysis of fluid flow involves adiabatic flow through a valve. This assumption of no heat flow is valid if the valve in question is insulated, or if the flow through it is so rapid that any heat transferred is negligible.

Consider the system in Figure 1.6. If we define the system as the valve, and consider it to be at steady state, then the First Law (neglecting kinetic and potential energy changes) becomes:

$$\underline{H}_i \delta m_i - \underline{H}_o \delta m_o + \delta Q + \delta W = dU$$

Because the system boundaries do not move, the work term is zero. (Recall that the flow work is contained in the $\underline{H}_i$ and $\underline{H}_o$ terms.) If we assume that the valve is adiabatic—that is, that no heat enters or leaves the system—then δQ is zero. Because the mass into the system is equal to the mass out of the system, we have:

$$\delta Q = 0; \qquad \delta W = 0; \qquad \delta m_i = \delta m_o$$

Therefore

$$\underline{H}_i = \underline{H}_o \tag{1.18}$$

This special case of isenthalpic expansion is called the Joule–Thomson expansion, and one often sees reference to the Joule–Thomson coefficient η_{JT}, the change of temperature with pressure at constant enthalpy.

$$\eta_{JT} \equiv \left(\frac{\partial T}{\partial P}\right)_{\underline{H}} \tag{1.19}$$

We demonstrate in the next section that the Joule–Thomson coefficient for an ideal gas is zero.

1.15 EQUATIONS OF STATE

Each of the thermodynamic quantities we have defined, such as internal energy, enthalpy, and the heat capacities, is a function of temperature and pressure. The variation of enthalpy with temperature is described by the constant pressure heat capacity C_P. To determine the variation of enthalpy (or internal energy) with pressure

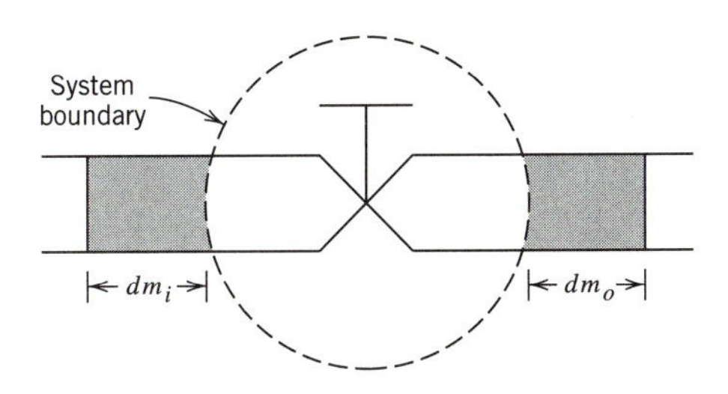

Figure 1.6 Flow through a valve.

requires a knowledge of the relationship among the physical variables that describe the condition of a material, that is, its equation of state.

For a gas, the equation of state relates the pressure (P), volume (V), temperature (T), and number of moles (n). The simplest equation of state for a gas is the ideal gas equation:

$$PV = nRT \tag{1.20a}$$

or

$$P\underline{V} = RT \tag{1.20b}$$

Based on the definition, an ideal gas does not condense into a liquid or solid. At constant temperature, there is a single value of specific volume for every pressure. There are no discontinuities in the P-V-T relationship, and a three-dimensional plot of the variables is a smooth surface (Figure 1.7).

In Eq. 1.20, the pressure is measured in pascals (newtons per square meter). The volume is measured in cubic meters. The term n represents the number of moles: one mole (mol) is 6.022×10^{23} molecules. R is the universal gas constant [8.314 J/(mol·K)], and T is the absolute temperature in kelvin. For the purpose of illustration let us calculate the volume of one mole of gas at 0°C (273.15 K) and one atmosphere pressure.

$$P = 1 \text{ atm} = 1.013 \times 10^5 \text{ N/m}^2$$

$$(1.013 \times 10^5)\,(\underline{V}) = 1\,(8.314)(273.15)$$

$$\underline{V} = 0.0224 \text{ m}^3\text{/mol} = 22.4 \text{ L/mol}$$

Another characteristic of an ideal gas is that its internal energy does not change with volume at constant temperature.

$$\left(\frac{\partial \underline{U}}{\partial \underline{V}}\right)_T = 0 \tag{1.21}$$

In this chapter, we use this concept as part of the definition of an ideal gas. In Chapter 3 we will prove that it follows as a consequence of the earlier definition (Eq. 1.20).

If the internal energy does not change with volume at constant temperature, it can be shown that the enthalpy does not change with pressure at constant temperature.

$$\underline{H} = \underline{U} + P\underline{V}$$

$$\left(\frac{\partial \underline{H}}{\partial P}\right)_T = \left(\frac{\partial \underline{U}}{\partial P}\right)_T + P\left(\frac{\partial \underline{V}}{\partial P}\right)_T + \underline{V}$$

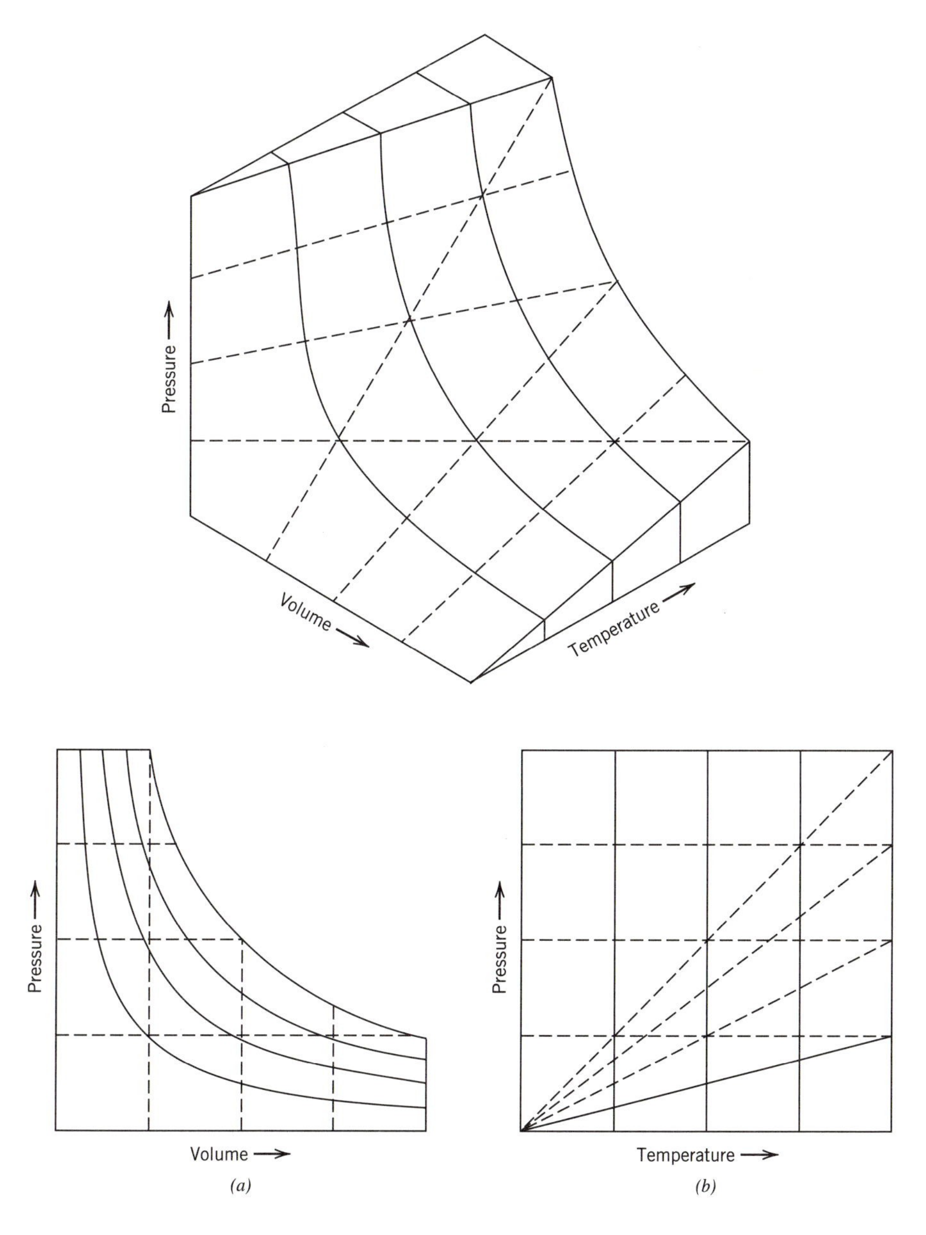

Figure 1.7 Projections of ideal gas *P*-*V*-*T* surface on (*a*) the *P*-*V* plane and (*b*) the *P*-*T* plane.

Using the chain rule and Eq. 1.20,

$$\left(\frac{\partial \underline{H}}{\partial P}\right)_T = \underbrace{\left(\frac{\partial \underline{U}}{\partial \underline{V}}\right)_T}_{0} \left(\frac{\partial \underline{V}}{\partial P}\right)_T + P\left(-\frac{RT}{P^2}\right) + \underline{V};$$

For an ideal gas, $\left(\frac{\partial \underline{U}}{\partial \underline{V}}\right)_T = 0$, hence,

$$\left(\frac{\partial \underline{H}}{\partial P}\right)_T + 0 - \underline{V} + \underline{V} = 0 \tag{1.22}$$

One conclusion that can be drawn from Eq. 1.22 is that the Joule–Thomson coefficient for an ideal gas (i.e., the change of temperature with pressure as the gas passes through a steady state, adiabatic valve) is zero.

$$\eta_{JT} = \left(\frac{\partial T}{\partial P}\right)_{\underline{H}} = -\frac{\left(\frac{\partial \underline{H}}{\partial P}\right)_T}{\left(\frac{\partial \underline{H}}{\partial T}\right)_P} \quad \text{using Eq. A.14 in Appendix A}$$

$$\eta_{JT} = 0 \quad \text{for an ideal gas} \tag{1.23}$$

We can show also that the difference in heat capacities $(C_P - C_V)$ of an ideal gas is the gas constant, R. Using the definitions of C_P and C_V:

$$C_P - C_V = \left(\frac{\partial \underline{H}}{\partial T}\right)_P - \left(\frac{\partial \underline{U}}{\partial T}\right)_V$$

From the definition of enthalpy (Eq. 1.11b)

$$\underline{H} = \underline{U} + P\underline{V}$$

$$\left(\frac{\partial \underline{H}}{\partial T}\right)_P = \left(\frac{\partial \underline{U}}{\partial T}\right)_P + P\left(\frac{\partial \underline{V}}{\partial T}\right)_P$$

Hence:

$$C_P - C_V = \left(\frac{\partial \underline{U}}{\partial T}\right)_P + P\left(\frac{\partial \underline{V}}{\partial T}\right)_P - \left(\frac{\partial \underline{U}}{\partial T}\right)_V \tag{1.24a}$$

Our problem at this point is to find the relationship between the partial derivatives $(\partial \underline{U}/\partial \underline{T})_P$ and $(\partial \underline{U}/\partial \underline{T})_V$. Recognizing that the specific internal energy ($\underline{U}$) of a gas is a function of its temperature and specific volume:

$$\underline{U} = \underline{U}\,(\mathrm{T},\underline{V})$$

$$d\underline{U} = \left(\frac{\partial \underline{U}}{\partial T}\right)_{\underline{V}} dT + \left(\frac{\partial \underline{U}}{\partial \underline{V}}\right)_T d\underline{V} \quad \textbf{(1.24b)}$$

But we also know that the specific volume of a gas is a function of its temperature and pressure (see Eq. 1.20 for an ideal gas):

$$\underline{V} = \underline{V}\,(T,\mathrm{P})$$

$$d\underline{V} = \left(\frac{\partial \underline{V}}{\partial T}\right)_P dT + \left(\frac{\partial \underline{V}}{\partial P}\right)_T dP \quad \textbf{(1.25)}$$

Substituting the value of $d\underline{V}$ from Eq. 1.25 in Eq. 1.24b yields:

$$d\underline{U} = \left(\frac{\partial \underline{U}}{\partial T}\right)_V dT + \left(\frac{\partial \underline{U}}{\partial \underline{V}}\right)_T \left[\left(\frac{\partial \underline{V}}{\partial T}\right)_P dT + \left(\frac{\partial \underline{V}}{\partial P}\right)_T dP\right]$$

This equation is valid in general, and also at constant pressure, hence:

$$(d\underline{U})_P = \left(\frac{\partial \underline{U}}{\partial T}\right)_V (dT)_P + \left(\frac{\partial \underline{U}}{\partial \underline{V}}\right)_T \left[\left(\frac{\partial \underline{V}}{\partial T}\right)_P (dT)_P\right]$$

Rewriting in the more familiar partial differential form:

$$\left(\frac{\partial \underline{U}}{\partial T}\right)_P = \left(\frac{\partial \underline{U}}{\partial T}\right)_V + \left(\frac{\partial \underline{U}}{\partial \underline{V}}\right)_T \left(\frac{\partial \underline{V}}{\partial T}\right)_P \quad \textbf{(1.26)}$$

Combining Eq. 1.26 with Eq. 1.24 yields:

$$C_P - C_V = \left(\frac{\partial \underline{U}}{\partial T}\right)_V + \left(\frac{\partial \underline{U}}{\partial \underline{V}}\right)_T \left(\frac{\partial \underline{V}}{\partial T}\right)_P + P\left(\frac{\partial \underline{V}}{\partial T}\right)_P - \left(\frac{\partial \underline{U}}{\partial T}\right)_V$$

$$C_P - C_V = \left[P + \left(\frac{\partial \underline{U}}{\partial \underline{V}}\right)_T\right]\left(\frac{\partial \underline{V}}{\partial T}\right)_P$$

Recall that for an ideal gas the specific internal energy ($\underline{U}$) does not change with volume at constant temperature (Eq. 1.21), namely, $(\partial \underline{U}/\partial \underline{V})_T = 0$. Therefore:

$$C_P - C_V = P\left(\frac{\partial \underline{V}}{\partial T}\right)_P \qquad \text{for an ideal gas}$$

Introducing the equation of state for an ideal gas:

$$P\underline{V} = RT$$

$$\left(\frac{\partial \underline{V}}{\partial T}\right)_P = \frac{R}{P}$$

Therefore for an ideal gas:

$$C_P - C_V = R \tag{1.27}$$

Equations for the heat capacity of some gases as a function of temperature are given in Table 1.1. It is useful to note that there are useful approximations of the heat capacity of elemental gases at low pressure. For a *monatomic* ideal gas, such as helium:

$$C_V = \tfrac{3}{2} R$$

$$C_P = \tfrac{5}{2} R$$

For a *diatomic* ideal gas, such as oxygen:

$$C_V = \tfrac{5}{2} R$$
$$C_P = \tfrac{7}{2} R$$

To gain an appreciation for these approximations, the heat capacity of nitrogen (N_2) at 298 K from the data in Table 1.1 is:

$$C_P\ (N_2) = 27.87 + 4.27 \times 10^{-3}\ T\ \text{J/(mol·K)}$$

$$C_P = 27.87 \times 4.27 \times 10^{-3}\ (298)$$

$$C_P = 29.14\ \text{J/(mol·K)}$$

The value using $C_P = \frac{7}{2} R$ is 29.1 J/(mol·K). As another comparison, the heat capacity of oxygen at 298 K is 29.32 J/(mol·K).

1.16 NONIDEAL GASES

As discussed in Section 1.15, the *P-V-T* relationship for an ideal gas has no discontinuities, and the three-dimensional surface representing the *P-V-T* behavior of an ideal gas is smooth (Figure 1.7). As long as a gas behaves ideally (Eq. 1.20), it does not change its state into a liquid or solid. This is, of course, not true of real (nonideal) gases. Real gases condense into liquids in certain pressure–temperature regimes and change into solids in others. For example, water vapor when compressed isothermally at 298 K will eventually condense into liquid water at about 3170 Pa. These

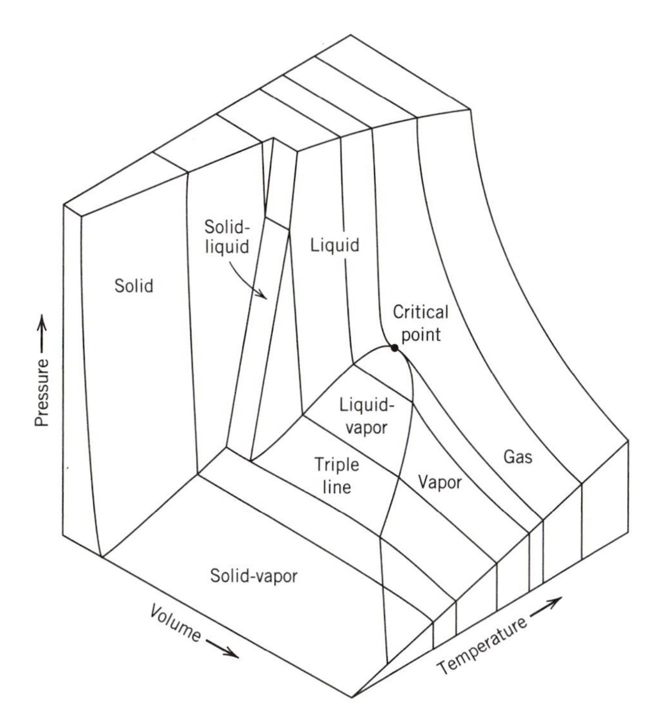

Figure 1.8 *P-V-T* surface for a substance that contracts on freezing.

regimes can be represented in diagrammatic form as in Figure 1.8. The projections of the three-dimensional (*P-V-T*) behavior of materials can be projected on the *P-T* and *P-V* planes as in Figure 1.9.

At temperatures below its critical temperature, a gas can be condensed into a liquid by compressing it at constant temperature (isothermally). At higher temperatures, the difference between liquids and gases no longer exists. There is no discontinuity as a gas is compressed. As a gas moves into a temperature–pressure range far above its critical temperature and pressure, its *P-V-T* relationship approaches that of an ideal gas. At times, however, it is necessary to describe the *P-V-T* behavior of a real gas with an equation of state that is more complex than the ideal gas equation. There are many forms for the equations of state of nonideal gases (Refs. 1 and 2).[7]

When it is important to know the properties precisely, the use of an equation with many terms is justified (Ref. 2). For most purposes, however, it is sufficient to know whether the gas in question is ideal and, if it is not, by what factor the ideal gas

[7]A slightly more complex equation of state is the van der Waals equation: $(P + a/\underline{V}^2)(\underline{V} - b) = RT$, where the b term serves as a correction to account for the volume of the atoms or molecules of the gas. The a term corrects for interactions among the molecules.

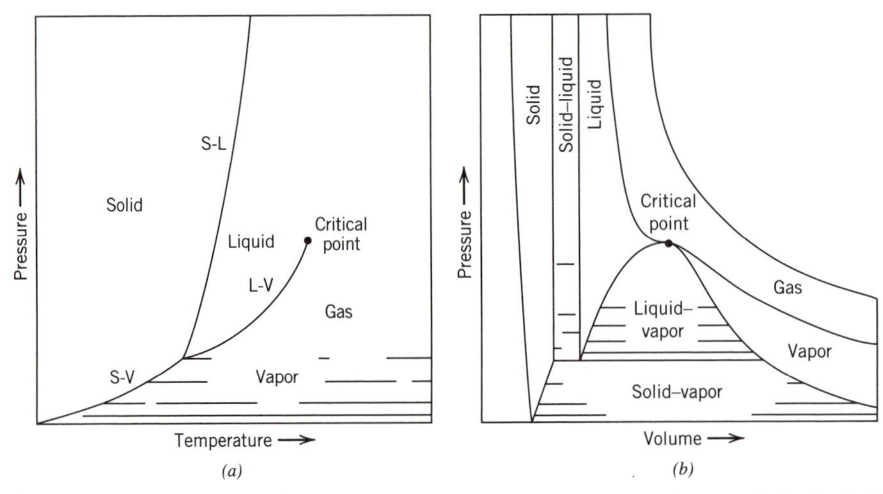

Figure 1.9 Projections of the surface in Figure 1.8 (*a*) on the *P-T* planes and (*b*) the *P-V* plane.

equation must be corrected. For this purpose, the ideal gas equation can be modified to include a correction term, *Z*, called the compressibility factor.

$$P\underline{V} = ZRT \tag{1.28}$$

The value of *Z* can be estimated using the "corresponding states" method. This method is based on the assumption that gases behave similarly with respect to non-ideality; that is, they have the same value of *Z* when they are in corresponding states relative to their respective critical pressures and critical temperatures.

In quantitative terms, the term *Z* in Eq. 1.28 can be determined by knowing the reduced pressure (P_r) and reduced temperature (T_r) of the gas. The reduced pressure is the pressure of the gas divided by its critical pressure (P_c). The reduced temperature is its temperature divided by its critical temperature (T_c), measured on an absolute scale.

$$P_r = \frac{P}{P_c} \quad \text{and} \quad T_r = \frac{T}{T_c} \tag{1.29}$$

Figure 1.10 plots *Z* versus reduced pressure for various values of reduced temperature. Values of P_c and T_c for several gases are listed in Table 1.2.

As an example, let us use Figure 1.10 to estimate *Z* for nitrogen at one atmosphere pressure and 298 K. For nitrogen, the critical pressure is 3.40×10^6 Pa and the critical temperature is 126 K. Thus the reduced pressure at one atmosphere (1.013×10^5 Pa) is $1.013 \times 10^5/3.40 \times 10^6 = 2.97 \times 10^{-2}$, and the reduced temperature is $298/126 = 2.37$. From Figure 1.10, *Z* is very nearly 1. That means that the assumption of ideality for nitrogen at one atmosphere and 298 K is justified.

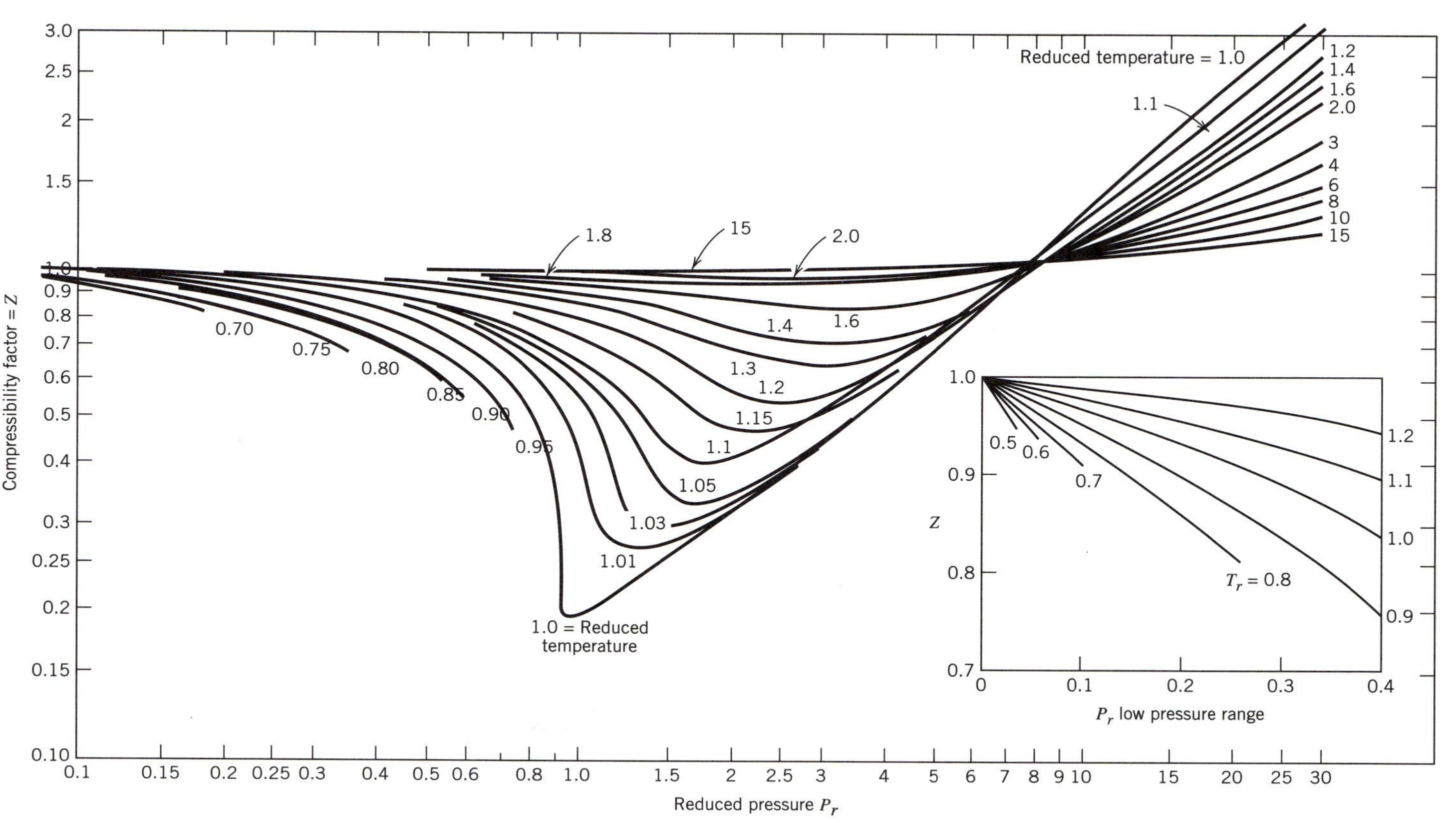

Figure 1.10 Experimental data showing generalized compressibility relation.

Table 1.2 Critical Constants for Gases

Gas	P_c (MPa)	T_c (K)
He	0.229	5.25
N_2	3.40	126
H_2	1.30	33.2
O_2	5.10	154
CO_2	7.40	304
SO_2	7.80	430
H_2O	22.1	647

By contrast, the critical temperature of carbon dioxide gas, namely, 304 K, is near room temperature. Its critical pressure (7.4×10^6 Pa) is near the pressures achieved in cylinders of gas as supplied in laboratories. These properties mean that the assumption of ideality is often not justified. For example, let us calculate the compressibility factor of carbon dioxide in a tank at about 0°C (273 K) and 650 pounds per square inch (psi) pressure (4.48×10^6 Pa):

$$T_r = \frac{273}{304} = 0.90$$

$$P_r = \frac{4.48 \times 10^6}{7.4 \times 10^6} = 0.61$$

Under these conditions, the compressibility factor, Z, is about 0.70. If the volume of the tank containing the carbon dioxide is known, then the mass of carbon dioxide can be determined using the equation:

$$PV = 0.70nRT$$

1.17 ADIABATIC COMPRESSION OR EXPANSION

As the pressure on a gas increases or decreases adiabatically (no heat flow), the temperature of the gas changes, increasing when the pressure increases. This behavior can be observed in nature as air rises rapidly (adiabatically) up a mountainside. At the higher altitude, the pressure is lower, and the temperature is also lower. At this point we have enough information to derive a quantitative relationship between the pressure and temperature of an ideal gas during an adiabatic pressure change.

Consider a volume of gas at a given temperature and pressure. Take a portion of the gas to be a closed system (Figure 1.11). If the process to be conducted is adiabatic, then δQ is zero. Because the boundary separating the system and surroundings is drawn arbitrarily through the gas, the pressure difference across the boundary is

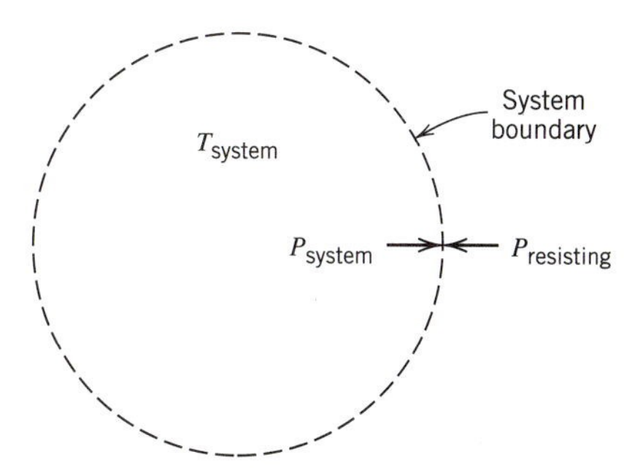

Figure 1.11 Closed system boundary in a gas phase (with $P_{resisting} = P_{system}$).

zero. Any work done can be considered to be reversible work because the resisting pressure is equal to the pressure of the system.

From the First Law (Eq. 1.7), and the definition of C_V (Eq. 1.16a):

$$\delta Q + \delta W = dU$$

$$-PdV = dU = nC_V\, dT$$

Because the gas is ideal:

$$PV = nRT$$

$$P\,dV + V\,dP = nR\,dT$$

Therefore:

$$-P\,dV = -nR\,dT + V\,dP = nC_V\,dT$$

$$V\,dP = n(C_V + R)dT$$

$$nRT\frac{dP}{P} = n(C_V + R)dT$$

$$\frac{dP}{P} = \left(\frac{C_V + R}{R}\right)\frac{dT}{T}$$

From Eq. 1.27 ($C_p - C_v = R$)

$$\frac{dP}{P} = \left(\frac{C_P}{R}\right)\frac{dT}{T}$$

If C_p may be assumed to be constant during the process:

$$\int_{P_1}^{P_2} \frac{dP}{P} = \frac{C_P}{R} \int_{T_1}^{T_2} \frac{dT}{T}$$

$$\frac{P_2}{P_1} = \left(\frac{T_2}{T_1}\right)^{C_p/R} \quad \text{or} \quad \frac{T_2}{T_1} = \left(\frac{P_2}{P_1}\right)^{R/C_P} \qquad \textbf{(1.30)}$$

Alternative algebraic forms of this relationship (for a reversible, adiabatic process in an ideal gas) are:

$$TP^{-R/C_P} = \text{constant}$$

$$TP^{(\gamma-1)/\gamma} = \text{constant}$$

$$PV^{\gamma} = \text{constant}$$

where $\gamma = C_P/C_V$.

The relationship derived above, combined with the principle discussed in Section 1.14 (Joule–Thomson expansion), allows us to gain an interesting insight into a typical laboratory situation. Gaseous helium is to be used to quench (rapidly cool) a hot experimental sample. The helium is stored in an insulated 50-liter tank at a temperature of 25°C. The tank pressure is initially 20 atm. What will be the temperature of the first gas to hit the sample when the valve on the tank is opened and the gas allowed to escape into the quench chamber? For the purposes of this problem, we can assume that helium behaves as an ideal gas.

To solve this problem, take the valve as the system (an open system). If we can assume that the properties of the valve do not change during the experiment, and that we can neglect changes in potential and kinetic energy of the gas, the First Law balance can be written as:

$$\underline{H}_i \, \delta m_i - \underline{H}_o \, \delta m_o + \delta Q + \delta W = 0$$

There is no work done by the system. Assume that the flow is so rapid that no appreciable heat flows. An alternative assumption is an insulated valve. Since for the valve $\delta m_i = \delta m_o$,

$$\underline{H}_i = \underline{H}_o$$

But for an ideal gas, the enthalpy is not a function of pressure; that is, it is a function of temperature only. Therefore, the temperature of the first helium leaving the valve (striking the sample) is the same as the temperature of the gas in the tank (25°C):

$$H(T,P) = H(T) \qquad \text{for an ideal gas}$$

Therefore,

$$T_i = T_o$$

As the helium flows, the pressure in the tank drops. What will be the temperature of the helium entering the quench chamber when the pressure in the tank has fallen to 10 atm?

To approach this problem, we must determine what happens to the temperature in the tank as the pressure drops. To solve this problem, define the system as the quantity of gas remaining in the tank when the pressure reaches 10 atm. This is a closed system. The system expands, but no mass enters or leaves. The gas undergoes a reversible expansion, because the boundary between the system and the rest of the gas is arbitrarily drawn: that is, there is no pressure difference between the system and surroundings. Furthermore, the expansion is adiabatic, because the tank is insulated. Therefore, the pressure–temperature trajectory of the system can be described by Eq. 1.30. Using $\frac{5}{2}R$ as the constant pressure heat capacity of helium, we can solve for the temperature of the gas remaining in the tank:

$$\frac{T_2}{T_1} = \left(\frac{P_2}{P_1}\right)^{R/C_P} = \left(\frac{10}{20}\right)^{R/(5/2)R} = (0.5)^{0.40}$$

$$T_2 = T_1(0.758) = 298(0.758) = 226 \text{ K}$$

The temperature of the gas passing through the valve does not change, as illustrated in the first part of this problem. Therefore, the temperature of the gas striking the sample when the pressure in the tank has fallen to 10 atm is 226 K or −47°C.

1.18 ENTHALPIES OF FORMATION

Earlier we noted that internal energy is a point or state function. The same is true of enthalpy, because it is composed of terms (U, P, and V) that are each point functions. The enthalpy change of a material between states 1 and 2 is given as:

$$\Delta H = H_2 - H_1 \tag{1.31}$$

Neither internal energy nor enthalpy can be defined in absolute terms. There is no absolute zero of energy. Even at a temperature of absolute zero, a material may possess energy, depending on how we define it. We could consider the mass itself to be a form of energy. However, for convenience in tabulation, it is useful (and customary) to define a *reference* state for a material and to assign a value of zero to enthalpy for certain materials in that reference state. We take as such reference conditions the *elements* in their equilibrium states at 298 K and one atmosphere pressure. For example, the enthalpy of diatomic oxygen at 298 K and one atmosphere pressure is zero. The enthalpy of monatomic oxygen is not zero, because the monatomic form is not the equilibrium form at 298 K and one atmosphere.[8] As another

---

[8]Some very small amount of monatomic oxygen can exist at equilibrium at 298 K and one atmosphere pressure, as will be shown in Chapter 5. But the diatomic form of oxygen is predominant in nature.

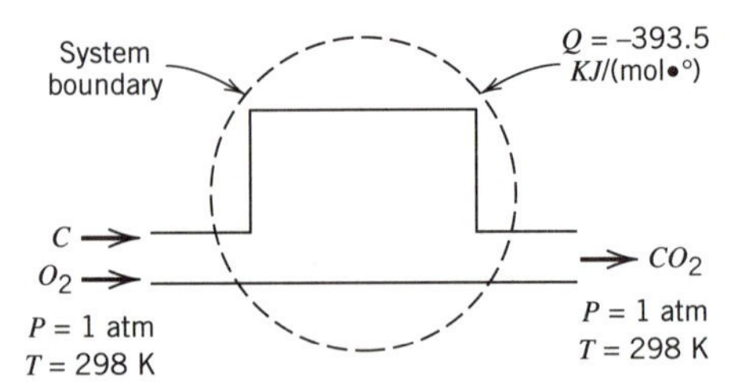

Figure 1.12 Combustion of carbon and oxygen.

example, the enthalpy of carbon as graphite (not diamond) at 298 K and one atmosphere pressure is taken as zero.

If the *elements* are assumed to have zero enthalpy when in their reference states, then *compounds* must have some other value of enthalpy at the reference conditions. To illustrate, consider the system in Figure 1.12 in which carbon (as graphite) and oxygen at a temperature of 298 K and a pressure of one atmosphere are introduced into a steady state system in which they will react to form carbon dioxide at 298 K and one atmosphere pressure. In terms of molar quantities, the First Law for the steady state system is

$$Q + W = \sum_{m_o} \underline{H}_o\, m_o - \sum_{m_i} \underline{H}_i\, m_i \tag{1.15}$$

Because no work is done:

$$Q = \sum_{m_o} \underline{H}_o\, m_o - \sum_{m_i} \underline{H}_i\, m_i$$

For the reaction that took place (i.e., $C + O_2 = CO_2$):

$$Q = \underline{H}_{CO_2}\, n_{CO_2} - \underline{H}_C\, n_C - \underline{H}_{O_2}\, n_{O_2} = \Delta H$$

If such an experiment were to be carried out, the heat transferred *from* the system would be 393.5 kJ per mole of carbon introduced into the system (or mole of carbon dioxide leaving the system). Applying this information to the First Law statement, we have

$$Q = -393.5 \text{ kJ} = \underline{H}_{CO_2} - \underline{H}_C - \underline{H}_{O_2}$$

But $\underline{H}_C = 0$, and $\underline{H}_{O_2} = 0$ because these elements are in their reference states. Therefore:

$$\underline{H}_{CO_2} = -393.5 \text{ kJ (at 298 K and } P = 1 \text{ atm)}$$

Thus the specific enthalpy of carbon dioxide, called the *heat of formation* of carbon dioxide (ΔH_f), is -393.5 kJ/mol at 298 K and one atmosphere pressure. Heats of formation for several compounds at 298 K are given in Table 1.3. Note that the heat of formation changes with temperature.

If the enthalpy change in a process is negative—that is, if heat is *evolved* from the change—the process is called *exothermic.* If the enthalpy change is *positive,* then heat must be *added* to the process, and it is called *endothermic.*

1.19 ENTHALPY CHANGES IN CHEMICAL REACTIONS

The enthalpy change for a chemical reaction may be calculated from the heats of formation of the compounds and elements involved in the reaction. As an example, let us calculate the enthalpy change (at 298 K) for the oxidation of methane (CH_4). The chemical equation describing the oxidation is:

$$CH_4 + 2O_2 \rightarrow CO_2 + 2H_2O(g)$$

The letter g is appended to the chemical symbol for water to indicate that the reaction product is water in the gaseous state.

The equation above can be written as the sum of three other equations:

$$\begin{array}{ll} \text{I. } C + O_2 \rightarrow CO_2 & \Delta H = \Delta \underline{H}_f\,(CO_2) \\ \text{II. } 2H_2 + O_2 \rightarrow 2H_2O(g) & \Delta H = 2\Delta \underline{H}_f\,(H_2O(g)) \\ \text{III. } C + 2H_2 \rightarrow CH_4 & \Delta H = \Delta \underline{H}_f\,(CH_4) \end{array}$$

As a result, the ΔH of reaction (at 298 K) in terms of the enthalpy changes of reactions I–III as follows (I + II − III):

$$\begin{aligned} \Delta H \text{ reaction} &= \Delta \underline{H}_{f(CO_2)} + 2\Delta \underline{H}_{f(H_2O(g))} - \Delta \underline{H}_{f(CH_4)} \\ &= -393.5 + 2(-241.8) - (-74.8) \\ &= -802.3 \text{ kJ} \qquad \text{at 298 K} \end{aligned}$$

In general, the enthalpy change for a chemical reaction at a specified temperature may be expressed as the sum of the enthalpies of formation of the products less the enthalpies of formation of the reactants at that temperature.

$$\Delta H_T = \sum_{\text{products}} n_p\, \Delta \underline{H}_{f,T} - \sum_{\text{reactants}} n_r\, \Delta \underline{H}_{f,T} \tag{1.32}$$

where the n's are the stoichiometric coefficients of the reaction.

The enthalpy change for methane when it is reacted with oxygen is called the enthalpy of combustion, or the heat of combustion of methane. Note that this value is called the *low* heat of combustion when the water produced is in the gaseous state. When the water produced is liquid, more heat is released in the process (the heat released upon liquefying the gaseous water), and the corresponding heat of combustion is higher, hence is referred to as the *high* heat of combustion.

Table 1.3 Enthalpies of Formation ($\Delta H^\circ_{f,298}$) and Standard Entropies (S°_{298}) for Selected Elements and Compounds

Material	ΔH°_{298} (kJ/mol)	S°_{298} [J/(mol · K)]
Ag	0	42.68
Ag_2O	−30.5	121.8
Al	0	28.33
Al_2O_3	−1674	51.0
Aa	0	47.4
C (graphite)	0	5.694
C (diamond)	+1.900	2.439
CO	−110.5	197.9
CO_2	−393.5	213.8
CH_4	−62.3	186.2
Ca	0	41.6
CaO	−635.6	39.8
$CaCO_3$	−1207	88.7
Cr	0	23.8
Cr_2O_3	−1130	81.2
Cu	0	33.35
Cu_2O	−167.4	93.9
CuO	−155.2	42.7
Fe	0	27.2
$Fe_{.95}O$	−265.7	59.4
Fe_2O_3	−824.3	90.0
Fe_3O_4	−1121.3	146.4
H_2	0	130.6
H_2O (l)	−285.9	70.1
H	+218.0	114.6
Li	0	28.0
LiCl	−405	58.1
Mg	0	32.5
MgO	−601.2	27.4
$MgCl_2$	−641.8	89.5
$MgCO_3$	−1096.2	65.7
Ni	0	29.8
NiO ($NiO_{1.005}$)	−240.6	38.5
Pb	0	64.9
PbO	−219.2	67.8
Si	0	18.83
SiO_2 (quartz)	−879.5	41.8

Source: Values from O. Kubaschewski and E. L. L. Evans, *Metallurgical Thermochemistry,* Wiley, New York, 1961, which lists values for many more substances, as well.

To calculate enthalpy changes for chemical reactions at temperatures other than 298 K (the reference temperature) requires some information about the temperature dependence of the enthalpies of the compounds involved (i.e., their constant pressure heat capacities).[9]

$$\left(\frac{\partial \underline{H}}{\partial T}\right) = C_P$$
$$d\underline{H} = C_P dT \tag{1.33}$$
$$\int_{298}^{T} d\underline{H} = \int_{298}^{T} C_P dT$$
$$\underline{H}_T - \underline{H}_{298} = \int_{298}^{T} C_P dT$$

If the heat capacity of the material is constant, Eq. 1.33 becomes:

$$\underline{H}_T - \underline{H}_{298} = C_P(T - 298)$$

If, however, the heat capacity varies with temperature (see Table 1.1), then C_p can be expressed using a polynomial expansion. Usually three terms in a polynomial are sufficient to describe the heat capacity with sufficient accuracy. Sometimes a fourth is used:

$$C_P = A + BT + CT^{-2} + DT^2$$

Then:

$$\underline{H}_T - \underline{H}_{298} = \int_{298}^{T} (A + BT + CT^{-2} + DT^2)dT$$

$$\underline{H}_T - \underline{H}_{298} = A\,(T - 298) + \frac{B}{2}\,(T^2 - 298^2) - C\left(\frac{1}{T} - \frac{1}{298}\right) + \frac{D}{3}\,(T^3 - 298^3)$$

[9]In some tables of thermochemical information, the JANAF tables for example (Ref. 3), the values of $\Delta \underline{H}_f^\circ$ for many compounds are given at various temperatures, in 100 K intervals starting at 0 K, as well as at 298 K. These values of $\Delta \underline{H}_f^\circ$ are based on assigning a value of zero to the heat of formation of each of the elements in a standard state, the equilibrium state, and one atmosphere at the temperature for which the value is provided. If the values of ΔH_f are known at the temperature of interest for all the compounds in a reaction, the ΔH for the overall reaction may be calculated using Eq. 1.32. Of course, the availability of these data considerably simplifies the calculation of ΔH at temperatures other than 298 K.

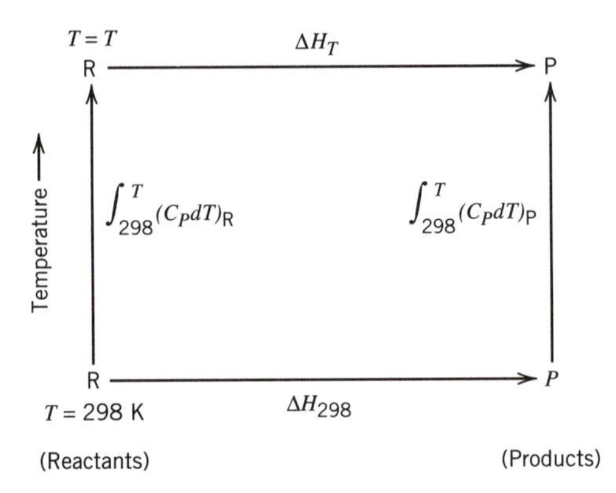

Figure 1.13 Change of enthalpy of reaction with temperature.

Because enthalpy is a point function, the enthalpy change for a reaction at temperature T (other than 298 K) is (Figure 1.13):

$$\int_{298}^{T} \Sigma(nC_P dT)_{\text{reactants}} + \Delta H_T = \Delta H_{298} + \int_{298}^{T} \Sigma(nC_P dt)_{\text{products}}$$

$$\Delta H_T = \Delta H_{298} + \int_{298}^{T} \left[\Sigma(nC_P)_{\text{products}} - \Sigma(nC_P)_{\text{reactants}} \right] dT \tag{1.34}$$

1.20 ADIABATIC TEMPERATURE CHANGE IN CHEMICAL REACTIONS

We can use the foregoing principles and tabulated information to calculate the temperature change when a chemical reaction is conducted under adiabatic conditions. The method can be used to calculate the highest temperature that can be achieved in a combustion reaction (adiabatic flame temperature). It can also be used to calculate the temperature that can be reached in polymerization reactions if no cooling is provided during the process.

As an example of an adiabatic flame temperature calculation, consider a burner in which a fuel gas at 298 K is burned with the stoichiometric amount of dry air (21% oxygen and 79% nitrogen), also at 298 K. The term "stoichiometric amount" means that just enough oxygen is used to complete the combustion. Ordinarily, a modest amount of air in excess of this quantity (10–20%) is used in most burners to ensure complete combustion. Incomplete combustion can be caused by improper mixing of the fuel and air. The reaction takes place at one atmosphere pressure. The composition of the fuel gas being burned is (by volume):

20% Carbon monoxide (CO)

30% Carbon dioxide (CO_2)

50% Nitrogen (N_2)

The nitrogen and carbon dioxide do not react. The combustion reaction for the carbon monoxide is:

$$CO + \tfrac{1}{2} O_2 \rightarrow CO_2$$

To determine the composition of the burner exit gas, it is useful to construct a table of the gases entering and leaving the burner. Note that nitrogen enters from two sources: the fuel gas and the air used for combustion. Take as the basis for the calculation one mole of fuel gas entering the burner. The stoichiometric amount of oxygen needed is 0.5 mol per mole of CO (i.e., 0.5 × 0.20 = 0.10 mol). The amount of nitrogen accompanying this oxygen is (79/21) × 0.10 mol. The nitrogen in the fuel gas is 0.50 mol.

Gas	Moles in	Moles out	Exit Gas Composition (%)
CO	0.20	0	
CO_2	0.30	0.50	36
O_2	0.10	0	
N_2	0.50 + (79/21) × 0.10	0.88	64
Total moles in exit gas		1.38	100

Take the burner as the system. The First Law for a steady state burner is:

$$Q + W = \Sigma \underline{H}_o \, n_o - \Sigma \underline{H}_i \, n_i$$

No work is done on or by the burner. Because we want to find the highest temperature that can be reached, we assume that the burner loses no heat to the surroundings; that is, it is adiabatic ($Q = 0$). Hence:

$$\Sigma \underline{H}_i \, n_i = \Sigma \underline{H}_o \, n_o$$

The enthalpies of the products leaving the burner are equal to the sum of the enthalpies of the reactants. We express the enthalpies thus:

$$\Sigma \underline{H}_o \, n_o = 0.5 \, \underline{H}_{CO_2,T} + 0.88 \, \underline{H}_{N_2,T} \qquad \textbf{(1.35a)}$$

$$\Sigma \underline{H}_i\, n_i = 0.2\, \underline{H}_{CO,298} + 0.3\, \underline{H}_{CO_2,298} + 0.1\, \underline{H}_{O_2,298} + 0.88\, \underline{H}_{N_2,298} \quad \textbf{(1.35b)}$$

The expressions for the individual enthalpy terms are:

$$\underline{H}_{O_2,298} = 0 \quad \text{and} \quad \underline{H}_{N_2,298} = 0 \quad \text{(reference state)}$$

$$\underline{H}_{N_2,T} = \int_{298}^{T} C_{P,N_2} dT$$

$$\underline{H}_{CO,298} = \Delta \underline{H}_{f,CO,298}$$

$$\underline{H}_{CO_2,298} = \Delta \underline{H}_{f,CO_2,298}$$

$$\underline{H}_{CO_2,T} = \Delta \underline{H}_{f,CO_2,298} + \int_{298}^{T} C_{P,CO_2} dT$$

Substituting in Eqs. 1.35a and 1.35b:

$$\Sigma \underline{H}_o\, n_o = 0.5\, \Delta \underline{H}_{f,CO_2,298} + 0.5 \int_{298}^{T} C_{P,CO_2} dT + 0.88 \int_{298}^{T} C_{P,N_2} dT$$

$$\Sigma \underline{H}_i\, n_i = 0.2\, \Delta \underline{H}_{f,CO,298} + 0.3\, \Delta \underline{H}_{f,CO_2,298}$$

Combining:

$$0.5 \int_{298}^{T} C_{P,CO_2} dT + 0.88 \int_{298}^{T} C_{P,N_2} dT = 0.2\, (\Delta \underline{H}_{f,CO,298} - \Delta \underline{H}_{f,CO_2,298})$$

$$0.5 \int_{298}^{T} C_{P,CO_2} dT + 0.88 \int_{298}^{T} C_{P,N_2} dT = 0.2\, (-110{,}500 + 393{,}500)$$

$$0.5 \int_{298}^{T} C_{P,CO_2} dT + 0.88 \int_{298}^{T} C_{P,N_2} dT = 56{,}600 \text{ J}$$

For the purpose of this example, assume that the heat capacities are constant at 34.3 J/(mol·K) for nitrogen, and 57.3 J/(mol·K) for carbon dioxide. Then:

$$[0.5(57.3) + 0.88(34.3)][T - 298] = 56{,}600$$

Solving

$$T = 1260 \text{ K}$$

Thus the maximum temperature of these gases leaving the burner, referred to as the adiabatic flame temperature (AFT), is 1260 K. This AFT is a function of many variables, including the temperature of the materials fed into the burner and the composition of the combustion gases. We assumed, for example, that the air used for combustion contained no moisture. If we had taken into account the moisture content, the water vapor associated with the combustion air would have had to be heated to the final flame temperature, and the AFT would have been lower. If the combustion air had been preheated to a temperature higher than 298 K, the enthalpy of the reactants would have been greater than zero, and the AFT would have been higher.

A word of caution should be added at this point. The calculation above assumed implicitly that the combustion reaction has essentially proceeded to completion. This may not be the case at higher calculated combustion temperatures. For example, the calculated adiabatic flame temperature of methane (CH_4) burned with pure oxygen was calculated assuming that a complete combustion reaction to carbon dioxide and gaseous water would be in excess of 5000 K. Based on our knowledge of chemical reaction equilibria, we know that the combustion reaction would not proceed to completion at this temperature, and the actual adiabatic flame temperature would thus be lower. Chapter 5 discusses a method for handling such cases.

In the case of combustion reactions, we are usually trying to achieve a high flame temperature and are looking for the conditions that will maximize the AFT. In some other reactions we are concerned that the temperature may rise dangerously high, as happens in some polymerization reactions. Take as an example the polymerization of methyl methacrylate. We can represent the reactions as:

$$\mathrm{M(l)} = \frac{1}{n} P_n$$

where M(l) represents the liquid monomer and P_n represents the polymer with average degree of polymerization *n,* in the amorphous form.

The enthalpy of polymerization $\Delta \underline{H}^\circ_\mathrm{p}$ is the difference in enthalpy between the polymer and the monomers that formed it:

$$\Delta \underline{H}^\circ_\mathrm{p} = \frac{1}{n} \Delta \underline{H}^\circ_\mathrm{f}\,(P_n) - \Delta \underline{H}^\circ_\mathrm{f}\,(\mathrm{M})$$

In the case of methyl methacrylate, the value of $\Delta \underline{H}^\circ_\mathrm{p}$ is -56.1 kJ/mol. For the purpose of our calculation, let us take an extreme case, one in which one mole of monomer is completely polymerized under adiabatic conditions, starting at 298 K. As in the preceding example,

$$\Sigma \underline{H}_\mathrm{i}\, n_\mathrm{i} = \Sigma \underline{H}_\mathrm{o}\, n_\mathrm{o}$$

$$\Delta \underline{H}^\circ_\mathrm{f}\,(\mathrm{M}) = \frac{1}{n} \Delta \underline{H}^\circ_\mathrm{f}\,(P_n) + C_p\,(T - 298)$$

The heat capacity is approximately 2 J/(g·K), which is about 200 J/(mole·K) (molecular weight = 100 g per mole of monomer). Solving:

$$C_p\,(T - 298) = \Delta \underline{H}^\circ_{\rm f}\,({\rm M}) - \frac{1}{n}\,\Delta \underline{H}^\circ_{\rm f}\,(P_n) = -\Delta \underline{H}_{\rm p}$$

$$T = \frac{56{,}100}{200} + 298 = 578 \text{ K or } 305°\text{C}$$

To heat a polymer to this temperature is dangerous. In a practical sense, therefore, some cooling will have to be provided during the polymerization. The word of caution introduced after the discussion of adiabatic flame temperature applies in this case, too. At 305°C, the polymerization reaction will not proceed. To determine an actual maximum temperature, which may also be a dangerous one, we will have to consider the chemical equilibria involved. This is done in Chapter 5.

REFERENCES

1. Lewis, G. N., and Randall, M., *Thermodynamics,* 2nd rev. ed., K. S. Pitzer and L. Brewer, McGraw-Hill, New York, 1961.
2. Balzhiser, R. E., Samuels, Michael R., and Eliassen, J. D., *Chemical Engineering Thermodynamics,* Prentice-Hall, Englewood Cliffs, NJ, 1972.
3. Stull, D. R., and Prophet, H., *JANAF Thermochemical Tables,* 2nd ed., National Standards Research Data Service–National Bureau of Standards, Washington, DC, 1971.

PROBLEMS

1.1 A lead bullet is fired at a rigid surface. At what speed must it travel to melt on impact, if its initial temperature is 25°C and heating of the rigid surface is neglected? The melting point of lead is 327°C. The molar heat of fusion of lead is 4.8 kJ/mol. The molar heat capacity C_P of lead may be taken as 29.3 J/(mol·K).

1.2 What is the average power production in watts of a person who burns 2500 kcal of food in a day? Estimate the average additional power production of a 75 kg man who is climbing a mountain at the rate of 20 m/min.

1.3 One cubic decimeter (1 dm^3) of water is broken into droplets having a diameter of one micrometer (1 μm) at 20°C.

(a) What is the total area of the droplets?

(b) Calculate the minimum work required to produce the droplets. Assume that the droplets are at rest (i.e., have zero velocity).

Water has a surface tension of 72.75 dyn/cm at 20°C.

Note: The term *surface energy* (erg/cm^2) is also used for *surface tension* (dyn/cm)

1.4 Gaseous helium is to be used to quench a hot piece of metal. The helium is in storage in an insulated tank with a volume of 50 L and a temperature of 25°C. The pressure is 20 atm. Assume that helium is an ideal gas.

(a) When the valve is opened and the gas escapes into the quench chamber (pressure = 1 atm), what will be the temperature of the first gas to hit the specimen?

(b) As the helium flows, the pressure in the tank drops. What will be the temperature of the helium entering the quench chamber when the pressure in the tank has fallen to 20 atm?

1.5 An evacuated ($P = 0$), insulated tank is surrounded by a very large volume (assume infinite volume) of an ideal gas at a temperature T_0. The valve on the tank is opened and the surrounding gas is allowed to flow quickly into the tank until the pressure inside the tank equals the pressure outside. Assume that no heat flow takes place. What is the final temperature of the gas in the tank?

The heat capacities of the gas, C_P and C_V, each may be assumed to be constant over the temperature range spanned by the experiment. Your answer may be left in terms of C_P and C_V.

Hint: One way to approach the problem is to define the system as the gas that ends up in the tank.

1.6 Calculate the heat of reaction of methane (CH_4) with oxygen at 298 K, assuming that the products of reaction are CO_2 and H_2O (gas). [This heat of reaction is also called the low calorific power of methane].

Convert the answer into units of Btu/1000 SCF of methane. SCF means standard cubic feet, taken at 298 K and 1 atm.

Note: This value is a good approximation for the low calorific power of natural gas.

DATA

For	ΔH°_{298} [kcal/(g·mol)]
CH_4 (g)	−17.89
CO_2 (g)	−94.05
H_2O (g)	−57.80

1.7 Methane is delivered at 298 K to a glass factory, which operates a melting furnace at 1600 K. The fuel is mixed with a quantity of air, also at 298 K, which is 10% in excess of the amount theoretically needed for complete combustion. (Air is approximately 21% O_2 and 79% N_2).

(a) Assuming complete combustion, what is the composition of the flue gas (the gas following combustion)?

(b) What is the temperature of the gas, assuming no heat loss?

(c) The furnace processes 2000 kg of glass hourly, and its heat losses to the surroundings average 400,000 kJ/h. Calculate the fuel consumption at STP (in m³/h) assuming that for the glass $\underline{H}^\circ_{1600} - \underline{H}^\circ_{298} = 1200$ kJ/kg.

(d) A heat exchanger is installed to transfer some of the sensible heat of the flue gas to the combustion air. Calculate the decrease in fuel consumption if the combustion air is heated to 800 K.

DATA

STP means $T = 298$ K and $P = 1.0$ atm.

For	C_p [cal/(mol·°C)]
CH_4	16
CO_2	13.7
H_2O (g)	11.9
N_2	8.2
O_2	8.2

1.8 In an investigation of the thermodynamic properties of α-manganese, the following heat contents were determined:

$$\underline{H}^\circ_{700} - \underline{H}^\circ_{298} = 12{,}113 \text{ J/(g·atom)}$$

$$\underline{H}^\circ_{1000} - \underline{H}^\circ_{298} = 22{,}803 \text{ J/(g·atom)}$$

Find a suitable equation for $H^\circ_T - H^\circ_{298}$ and also for C_P as a function of temperature in the form $(a + bT)$. Assume that no structural transformation takes place in the given temperature range.

1.9 A fuel gas containing 40% CO, 10% CO_2, and the rest N_2 (by volume) is burnt completely with air in a furnace. The incoming and ongoing temperatures of the gases in the furnace are 500°C (773 K) and 977°C (1250 K), respectively. Calculate (a) the maximum flame temperature and (b) heat supplied to the furnace per cu. ft of exhaust gas.

DATA

$$\Delta\underline{H}^\circ_{f,298,CO} = -110{,}458 \text{ J/mol}$$

$$\Delta\underline{H}^\circ_{f,298,CO_2} = -393{,}296 \text{ J/mol}$$

$$C_{P,CO} = 28.45 + 3.97 \times 10^{-3}\, T - 0.42 \times 10^5\, T^{-2} \text{ (J/mol·K)}$$

$$C_{P,CO_2} = 44.35 + 9.20 \times 10^{-3}\, T - 8.37 \times 10^5\, T^{-2} \text{ (J/mol·K)}$$

$$C_{P,O_2} = 29.92 + 4.10 \times 10^{-3}\, T - 1.67 \times 10^5\, T^{-2} \text{ (J/mol·K)}$$

$$C_{P,N_2} = 29.03 + 4.184 \times 10^{-3}\, T \text{ (J/mol·K)}$$

1.10 **(a)** For the reaction:

$$CO + \tfrac{1}{2}\, O_2 \rightarrow CO_2$$

What is the enthalpy of reaction (ΔH°) at 298 K?

(b) A fuel gas, with composition 50% CO, 50% N_2 is burned using the stoichiometric amount of air. What is the composition of the flue gas?

(c) If the fuel gas and the air enter the burner at 298 K, what is the highest temperature the flame may attain (adiabatic flame temperature)?

DATA

Standard heats of formation ΔH_f at 298 K

$$CO = -110{,}000 \quad (J/mol)$$
$$CO_2 = -393{,}000 \ (J/mol)$$

Heat capacities [J/(mol·K)] to be used for this problem.

$$N_2 = 33$$
$$O_2 = 33$$
$$CO = 34$$
$$CO_2 = 57$$

1.11 A particular blast furnace gas has the following composition (by volume):

CO	12%
CO_2	24%
H_2	4%
N_2	60%

(a) If the gas at 298 K is burned with the stoichiometric amount of dry air at 298 K, what is the composition of the flue gas? What is the adiabatic flame temperature?

(b) Repeat the calculations for 30% excess combustion air at 298 K.

(c) What is the adiabatic flame temperature when the blast furnace gas is preheated to 700 K? (The dry air is at 298 K.)

(d) Suppose the combustion air is not dry (i.e., has a partial pressure of water of 15 mm Hg and a total pressure of 760 mm Hg). How will the flame temperature be affected?

DATA (kJ/mol)

For	**ΔH_f (kJ/mol)**
CO	−110.523
CO_2	−393.513

For	**C_P [J/(mol·k)]**
CO	33
CO_2	57
H_2O (g)	50
N_2, O_2	34

1.12 A bath of molten copper is supercooled to 5°C below its true melting point. Nucleation of solid copper then takes place, and the solidification proceeds under adiabatic conditions. What percentage of the bath solidifies?

DATA

Heat of fusion for copper is 3100 cal/mol al 1083°C (the melting point of copper)

C_P (copper, liquid) = 7.5 cal/(mol·°C)

C_P (copper, solid) = $5.41 + (1.5 \times 10^{-3}\ T)$ cal/(mol·°C)

1.13 Cuprous oxide (Cu_2O) is being reduced by hydrogen in a furnace at 1000 K.

(a) Write the chemical reaction for the reduction of one mole of Cu_2O.

(b) How much heat is released or absorbed per mole reacted? Give the quantity of heat and state whether heat is evolved (exothermic reaction) or absorbed (endothermic reaction).

DATA

Heats of formation of 1000 K in cal/mol

Cu_2O = −41,900

H_2O = −59,210

1.14 (a) What is the enthalpy of pure, liquid aluminum at 1000 K? (Use pure, solid aluminum at 298 K as the reference state.)

(b) An electric resistance furnace is used to melt pure aluminum at the rate of 100 kg/h. The furnace is fed with solid aluminum at 298 K. The liquid aluminum leaves the furnace at 1000 K. What is the minimum electric power rating (kW) of the furnace?

DATA

For aluminum:

Atomic weight = 27 g/mol

Heat capacity of solid (C_P) = 26 J/(mol·K)

Heat capacity of liquid (C_P) = 29 J/(mol·K)

Melting point = 932 K

Heat of fusion ($\Delta \underline{H}_m$) = 10,700 J/mol

1.15 A waste material (dross from the melting of aluminum) is found to contain 1 wt % metallic aluminum. The rest may be assumed to be aluminum oxide (Al_2O_3). The aluminum is finely divided and dispersed in the Al_2O_3; that is, the two materials are thermally connected.

If the waste material is stored at 298 K, what is the maximum temperature to which it may rise if all the metallic aluminum is oxidized by air? The entire mass may be assumed to rise to the same temperature.

DATA

Atomic weights: Al = 27 g/mol

O = 16 g/mol

Heat capacity of solid aluminum = 26 J/(mol·K)
Heat capacity of solid Al_2O_3 = 104 J/(mol·K)

For Al_2O_3:
Heat of formation ($\Delta \underline{H}_{f,298}$) = −1,676,000 J/mol

1.16 Metals exhibit some interesting properties when they are rapidly solidified from the liquid state. An apparatus for the rapid solidification of copper is cooled by water.

In the apparatus, liquid copper at its melting point (1356 K) is sprayed on a cooling surface, where it solidifies and cools to 400 K. The copper is supplied to the apparatus at the rate of one kilogram per minute.

Cooling water is available at 20°C, and is not allowed to rise above 80°C. What is the minimum flow rate of water in the apparatus, in cubic meters per minute?

DATA

For water:	C_P = 4.184 J/g·K
	Density = 1 g/cm^3
For copper:	molecular weight = 63.54 g/mol
	C_P (solid) = 7 cal/(mol·K)
	Heat of fusion = 3120 cal/mol

Clearly state the system and basis for your calculation.

1.17 Water flowing through an insulated pipe at the rate of 5 L/min is to be heated from 20°C to 60°C by an electrical resistance heater.

Calculate the minimum power rating of the resistance heater in watts. Specify the system and the basis for your calculation.

DATA

For water:	Heat capacity = 4.184 J/(g·K)
	Density = 1 g/cm^3

1.18 The heat of evaporation of water at 100°C and 1 atm is 2261 J/g.

(a) What percentage of that energy is used as work done by the vapor?
(b) If the density of water vapor at 100°C and 1 atm pressure is 0.597 kg/m^3, what is the internal energy change for the evaporation of water?

1.19 What is the minimum amount of steam (at 100°C and 1 atm pressure) required to melt a kilogram of ice (at 0°C)? Use data for Problem 1.20.

1.20 In certain parts of the world pressurized water from beneath the surface of the earth is available as a source of thermal energy. To make steam, the "geothermal water" at 180°C is passed through a "flash evaporator" that operates at 1 atm pressure. Two streams come out of the evaporator, liquid water and water vapor. How much water vapor is formed per kilogram of geothermal water? Is the process reversible? Assume that water is incompressible

The vapor pressure of water at 180°C is 1.0021 MPa (about 10 atm).

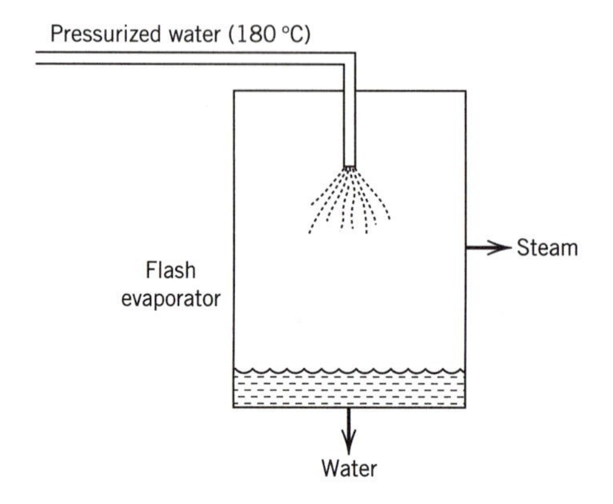

DATA

For Problems 1.19 and 1.20

C_P (water, liquid) $= 4.18$ J/(g·K)

C_P (water, vapor) $= 2.00$ J/(g·K)

$\Delta \underline{H}_{\text{vap}} = 2261$ J/g at 100°C

$\Delta \underline{H}_{\text{m}} = 334$ J/g at 0°C

Chapter 2

The Second Law

Chapter 1 dealt with the first law of thermodynamics, the conservation of energy and energy transfer in terms of heat and work. Critical to the discussion of the First Law is the postulate that energy is a state function. That is, the change in the energy of a system depends on the final and initial states of the system, not on the path between the two. This chapter deals with another thermodynamic property, *entropy,* which is also a state function.

The origins of the entropy function lie in the study of heat engines, devices that convert thermal energy (heat) into mechanical energy (work). The first law of thermodynamics accounts for the energies involved in such a conversion but places no limits on the amounts that can be converted. The second law is concerned with limits on the conversion of "heat" into "work" by heat engines. The principles governing heat engines were investigated as early as 1824 by a French engineer, Sadi Carnot. Through a consideration of an idealized heat engine operating between two heat reservoirs, Carnot found that not all the heat removed from a high temperature reservoir is converted into work. In fact, as discussed in Section 2.5, the amount that can be converted is governed by the temperatures of the two reservoirs.

The conclusion that Carnot reached can be derived using the entropy function, defined in Section 2.1.

2.1 ENTROPY AS A STATE FUNCTION

It is certainly not obvious, but it can be shown, that for a closed system the reversible heat flow divided by the absolute temperature of the system is a state or point function. In mathematical terms:

$$\frac{\delta Q_{\text{rev}}}{T} = dS \tag{2.1}$$

The proof that this property, S, is indeed a state function is beyond the scope of this text. It is presented as the theorem of Carathéodory in a classical text on mathematical physics (Ref. 1). The function S, called entropy, is a useful and powerful concept in the science of thermodynamics. The units of entropy are energy units divided by temperature: joules per kelvin in the SI system.

This chapter relates changes in entropy to transfers of heat and to irreversibilities in a macroscopic sense. We deal with the macroscopic properties of a system without reference to atomic or molecular structures or movements. In Chapter 10, on statistical thermodynamics, we consider entropy as a measure of the randomness of a system in terms of the atomic and molecular structure.

To explore some aspects of the entropy function, consider a closed system that undergoes a change in internal energy, dU. From the First Law:

$$dU = \delta Q + \delta W$$

Remembering that the change in the internal energy of the system is independent of path, we may choose a reversible path between the same starting and ending points U and $U + dU$, so that:

$$dU = \delta Q_{\text{rev}} + \delta W_{\text{rev}}$$

Because U is independent of the path, the two values of dU are equal. Hence we can write:

$$dU = \delta Q_{\text{rev}} + \delta W_{\text{rev}} = \delta Q + \delta W$$

$$\delta Q_{\text{rev}} = \delta Q - \delta W_{\text{rev}} + \delta W$$

The combination of the work terms $(-\delta W_{\text{rev}} + \delta W)$ is called lost work and will be noted as $-\delta\text{lw}$. It is the difference between the work that could have been done had the process been reversible and the actual work that was done. It is a measure of the irreversibilities involved in the transfers of energy between the system and surroundings.[1]

$$\delta Q_{\text{rev}} = \delta Q - \delta\text{lw} \tag{2.2}$$

[1]An exploration of the nature and interrelation of these irreversibilities is covered in a study called, appropriately, irreversible thermodynamics. A detailed treatment of the subject is beyond the scope of this text, but can be found in textbooks on the subject (Refs. 2 and 3).

One type of lost work is illustrated in Figure 2.1. A closed system contains a gas originally at P_1 and V_1. If the gas is expanded reversibly—that is, against a force that is always balanced against the pressure of the gas in the system—the work done by the system is the negative of the integral under the *P-V* curve followed by the system, as shown in Figure 2.1*a*. It is negative because the system is doing work on the surroundings. But, if the resisting force (or pressure) during the expansion is *constant,* and equal to the final pressure in the system, the expansion is irreversible. The work done is just $-P_2(V_2 - V_1)$. The difference between the *reversible* work that *could have been done* and the *actual* work that *was* done is the lost work (Figure 2.1*b*).

Combining Eqs. 2.2 and 2.1 yields an expression for the change in entropy of a system:

$$dS = \frac{\delta Q_{(\text{actual})}}{T} - \frac{\delta \text{lw}}{T}$$

The use of the lost work concept can be illustrated by a simple example. Consider the isothermal expansion of one mole of an ideal gas from a volume, V, to double the volume, $2V$. This expansion will be conducted two ways. First, the expansion will be conducted reversibly: that is, the resisting pressure will be equal to the pressure of the gas. Second, the gas will be expanded against no pressure under adiabatic conditions (no heat flow); this can be done by enclosing the gas in a cylinder of volume $2V$, but with the gas originally in a section of the cylinder of volume V (i.e., bounded by a membrane on one side). There is a vacuum on the other side of the membrane. The expansion is conducted by piercing the membrane and allowing the gas to fill the entire volume, $2V$. In this case there is no work done by the gas because

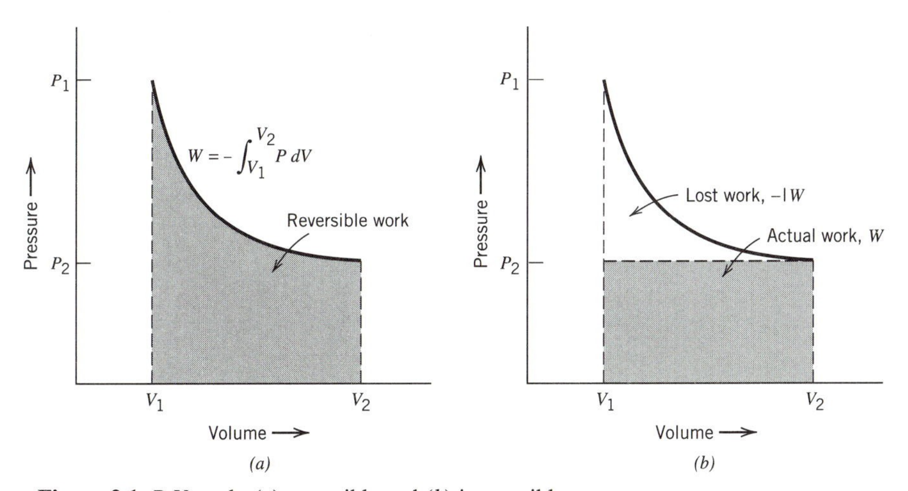

Figure 2.1 *P-V* work: (*a*) reversible and (*b*) irreversible.

the resisting pressure (the vacuum) is zero. To analyze the first case, we write the First Law,

$$\Delta U = Q + W$$

The change in internal energy, ΔU, is zero because the internal energy of an ideal gas is only a function of temperature (Eq. 1.21), and the expansion was isothermal. The work term can be evaluated as follows:

$$W = -\int_{V}^{2V} P\,dV = -\int_{V}^{2V} \left(\frac{RT}{V}\right) dV \qquad \text{ideal gas}$$

$$W = -RT \ln \left(\frac{2V}{V}\right) = -RT \ln 2$$

Considering the First Law ($Q + W = 0$), the amount of heat added to the system must have been $+RT \ln 2$, and that is Q_{rev} by the condition of the problem. The entropy change of the gas is thus:

$$\Delta \underline{S} = \frac{Q_{\text{rev}}}{T} = \frac{RT \ln 2}{T} = R \ln 2 \tag{2.3}$$

Now consider the second case, expansion into a vacuum under adiabatic conditions. The heat term is zero because the expansion is adiabatic. The work term is zero because there is no resisting pressure. Hence the change in internal energy (ΔU) is zero. Since, however, the internal energy for an ideal gas is a function of temperature alone, the initial temperature and the final temperature are identical. The final condition of the gas in the second case is same as the final condition of the gas in the first case. We may calculate the entropy change of the gas using Eq. 2.3. The heat transferred in the second case is zero, hence:

$$\Delta \underline{S} = -\frac{\text{lw}}{T}$$

But the lost work is the work that could have been done if the process had been conducted along a reversible path (i.e., $-RT \ln 2$). Hence:

$$\Delta \underline{S} = -\left(\frac{-RT \ln 2}{T}\right) = R \ln 2$$

Thus, the calculated entropy change of the gas is the same in both cases, as it should be, because the beginning and final states of the gas are the same in the two cases.

2.2 ENTROPY NOT CONSERVED

One of the unusual, and sometimes confusing, characteristics of the entropy function is that, in contrast to the energy function (U), entropy is *not* conserved in natural processes. To illustrate, consider the physical situation in Figure 2.2. The bodies I and II are at two different temperatures, T_I and T_{II}, and are connected by a thermal conductor, such as a bar of copper. The system is taken to be at steady state. The temperatures at various points in the copper bar do not change. If T_I is greater than T_{II}, heat will flow from I to II. Assuming no extraneous heat losses, all the heat that flows from I ends up in II; that is, $Q_I = -Q_{II}$.

Taking the whole apparatus (bodies I and II and the connector) as the system, the change in internal energy is zero. Energy is conserved.

$$\Delta U_{I+II} = \Delta U_I + \Delta U_{II} = Q_I + Q_{II} = 0$$

Consider now the change in entropy of this total system. The process of heat transfer at the boundaries of reservoirs I and II is considered to be reversible. That is, the temperature of the copper bar immediately adjacent to a reservoir is essentially at the temperature of the reservoir.

$$\Delta S_{I+II} = \Delta S_I + \Delta S_{II}$$

$$\Delta S_I = \frac{Q_I}{T_I} \qquad \text{and} \qquad \Delta S_{II} = \frac{Q_{II}}{T_{II}}$$

$$\Delta S_{I+II} = \frac{Q_I}{T_I} + \frac{Q_{II}}{T_{II}}$$

But

$$Q_{II} = -Q_I$$

$$\Delta S_{I+II} = -\frac{Q_{II}}{T_I} + \frac{Q_{II}}{T_{II}}$$

$$\Delta S_{I+II} = Q_{II}\left[\frac{T_I - T_{II}}{T_I\, T_{II}}\right]$$

The conclusion is, thus, that the entropy of this isolated system is not conserved. The heat flow was irreversible because the two temperatures, T_I and T_{II}, differed by

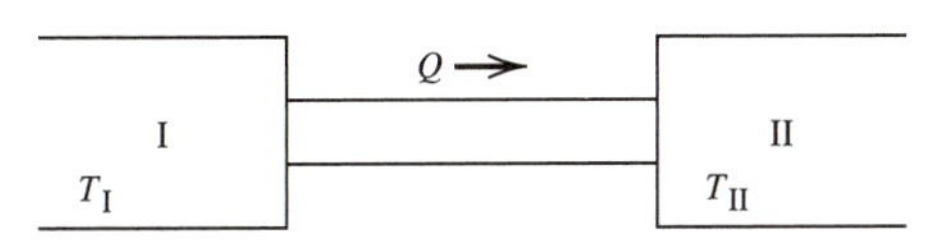

Figure 2.2 Heat conduction from T_I to T_{II}.

more than an infinitesimal amount. That irreversibility gave rise to an entropy increase. Because heat flows spontaneously only from higher to lower temperatures, it should be apparent that the entropy of this isolated system can never decrease; it can only increase or remain constant.

2.3 OPEN SYSTEM ENTROPY BALANCE

To state the Second Law for an open system, some terms must be added to account for the entropy of material entering or leaving the system:

$$\sum_{m_i} \underline{S}_i \, \delta m_i - \sum_{m_o} \underline{S}_o \, \delta m_o + \frac{\delta Q}{T} - \frac{\delta \text{lw}}{T} = dS \tag{2.4}$$

where the term $\Sigma_{m_i} \underline{S}_i \, \delta m_i$ represents the entropy of all the material entering the system. Each term in the sum is the product of the specific entropy of one of the materials multiplied by the mass of that material entering. Similarly, the term $\Sigma_{m_o} \underline{S}_o \, \delta m_o$ represents the entropy of all the material leaving the system.

2.4 ADIABATIC, REVERSIBLE, STEADY STATE SYSTEM

An illustration of the usefulness of the entropy function can be obtained by considering the flow of a material through an *adiabatic, steady state* machine in a *reversible* manner. In the statement of second law (Eq. 2.4):

(i) δQ is zero because the process is *adiabatic,*
(ii) δlw is zero because the process is *reversible,* and
(iii) dS is zero because the properties and condition of the machine do not change with time.

The last point, labeled iii, is sometimes difficult to understand. Remember that entropy is a state function. The entropy of each part of the machine (and its contents) is the product of the specific entropy of the material and its mass. At steady state, all the quantities that define the system (e.g., temperature and pressure) do not change with time. Hence the specific entropy does not change with time. The mass of each part of the system does not change with time at steady state. Hence, the entropy (the product of specific entropy and mass) of each part does not change. The total entropy, the sum of the entropies of the parts, also does not change. Hence, $dS = 0$.

The result of all of this is:

$$\sum_{m_i} \underline{S}_i \, \delta m_i = \sum_{m_o} \underline{S}_o \, \delta m_o \tag{2.5a}$$

or for one fluid

$$\underline{S}_i = \underline{S}_o \qquad \text{because} \quad \delta m_i = \delta m_o \tag{2.5b}$$

Thus the entropy of a fluid passing through an adiabatic, reversible (and steady state) machine is constant. The process is termed *isentropic*. We can, using the method discussed in Chapter 3, calculate the pressure–temperature relationship of a fluid as it passes through a process of this type. Such a baseline calculation provides the standard against which real expansions are compared to calculate efficiencies. The pressure–temperature relationship for an ideal gas in such an expression was calculated in Section 1.17.

2.5 HEAT ENGINES

It was shown in Section 2.2 that if two heat reservoirs, one at a high temperature, T_{I}, and one at a low temperature, T_{II}, are connected thermally so that heat flows from T_{I} to T_{II}, the net entropy change of the universe is positive. To explain, the system, consisting of the two reservoirs and their connection, was assumed to be isolated from the surroundings. The system showed an entropy increase, and the surroundings were unaffected. The universe, the sum of the system and surroundings, therefore underwent an entropy increase. Of course, the two heat reservoirs need not be connected in such an irreversible manner. The Second Law allows us to calculate how much work could be done if the irreversibilities were eliminated.

Suppose that a machine is placed between the high temperature (T_{H}) and low temperature (T_{L}) reservoirs, and that the machine (called a heat engine) is able to absorb heat reversibly from T_{H} and to discharge heat reversibly to T_{L}. In the process, the machine would do some work. Figure 2.3 schematically shows such a machine. Because we are not interested in the changes it undergoes, the machine will be assumed to be a steady state device ($\Delta U = 0$; $\Delta S = 0$). Because we want to calculate the maximum amount of work that can be done, it will also be assumed to be a reversible machine (lw $= 0$). Any irreversibilities would reduce its work output.

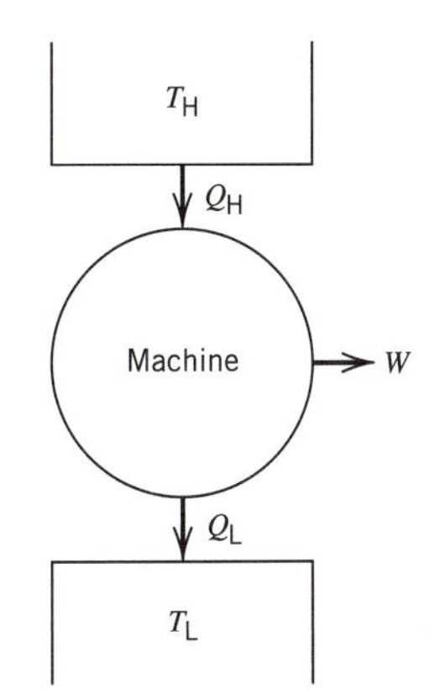

Figure 2.3 Heat engine (machine) operating between T_{H} and T_{L}.

Take the machine as the system. Because it is a closed system, the First Law for the reversible steady state machine can be written:

$$\Sigma\, \delta Q + \delta W_{rev} = 0$$
$$\delta Q_H + \delta Q_L + \delta W_{rev} = 0 \tag{2.6}$$

Because the machine is a closed system ($\delta m_i = \delta m_o = 0$), steady state ($dS = 0$), and reversible ($\delta lw = 0$), the Second Law (Eq. 2.4) is reduced to:

$$\Sigma \frac{\delta Q}{T} = 0$$
$$\frac{\delta Q_H}{T_H} + \frac{\delta Q_L}{T_L} = 0 \tag{2.7}$$
$$\delta Q_L = -\frac{T_L}{T_H}\,\delta Q_H$$

Combining with Eq. 2.6:

$$-\delta W_{rev} = \frac{\delta Q_H}{T_H}\,[T_H - T_L] \tag{2.8}$$

The efficiency of conversion of heat into work in this heat engine (HE) is defined[2] as the work done (δW_{rev}) divided by the quantity of energy that flowed as heat out of the high temperature source (δQ_H):

$$\eta_{HE} = \frac{-\delta W_{rev}}{\delta Q_H} = \frac{T_H - T_L}{T_H} \tag{2.9}$$

The conclusion is thus that the maximum amount of work δW_{rev} that can be obtained from a quantity of heat δQ_H extracted from a high temperature (T_H) source depends on the temperature of the low temperature reservoir (T_L) to which heat may be discharged. It is impossible to have a low temperature heat sink at absolute zero of temperature. Therefore some heat must always be discharged to the low temperature reservoir. This illustrates Lord Kelvin's statement of the Second Law: ''It is impos-

---

[2]Equations 2.6–2.9 were derived for an infinitesimal quantity of heat, δQ_H, recovered from a high temperature source. The same equations apply to finite quantities of heat when the heat sources and sinks at T_H and T_L may be considered to be infinite (i.e., when their temperatures do not change when heat is removed or added).

sible to perform a process whose sole result is the conversion of heat into an equivalent amount of work.''

As an example, let us calculate the maximum efficiency of conversion in a heat engine that operates between the boiling point of water 100°C (373.15 K) and room temperature, taken to be 20°C (293.15 K).

$$\eta_{\max} = \frac{T_{\mathrm{H}} - T_{\mathrm{L}}}{T_{\mathrm{H}}} = 1 - \frac{T_{\mathrm{L}}}{T_{\mathrm{H}}}$$

$$\eta_{\max} = 1 - \frac{293.15}{373.15} = \frac{80}{373.15} = 0.214$$

$$\eta_{\max} = 21.4\%$$

The efficiency of conversion of heat into work, η, is the work output of the machine divided by the heat input. The efficiency cannot be 100% because T_{L} cannot be zero. It should also be apparent that the maximum efficiency of conversion of heat into work depends on the temperature at which the heat is supplied, T_{H}. The higher the temperature, the higher the possible conversion efficiency. This relationship is the driving force behind a great deal of the research and development in the field of high temperature materials. By selecting fuels and oxidation conditions, very high gas (flame) temperatures may be attained, as discussed in Section 1.20. If the energy in these gases is to be converted into work (e.g., in the form of electricity) the gas must be passed through a heat engine, such as a turbine. The hot gases will have to expand against the blades of the turbine. The maximum temperature permissible in this heat engine is determined by the ability of the blade materials to withstand the stresses in the spinning turbine at that temperature. As we have seen (Eq. 2.9), higher temperatures mean higher possible efficiency. The promise of higher efficiency of conversion has stimulated many years of fruitful research and development in materials that retain useful mechanical properties at high temperatures, such as high alloy metallic materials, and is now stimulating similar work in ceramics.

It is important to note that the limitation imposed by the Second Law applies only to the conversion of heat into an equivalent amount of work, not to other conversion processes. For example, there is no limit on the efficiency of conversion between potential energy and work. All the potential energy change of a mass that is lowered from a height z_2 to z_1 $[mg(z_2 - z_1)]$ may be converted into work. All electrical work done on a system—for example, by having a current i flow through a potential V for a time t—may be converted into mechanical work equivalent to the product, $V \times i \times t$, by a 100% efficient electric motor. There is also no limit on the amount of work that can be converted into heat. All the electrical work described above could have been converted into heat by passing the current through a resistor. *The Second Law limitation on efficiency of conversion applies only to the conversion of heat into work.*

2.6 DIAGRAMMATIC REPRESENTATION

Energy conversion processes can be usefully displayed on a set of axes in which an intensive variable is displayed on the vertical axis and the corresponding (conjugate) extensive variable is displayed on the horizontal axis. For example, in the case of potential energy, the height multiplied by the acceleration of gravity would be the intensive variable, and *m* (mass) the extensive variable. For expansions and contractions of a fluid, the pressure is the intensive variable and volume the extensive. From Figure 2.4 it can be seen that the work that can be obtained by moving from point 1 to point 2 is $P_1(V_2 - V_1)$. No work is done on the 2–3 path (no volume change). The work done on the system to move it from point 3 to point 4 is $P_3(V_4 - V_3)$. Again no work is done from 4 to 1. The net work done by the fluid as it traverses the cycle from 1 to 4 and back to 1 is $(P_1 - P_3)(V_2 - V_1)$, that is, the cyclic integral of $P\,dV$, $\oint P\,dV$.

The same approach can be used for thermal energy, where T is the intensive variable of thermal energy and S is the extensive variable. Figure 2.5 shows the temperature–entropy plot of the working fluid in a thermodynamic machine such as the one discussed in Section 2.5. The ''working'' fluid circulates in the heat engine, absorbing heat from the high temperature reservoir, passing through a mechanical device to convert some of its thermal energy into work, and then discharging heat to the low temperature reservoir. The machine could be a very much simplified Rankine cycle device (Figure 2.6) in which water is boiled at a high temperature (T_H), as represented by path 1–2. The entropy of gaseous water (steam) is higher than that of liquid water. The steam is then passed through an adiabatic and reversible expander such as a turbine. The entropy change upon passing through the turbine is zero (see Section 2.4, Eq. 2.5a) Thus, path 2–3 is isentropic. The path from point 3 to point 4 represents heat removal in a condenser, where the low temperature steam is condensed into liquid water. The water is then pumped, isentropically, from 4 to 1, to return it to the boiler. The pump is run by part of the work output of the turbine. (Note that this is a simplified version of the actual Rankine cycle.)

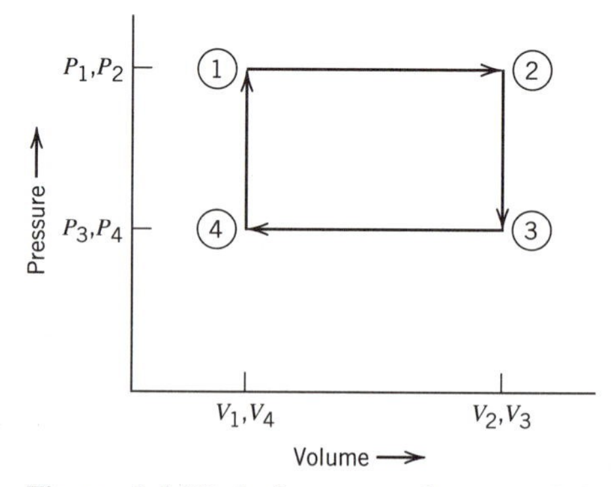

Figure 2.4 Work from a cycle on a *P-V* diagram.

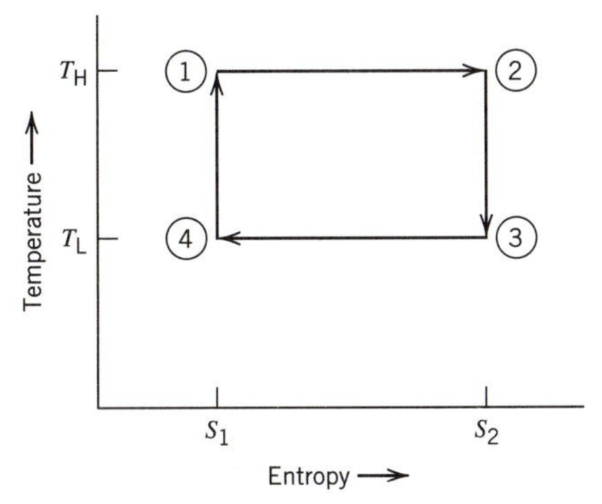

Figure 2.5 Work from a cycle on a temperature–entropy (T–S) diagram.

In Figure 2.7, the heat removed from the high temperature source is $Q_H = T_H(S_2 - S_1)$. The heat rejected to the low temperature sink is $T_L(S_1 - S_2)$. The work done is the area enclosed in the path 1–2–3–4–1: that is $(T_H - T_L) \times (S_2 - S_1)$. The quantity $S_2 - S_1$ is also Q_H/T_H. The work done is thus $Q_H(T_H - T_L)/T_H$, as in Eq. 2.8.

2.7 REFRIGERATORS

A refrigerator (or air conditioner) is a machine designed to remove heat from a low temperature region and pump it to a higher temperature by doing some work on it. The temperature inside a household refrigerator is lower than the ambient temperature (room temperature). Heat leaks into the refrigerator through its walls, sealing

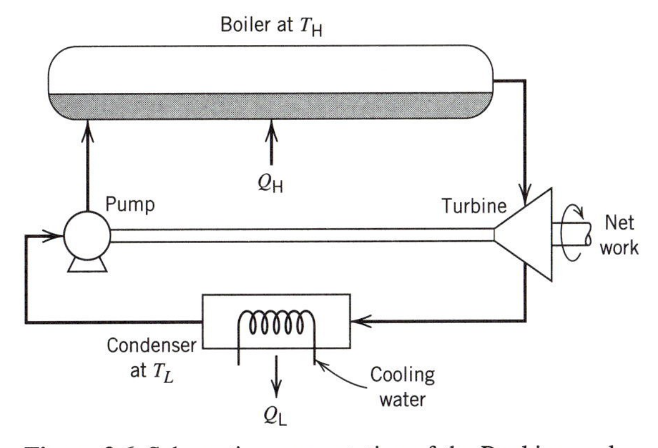

Figure 2.6 Schematic representation of the Rankine cycle.

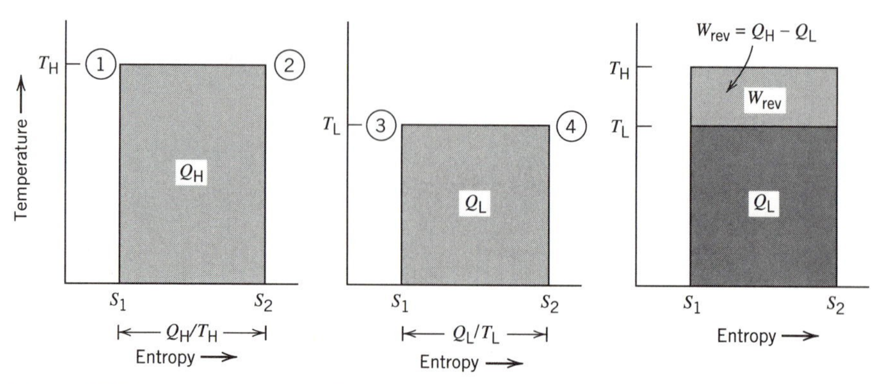

Figure 2.7 The Second Law on a *T–S* diagram for a heat engine.

gaskets, and openings. To maintain the inside of the refrigerator at the desired low temperature, this thermal energy must be "pumped" back to the higher temperature and released into the ambient. A machine to do this is similar to the machine described in Figure 2.7, but run in reverse.

Without reference to the details of a specific machine, the coefficient of performance of a refrigerator—that is, the heat removed from the low temperature reservoir divided by the work required (Q_L/W_{rev})—may be derived from a consideration of a temperature–entropy diagram in which the working fluid in the engine flows from 4 to 3 to 2 to 1, the reverse of the direction for a heat engine (Figure 2.8). The coefficient of performance is:

$$\eta_{refrig} = \frac{Q_L}{W_{rev}} = \frac{T_L}{T_H - T_L} \tag{2.10}$$

Note that this figure of merit (not an efficiency) can be greater than unity. For example, if the interior of the refrigerator is to be maintained at $-20°C$ (253.15 K) and room temperature is $+30°C$ (303.15 K), the maximum figure of merit is:

$$\eta = \frac{253.15}{303.15 - 253.15} = \frac{253.15}{50} = 5.06$$

$$5.06 = \frac{Q_L}{W_{rev}}$$

$$5.06 W_{rev} = Q_L$$

Thus, if the machine operates at maximum efficiency, for every joule of work done by the refrigerator, 5.06 J of heat is removed from the interior of the refrigerator. The heat released to the ambient is (based on the First Law) the sum of the two, 6.06 J. This use of one joule of electrical work to release 6.06 J of heat into a room (under ideal, maximum efficiency conditions) is the basis of heat pumps, as discussed in the next section.

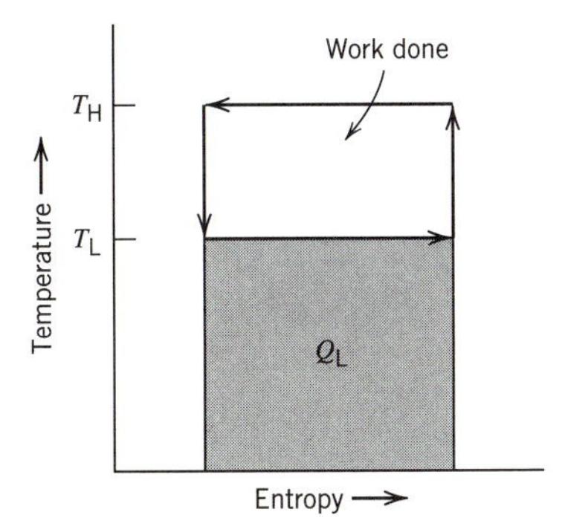

Figure 2.8 *T–S* diagram for a heat pump or refrigerator.

2.8 HEAT PUMPS

A refrigerator, as we have just seen, is a device for pumping thermal energy from a lower temperature region to a higher temperature region. In the case of a refrigerator, the machine is used to keep a low temperature compartment at a specified low temperature by removing the heat that leaks in from the surroundings. The same type of device could be used to supply heat to a room at a temperature higher than the outdoor temperature. From Figure 2.8 note that the heat supplied to the room (at T_H) is greater than the work done by the machine—that is, greater than the electrical energy supplied to the motor that runs the machine. This means that the system can be more ''efficient'' than one involving just an electrical resistor to heat the room. Because the heat pump is rated on its ability to deliver heat to the high temperature sink, its coefficient of performance is based on Q_H rather than Q_L. Its coefficient is

$$\eta_{\text{heat pump}} = \frac{Q_H}{W_{\text{rev}}} = \frac{T_H}{T_H - T_L} \tag{2.11}$$

Thus if the refrigerator discussed in Section 2.7 were to be evaluated as a heat pump, its figure of merit would be

$$\eta = \frac{303.15}{303.15 - 253.15} = \frac{303.15}{50} = 6.06$$

A heat pump operating between these two temperatures can thus deliver 6.06 J of heat into the high temperature region for every joule of electrical work done by the machine, under ideal maximum efficiency conditions.

2.9 ENTROPY CHANGES

The specific entropy change for a material heated between two temperatures, T_1 and T_2, at constant pressure (assuming there is no change of state between the two temperatures) is

$$d\underline{S} = \frac{\delta Q_{\text{rev}}}{T} = \frac{C_P dT}{T}$$
$$\underline{S}_2 - \underline{S}_1 = \int_{S_1}^{S_2} d\underline{S} = \int_{T_1}^{T_2} \frac{C_P dT}{T} \tag{2.12}$$

If there is a change of state (a phase change) between the two temperatures, such as melting of a solid, the entropy change must be calculated in steps: first we find

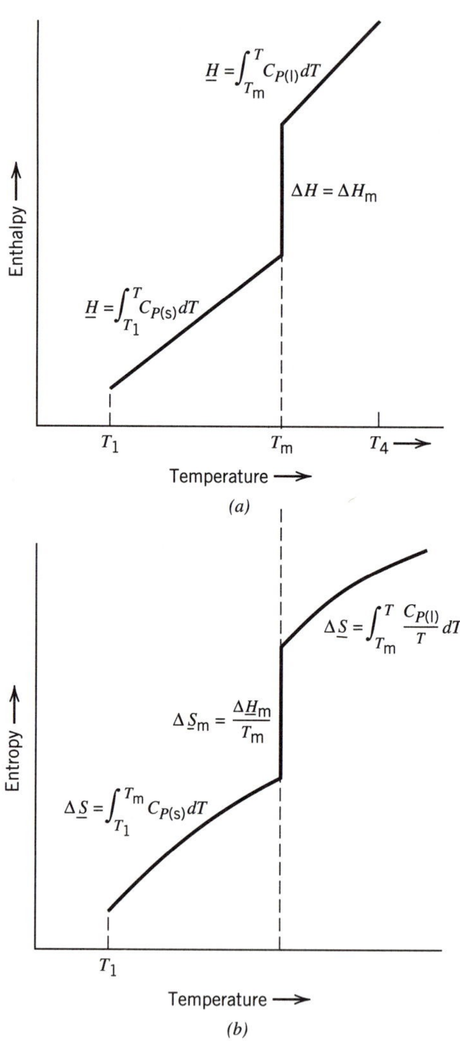

Figure 2.9 Changes during melting in (*a*) enthalpy and (*b*) entropy.

the entropy change from T_1 to the melting temperature T_m, then the entropy of melting (see Figure 2.9),

$$\Delta \underline{S}_{\text{melting}} = \frac{\Delta \underline{H}_{\text{melting}}}{T_{\text{melting}}} \tag{2.13}$$

and finally the entropy change of the liquid from the melting temperature, T_m, to T_2. Note that in each case the entropy change is calculated for a *reversible* process. The entropy change upon melting is the heat of fusion divided by the melting temperature, because the change from solid to liquid is *reversible* only at the melting temperature. The calculation would be invalid at temperatures other than the melting temperature, because there the process is irreversible.

The need to calculate entropy changes for *irreversible* reactions often arises in the study of transformations. In this case it is necessary to define a *reversible* path between the initial and final state of the system, and to calculate the entropy changes along that path. As an example, let us calculate the entropy change for the solidification of copper at 1300 K. The melting temperature of copper is 1356 K. The situation is described in Figure 2.10. It is necessary to calculate the entropy change for liquid copper from 1300 K to 1356 K, then the entropy of solidification ($\Delta \underline{H}_S/T_S$), then the entropy change for solid copper from 1356 K to 1300 K.

$$\Delta \underline{S}_{1300\text{ K}} = \int_{1300}^{1356} \frac{C_{P(l)}}{T}\, dT + \frac{\Delta \underline{H}_S}{1356} + \int_{1356}^{1300} \frac{C_{P(s)}}{T}\, dT \tag{2.14}$$

If we may make the simplifying assumptions that the heat capacities of the liquid and solid are constant and equal ($C_{P(l)} = C_{P(s)}$), the change of entropy upon solidification is the same at 1300 and 1356 K. Note carefully that the entropy change, ΔS_{1300}, is *not* the heat transferred at 1300 K (ΔH_{1300}) divided by 1300 K.

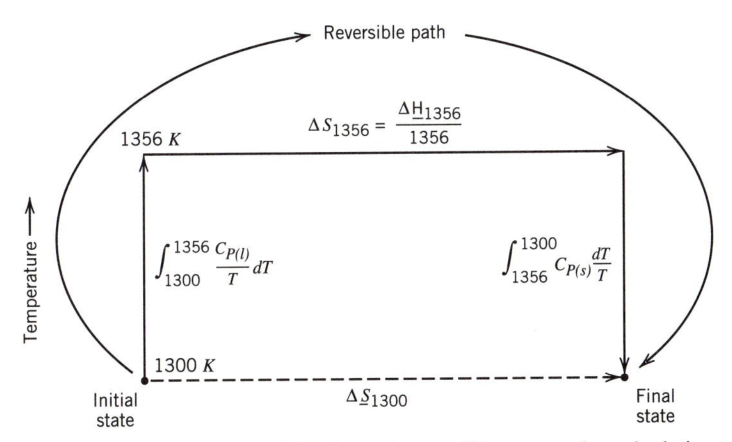

Figure 2.10 Calculation of ΔS for an irreversible process by calculating ΔS for a reversible path with the same initial and final states.

2.10 ENTROPY CHANGES IN CHEMICAL REACTIONS AND THE THIRD LAW

The technique for calculating the entropy changes for a chemical reaction is similar to the technique used in the case of enthalpy changes, with one very important difference. That difference is embodied in a relationship often referred to as the Third Law. The third law of thermodynamics, enunciated by Nernst in 1906, states that in any chemical reaction involving only *pure, crystalline* substances, the change of entropy is zero at the absolute zero of temperature. The term "pure" refers to pure elements and stoichiometrically balanced compounds that are perfectly crystalline. A stronger statement of the law by Planck asserts that the entropy of a pure, perfectly crystalline substance is zero at the absolute zero of temperature. We explore the reasons for these statements in Chapter 10. For our purposes at this point, the two are equivalent and lead to a simplification in the calculation of entropies of reaction. The Third Law is simply:

$$\Delta S_0^\circ = 0 \qquad \textbf{(2.15)}$$

A comment on notation may be useful. The superscript zero on a term representing a thermodynamic property is generally used to indicate that we are dealing with the thermodynamic properties of materials in some standard state. In this case, the designation S° means pure, perfectly crystalline substances at standard pressure (one atmosphere). The subscript, zero in Eq. 2.15, is the temperature.

Thus the term ΔS_0° means the change in entropy of a reaction at zero kelvin in which each of the chemical entities is in its "standard" state. To cite another example, the designation ΔH_{298}° means the enthalpy change of a reaction at 298 K in which each of the chemical entities is in its "standard state."

The Third Law provides a great simplification in the calculation of entropies of reaction. Whereas in enthalpy calculations, values of enthalpies of formation must be tabulated for each chemical compound at a reference temperature (298 K), the entropy of formation for *all pure, perfectly crystalline compounds* at absolute zero kelvin is zero. To calculate the entropy change for a chemical reaction at a specified temperature (T) requires only a knowledge of the entropy of each of the substances involved above 0 K ($S_T^\circ - S_0^\circ$). This quantity can be calculated[3] from heat capacity information:

$$\underline{S}_{298}^\circ = \int_0^{298} \frac{C_P}{T}\,dT \qquad \textbf{(2.16)}$$

[3]The integral in Eq. 2.16 can be evaluated because the heat capacity of substances at very low temperatures decreases rapidly as a function of temperature. As the temperature of zero kelvin is approached, the heat capacity of solids is approximately proportional to the third power of the absolute temperature. This relationship is discussed in Chapter 10. There is also a linear contribution to the heat capacity from free conduction electrons in the case of metals. In any event, the integral may be evaluated.

Because entropy changes near room temperature are often needed in thermodynamic calculations, the integral in Eq. 2.16 has been evaluated for many substances at 298 K. These values are generally called the "standard" entropies for the compounds. The integration must take phase changes into account. The form in Eq. 2.16 assumes that no phase changes take place between 0 and 298 K, as in the case of pure copper. However, in the case of oxygen, which is solid near 0 K and gaseous at 298 K, the integration must take into account the melting of solid oxygen at 54.36 K and the boiling of the liquid at 90.19 K. In that case the integral is

$$\underline{S}^\circ_{298} = \int_0^{T_\mathrm{m}} \frac{C_P}{T}\,dT + \frac{\Delta\underline{H}_\mathrm{m}}{T_\mathrm{m}} + \int_{T_\mathrm{m}}^{T_\mathrm{b}} \frac{C_P}{T}\,dT + \frac{\Delta\underline{H}_\mathrm{b}}{T_\mathrm{b}} + \int_{T_\mathrm{b}}^{298} \frac{C_P}{T}\,dT \qquad \textbf{(2.17)}$$

where $\Delta\underline{H}_\mathrm{m}$ and $\Delta\underline{H}_\mathrm{b}$ are the molar enthalpies of melting and boiling, respectively, and T_m and T_b are the melting and boiling temperatures.

Because the entropy change for a chemical reaction among pure, crystalline substances is zero at the absolute zero of temperature (Third Law), the entropy change for a chemical reaction at 298 K is simply the sum of the standard entropies of the products minus the sum of the standard entropies of the reactants (the underscore signifies standard *molar* entropies)

$$\Delta S^\circ_{298} = \sum_{\text{products}} n_p \underline{S}^\circ_{298,p} - \sum_{\text{react.}} n_r \underline{S}^\circ_{298,r} \qquad \textbf{(2.18)}$$

where the subscripts p and r refer to the products and the reactants of the reaction, respectively. For example, the standard entropy change for the formation of silver chloride, AgCl, from its elements at 298 K is:

$$Ag + \tfrac{1}{2}Cl_2 = AgCl$$

$$\Delta S^\circ_{298} = \underline{S}^\circ_{298,\mathrm{AgCl}} - \underline{S}^\circ_{298,\mathrm{Ag}} - \tfrac{1}{2}\underline{S}^\circ_{298,\mathrm{Cl_2}}$$

$$\Delta S^\circ_{298} = 96.53 - 42.73 - \tfrac{1}{2}(223.0)$$

$$\Delta S^\circ_{298} = -57.7 \text{ J/(mol·K)}$$

To calculate entropy changes at other temperatures, a procedure similar to the one for calculating enthalpy changes is used (see Eq. 1.34):

$$\Delta S^\circ_T = \Delta S^\circ_{298} + \int_{298}^{T} \frac{n_p C_{P,\mathrm{prod}} - n_r C_{P,\mathrm{react}}}{T}\,dT \qquad \textbf{(2.19)}$$

Care must be taken to evaluate the integral in Eq. 2.19 when changes in state, such as melting or boiling, are encountered. In such cases the entropy changes must be evaluated using the technique illustrated in Eq. 2.17.

REFERENCES

1. Margenau, Henry, and Murphy, George Moseley, *The Mathematics of Physics and Chemistry,* Van Nostrand, New York, 1956, p 26.
2. Denbigh, K. G., *Thermodynamics of the Steady State,* John Wiley, New York, 1951.
3. Forland, K. S., Forland T., and Ratkje, S. K. *Irreversible Thermodynamics,* John Wiley, New York, 1988.

PROBLEMS

2.1 The solar energy flux is about 4 J cm²/min. In a nonfocusing collector the surface temperature can reach a value of about 90°C. If we operate a heat engine using the collector as the heat source and a low temperature reservoir at 25°C, calculate the area of collector needed if the heat engine is to produce 1 horsepower (hp). Assume that the engine operates at maximum efficiency.

2.2 A refrigerator is operated by a 0.25 hp motor. If the interior of the box is to be maintained at −20°C against a maximum exterior temperature of 35°C, what is the maximum heat leak (in watts) into the box that can be tolerated if the motor runs continuously? Assume that the coefficient of performance is 75% of the value for a reversible engine.

2.3 Suppose an electrical motor supplies the work to operate a Carnot refrigerator. The interior of the refrigerator is at 0°C. Liquid water is taken in at 0°C and converted to ice at 0°C. To convert 1 g of ice to 1 g of liquid, $\Delta H_{\text{fus}} = 334$ J/g is required. If the temperature outside the box is 20°C, what mass of ice can be produced in one minute by a 0.25 hp motor running continuously? Assume that the refrigerator is perfectly insulated and that the efficiencies involved have their largest possible values.

2.4 Under 1 atm pressure, helium boils at 4.216 K. The heat of vaporization is 84 J/mol. What size motor (in horsepower) is needed to run a refrigerator that must condense 2 mol of gaseous helium at 4.216 K to liquid at the same temperature in one minute? Assume that the ambient temperature is 300 K and that the coefficient of performance of the refrigerator is 50% of the maximum possible.

2.5 If a fossil fuel power plant operating between 540 and 50°C provides the electrical power to run a heat pump that works between 25 and 5°C, what is the amount of heat pumped into the house per unit amount of heat extracted from the power plant boiler?

(a) Assume that the efficiencies are equal to the theoretical maximum values.

(b) Assume that the power plant efficiency is 70% of maximum and that the coefficient of performance of the heat pump is 10% of maximum.

(c) If a furnace can use 80% of the energy in fossil fuel to heat the house, would it be more economical in terms of overall fossil fuel consumption to use a heat pump or a furnace? Do the calculations for cases a and b.

2.6 Calculate ΔU and ΔS when 0.5 mol of liquid water at 273 K is mixed with 0.5

mol of liquid water at 373 K, and the system is allowed to reach equilibrium in an adiabatic enclosure. Assume that C_P is 77 J/(mol·K) from 273 K to 373 K.

2.7 A modern coal-burning power plant operates with a steam outlet from the boiler at 540°C and a condensate temperature of 30°C.

(a) What is the maximum electrical work that can be produced by the plant per joule of heat provided to the boiler?

(b) How many metric tons (1000 kg) of coal per hour is required if the plant output is to be 500 MW (megawatts). Assume the maximum efficiency for the plant. The heat of combustion of coal is 29.0 MJ/kg.

(c) Electricity is used to heat a home at 25°C when the outdoor temperature is 10°C by passing a current through resistors. What is the maximum amount of heat that can be added to the home per kilowatt-hour of electrical energy supplied?

2.8 An electrical resistor is immersed in water at the boiling temperature of water (100°C). The electrical energy input into the resistor is at the rate of one kilowatt.

(a) Calculate the rate of evaporation of the water in grams per second if the water container is insulated: that is, no heat is allowed to flow to or from the water except for that provided by the resistor.

(b) At what rate could water could be evaporated if electrical energy were supplied at the rate of 1 kW to a heat pump operating between 25 and 100°C?

DATA

For water: Enthalpy of evaporation is 40,000 J/mol at 100°C.
Molecular weight is 18 g/mol
Density is about 1 g/cm^3.

2.9 Some aluminum parts are being quenched (cooled rapidly) from 480°C to −20°C by immersing them in a brine, which is maintained at −20°C by a refrigerator. The aluminum is being fed into the brine at a rate of one kilogram per minute. The refrigerator operates in an environment at 30°C; that is, the refrigerator may reject heat at 30°C.

What is the minimum power rating, in kilowatts, of the motor required to operate the refrigerator?

DATA

For aluminum: Heat capacity is 28 J/(mol·K)
Molecular weight is 27 g/mol

2.10 An electric power generating plant has a rated output of 100 MW. The boiler of the plant operates at 300°C. The condenser operates at 40°C.

(a) At what rate (joules per hour) must heat be supplied to the boiler?

(b) The condenser is cooled by water, which may undergo a temperature rise of no more than 10°C. What volume of cooling water, in cubic meters per hour, is required to operate the plant?

(c) The boiler temperature is to be raised to 540°C, but the condenser temperature and the electric output will remain the same. Will the cooling water requirement be increased, decreased, or remain the same? Justify your answer.

DATA

For water: Heat capacity is 4.184 J/(g·K)

Density is 1 g/cm^3.

2.11 (a) Heat engines convert heat that is available at different temperatures to work. There have been several proposals to generate electricity by using a heat engine that operates on the temperature differences available at different depths in the oceans.

Assume that surface water is at 20°C, that water at a great depth is at 4°C, and that both may be considered to be infinite in extent. How many joules of electrical energy may be generated for each joule of energy absorbed from the surface water?

(b) The hydroelectric generation of electricity uses the drop in height of water as the energy source. In a particular region the level of a river drops from 100 m above sea level to 70 m above sea level. What fraction of the potential energy change between those two levels may be converted into electrical energy?

How much electrical energy, in kilowatt-hours, may be generated per cubic meter of water that undergoes such a drop?

2.12 A sports facility has both an ice rink and a swimming pool. To keep the ice frozen during the summer requires the removal from the rink of 10^5 kJ of thermal energy per hour.

It has been suggested that this task be performed by a thermodynamic machine (a heat engine, a refrigerator, or a heat pump), which would use the swimming pool as the high temperature reservoir (heat sink). The ice in the rink is to be maintained at a temperature of −15°C, and the swimming pool operates at 20°C.

(a) What is the theoretical minimum power, in kilowatts, required to run the machine?

(b) How much heat, in joules per hour, would be supplied to the pool by this machine?

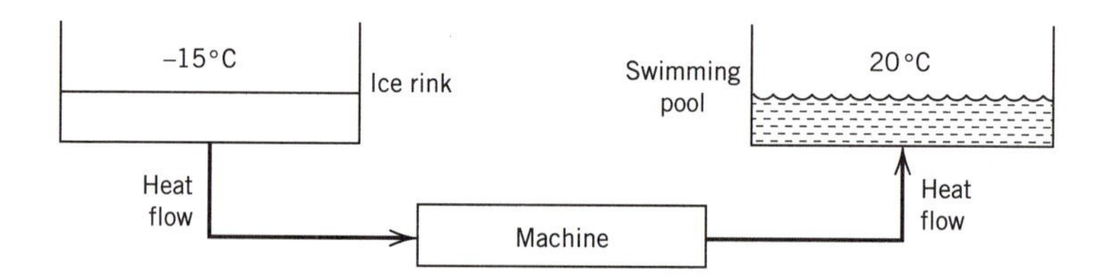

2.13 Aluminum nitride (AlN), a material that is sometimes used in electronic "chip carriers," may be formed by reacting metallic aluminum with pure, gaseous nitrogen (N_2).

(a) Write the chemical reaction for the formation of one mole of aluminum nitride.

(b) What is the change in enthalpy for the reaction (ΔH) if all the reactants and products are at 298 K and 1 atm pressure?

(c) What is the change in entropy (ΔS) for the reaction if all the reactants and products are at 298 K and one atm pressure?

(d) Repeat parts (b) and (c) when the reactants and products are at 298 K and 10 atm pressure.

DATA

$$\Delta \underline{H}^\circ_{f,298} \text{ for AlN} = -76{,}470 \text{ cal/mol}$$

For	**Standard Entropies S° at 298 K [cal/(mol·K)]**
Aluminum	6.77
Nitrogen	45.77
Aluminum nitride	4.82

2.14 Compute the entropy difference between 12 kg of water at 40°C and 12 kg of ice at −10°C.

DATA

$$C_{P,\text{water}} = 4.184 \text{ J/g·K}$$

$$C_{P,\text{ice}} = 2.1 \text{ J/g·K}$$

Heat of melting of ice is = 336 kJ/kg

2.15 A great deal of effort has been expended to find "high temperature superconductors": materials that are superconductors at temperatures higher than the boiling point of liquid nitrogen (77 K). Most of the older superconductors had to be operated with liquid helium (boiling point 4.2 K) as the cooling fluid. To estimate the savings possible in operating costs through the use of the "high temperature" superconductors, calculate the minimum work needed to compensate for a heat leak of 1 kJ into the superconductor for both "high temperature" superconductors and the older ones. Assume that the ambient temperature is 300 K.

2.16 **Note:** In this problem, part b is more difficult than part a. The low temperature (T_L) varies as the helium is cooled.

(a) The boiling point of helium is 4.2 K. At this temperature, the enthalpy of evaporation of helium is 83.3 J/mol. What is the minimum work required to liquefy one mole of helium starting with gaseous helium at 4.2 K? Assume that the ambient temperature is 300 K.

(b) What is the minimum work required to cool one mole of gaseous helium at 1 atm pressure from 300 K to 4.2 K? The ambient temperature is 300 K. The heat capacity may be taken as $\frac{3}{2}R$ J/(mol·K).

2.17 At 25°C, one mole of ideal gas undergoes a reversible, isothermal expansion during which its volume increases by a factor of 10.

(a) Compute the total amount of heat released (or absorbed) during the expansion.

(b) What is the entropy change of the gas?

(c) Suppose this expansion is accomplished in a very irreversible way so that no work is done by the system (but the expansion is still isothermal). What is the total amount of heat released (or absorbed) now?

(d) Water is stirred in a bucket that is thermally insulated (adiabatic), but under constant pressure. Does the enthalpy of the water change?

2.18 A nuclear power plant is designed to produce 500 MW of electrical power. The high temperature heat absorption in the boiler occurs at 360°C. Condensation will be done at 25°C by using cooling water.

What is the minimum amount of cooling water needed per second if the Environmental Protection Agency regulations forbid a temperature rise of the cooling water greater than 2°C? Assume maximum thermodynamic efficiency for the plant (no mechanical losses).

DATA

$$C_P \text{ for water is } 4.184 \text{ J/g·K}$$

2.19 A refrigerator driven by a 500 W electrical motor operates in a room where the temperature is kept constant at 22°C. The inside of the refrigerator is at 0°C.

How much ice can this refrigerator produce per minute from water at 0°C? Assume maximal thermal efficiency for the refrigerator (no mechanical losses).

DATA

$$C_{P(\text{water})} = 4.184 \text{ kJ/(kg·K)}$$

$$C_{P(\text{ice})} = 2.1 \text{ kJ/(kg·K)}$$

$$\text{Heat of melting for ice} = 335 \text{ J/g at } 0°\text{C}$$

Chapter 3

Property Relations

Chapters 1 and 2 discussed the first and second laws of thermodynamics and demonstrated the usefulness of the energy and entropy functions. This chapter first explores how the values of these energy and entropy functions are influenced by changes in several physical variables. We then define some additional functions used in thermodynamic calculations concerning physical and chemical equilibrium and derive some of the consequences of these definitions.

Part of the first objective of the chapter, to explore the variation of the energy functions (U and H) with temperature and pressure, has already been met for the case of an ideal gas. The heat capacities (C_V and C_P) give the change of $\underline{U}$ and $\underline{H}$ with temperature, at constant volume and constant pressure, respectively. We also discussed how the specific internal energy of an ideal gas changes with volume, and postulated that the variation is zero: that is, $(\partial \underline{U}/\partial \underline{V})_T = 0$.

In a sense, we have solved the problem of defining the specific internal energy as a function of temperature and specific volume for ideal gases. To illustrate using a diagram, visualize a three-dimensional plot with temperature (T) on one axis, specific volume ($\underline{V}$) on another, and specific internal energy ($\underline{U}$) on the third. Figure 3.1*a* is a two-dimensional representation of the diagram using the T and $\underline{V}$ axes. The $\underline{U}$ axis is perpendicular to the page. The specific internal energy at point B can be determined relative to point A if the values of the two partial derivatives are known everywhere. If the heat capacity at constant volume (C_V) is known at the specific

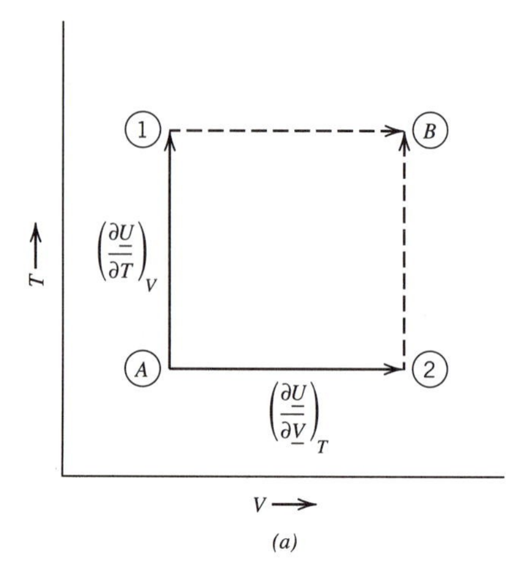

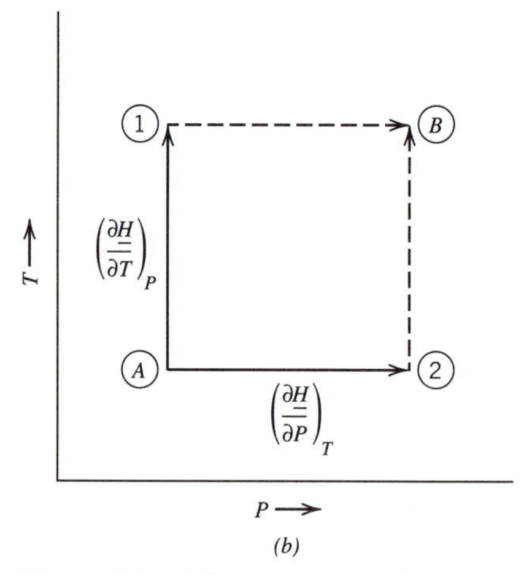

Figure 3.1 (*a*) Internal energy changes as a function of temperature T and specific volume $\underline{V}$. (*b*) Enthalpy as a function of temperature T and pressure P.

volume corresponding to point A, then the specific internal energy at intermediate point 1 can be determined through integration from T_A to T_B. For an ideal gas the variation of specific internal energy from point 1 to point B is zero. Hence, we can find the specific internal energy at any point on the T–$\underline{V}$ axes by knowing the constant volume heat capacity as a function of temperature.

It should be apparent that this heat capacity at constant volume is dependent only on temperature, because the specific internal energy does not depend on specific volume. The specific internal energy changes along lines A–2 and 1–B are zero. Later in the chapter we will prove that $(\partial C_V/\partial \underline{V})_T = 0$ for an ideal gas.

Using similar reasoning, we can show that the changes in specific enthalpy for an ideal gas can be determined as a function of temperature and pressure by knowing the constant pressure heat capacity as a function of temperature (Figure 3.1*b*). These are very useful simplifications for the ideal gas. Unfortunately, not all materials are ideal gases. This chapter, therefore, examines, in general, the dependence of properties, such as the internal energy, enthalpy, entropy, and other thermodynamic functions, on temperature, pressure, specific volume, and other intensive variables.

3.1 THE PROPERTY RELATION

Consider a *closed* system consisting of a *homogeneous* material with uniform properties throughout. Writing the First and Second laws for the material (neglecting potential and kinetic energy terms):

$$\text{First Law:} \qquad dU = \delta Q + \delta W \tag{3.1}$$

$$\text{Second Law:} \qquad dS = \frac{\delta Q}{T} - \frac{\delta \text{lw}}{T}$$

Multiplying both sides by T:

$$T\,dS = \delta Q - \delta\text{lw} \tag{3.2}$$

Subtracting Eq. 3.2 from Eq. 3.1 yields:

$$dU = T\,dS + \delta W + \delta\text{lw}$$

Noting that $\delta W + \delta\text{lw} = \delta W_{\text{rev}}$,

$$dU = T\,dS + \delta W_{\text{rev}} \tag{3.3}$$

The reversible work term, δW_{rev}, represents all the work terms including surface, electrical, stress effects, etc.

$$\delta W_{\text{rev}} = -P\,dV + \gamma\,dA + \Sigma\varepsilon\,dq + F\,dl \cdots$$

For convenience at this point, consider the P-V term to represent all the work terms. Then:

$$dU = T\,dS - P\,dV \tag{3.4}$$

This, the basic property equation, relates the change of internal energy of a material to changes in entropy and volume. By combining it with the definition of enthalpy ($H = U + PV$), and noting that $dH = dU + P\,dV + V\,dP$, we derive:

$$dH = T\,dS + V\,dP \tag{3.5}$$

3.2 THE FUNCTIONS *F* AND *G*

We can now proceed to define two other thermodynamic functions: F, the Helmholtz free energy, and G, the Gibbs free energy.[1]

These two terms are defined as measures of the work required to change a system from one state to another (state 1 to state 2) through *reversible* processes under *isothermal* conditions. Their utility will become apparent when, in a later chapter, we consider the thermodynamic conditions of equilibrium.

Rewriting Eq. 3.3,

$$\delta W_{\text{rev}} = dU - T\,dS \tag{3.6}$$

Integrating Eq. 3.6 at constant temperature:

$$W_{\text{rev}} = \Delta U - T\,\Delta S$$

$$W_{\text{rev}} = \Delta U - \Delta(TS) = \Delta(U - TS)$$

The function $U - TS$ is defined as the Helmholtz free energy, F. It is sometimes also called the work function because of the relationship in Eq. 3.7.

$$F \equiv U - TS \tag{3.7a}$$

$$W_{\text{rev}} = \Delta F \tag{3.7b}$$

In many situations, we will want to calculate the reversible work required to change a system from state 1 to state 2 at constant temperature and pressure, *exclusive of the P-V work.* Typically, these conditions arise when we are dealing with the thermodynamics of chemical changes at constant pressure. As discussed later in Chapters 4 and 5, equilibrium between two states is defined in terms of the reversible work required to transform a system from one state to the other. If the system can

[1]The notation for Helmholtz free energy and for Gibbs free energy varies among publications relating to thermodynamics. In some of the older, classic texts (Refs. 1 and 2), and in older collections of thermodynamic data (Refs. 3 and 4), A is used to denote the Helmholtz free energy, and F to denote Gibbs free energy.

freely expand or contract against constant pressure (e.g., atmospheric pressure), then the useful work that can be done by the system must be calculated exclusive of the P-V term. To proceed with the argument, let us define δW^* as that "useful" work.

$$\delta W_{rev} = \delta W^*_{rev} - P\,dV = dU - T\,dS$$

$$\delta W^*_{rev} = dU - T\,dS + P\,dV$$

Integrating at constant temperature and pressure:

$$\begin{aligned} W^*_{rev} &= \Delta U + P\,\Delta V - T\,\Delta S = \Delta(U + PV - TS) \\ W^*_{rev} &= \Delta(H - TS) = \Delta G \end{aligned} \tag{3.8}$$

The function $H - TS$ is defined as the Gibbs free energy G. To summarize:

$$F \equiv U - TS \tag{3.9a}$$

and

$$G \equiv H - TS \tag{3.10a}$$

Note that the two functions, F and G, are each point or state functions because all the terms in them are state functions. The units of F and G are energy units, joules in the SI system.

By differentiating the F and G functions (Eqs. 3.9 and 3.10) and making the appropriate substitutions,

$$dF = -S\,dT - P\,dV \tag{3.9b}$$

$$dG = -S\,dT + V\,dP \tag{3.10b}$$

3.3 CHEMICAL POTENTIALS

The relationships developed thus far in this chapter (Eqs. 3.4, 3.5, 3.9b, and 3.10b) deal with changes of various thermodynamic variables (U, H, F, and G) in *closed* systems consisting of a *homogeneous* material with uniform properties throughout. For example, Eq. 3.4 expressed the change of internal energy (U) as a function of entropy and volume changes for a closed system. We could have dealt with this as follows:

$$\begin{aligned} U &= U(S, V) \\ dU &= \left(\frac{\partial U}{\partial S}\right)_V dS + \left(\frac{\partial U}{\partial V}\right)_S dV \end{aligned} \tag{3.11}$$

Comparing Eq. 3.11 with Eq. 3.4, $dU = T\,dS - P\,dV$:

$$\left(\frac{\partial U}{\partial S}\right)_V = T \qquad \text{and} \qquad \left(\frac{\partial U}{\partial V}\right)_S - P \tag{3.12}$$

To be more useful, the thermodynamic equations will be extended to deal with changes in composition through additions of mass. In these equations we will measure mass in moles because the equations are used especially for the treatment of solutions and chemical equilibria, where molar compositions are important. We will denote mass by the letter n, number of moles. Proceeding as in the equations above, we write that internal energy as a function of entropy (S), volume (V), and the masses of the various constitutents of the system (n_i), measured in moles.

$$U = U(S, V, n_i)$$

$$dU = \left(\frac{\partial U}{\partial S}\right)_{V,n_i} dS + \left(\frac{\partial U}{\partial V}\right)_{S,n_i} dV + \sum_i \left(\frac{\partial U}{\partial n_i}\right)_{S,V,n_j \neq n_i} dn_i \tag{3.13}$$

$$dU = T\,dS - P\,dV + \sum_i \left(\frac{\partial U}{\partial n_i}\right)_{S,V,n_j \neq n_i} dn_i$$

Similarly, $H = H(S, P, n_i)$

$$dH = \left(\frac{\partial H}{\partial S}\right)_{P,n_i} dS + \left(\frac{\partial H}{\partial P}\right)_{S,n_i} dP + \sum_i \left(\frac{\partial H}{\partial n_i}\right)_{S,P,n_j \neq n_i} dn_i$$

$$dH = T\,dS + V\,dP + \sum_i \left(\frac{\partial H}{\partial n_i}\right)_{S,P,n_j \neq n_i} dn_i \tag{3.14}$$

And, $F = F(T, V, n_i)$

$$dF = \left(\frac{\partial F}{\partial T}\right)_{V,n_i} dT + \left(\frac{\partial F}{\partial V}\right)_{T,n_i} dV + \sum_i \left(\frac{\partial F}{\partial n_i}\right)_{T,V,n_j \neq n_i} dn_i$$

$$dF = -S\,dT - P\,dV + \sum_i \left(\frac{\partial F}{\partial n_i}\right)_{T,V,n_j \neq n_i} dn_i \tag{3.15}$$

Also, $G = G(T, P, n_i)$

$$dG = \left(\frac{\partial G}{\partial T}\right)_{P,n_i} dT + \left(\frac{\partial G}{\partial P}\right)_{T,n_i} dP + \sum_i \left(\frac{\partial G}{\partial n_i}\right)_{T,P,n_j \neq n_i} dn_i$$

$$dG = -S\,dT + V\,dP + \sum_i \left(\frac{\partial G}{\partial n_i}\right)_{T,P,n_j \neq n_i} dn_i \tag{3.16}$$

By noting that $G = F + PV$ (Eqs. 1.11a, 3.8, 3.9a, and 3.10a), we can write

$$dG = dF + P\,dV + V\,dP$$

Combining with the equations above:

$$dG = -S\,dT + V\,dP + \sum_i \left(\frac{\partial G_i}{\partial n_i}\right)_{T,P,n_j \neq n_i} dn_i$$

$$= -S\,dT + V\,dP + \sum_i \left(\frac{\partial F}{\partial n_i}\right)_{T,V,n_j \neq n_i} dn_i$$

$$\sum_i \left(\frac{\partial G_i}{\partial n_i}\right)_{T,P,n_j \neq n_i} dn_i = \sum_i \left(\frac{\partial F}{\partial n_i}\right)_{T,V,n_j \neq n_i} dn_i$$

Thus:

$$\left(\frac{\partial G_i}{\partial n_i}\right)_{T,P,n_j \neq n_i} = \left(\frac{\partial F}{\partial n_i}\right)_{T,V,n_j \neq n_i}$$

By repeating the procedure for each of the defined thermodynamic variables, we can show that:

$$\left(\frac{\partial G_i}{\partial n_i}\right)_{T,P,n_j \neq n_i} = \left(\frac{\partial F}{\partial n_i}\right)_{T,V,n_j \neq n_i} = \left(\frac{\partial H}{\partial n_i}\right)_{P,S,n_j \neq n_i} = \left(\frac{\partial U}{\partial n_i}\right)_{V,S,n_j \neq n_i} = \mu_i \quad \textbf{(3.17)}$$

where μ_i is called the chemical potential of the component i.

The chemical potential concept is used extensively in the treatment of the thermodynamics of solutions and of chemical reactions.

3.4 PARTIAL MOLAR QUANTITIES

A partial molar quantity is the partial derivative of that quantity with respect to mass (number of moles) at constant temperature and constant pressure, and the mass of all other materials in the system. Stated another way, it is the rate of change of that quantity as mass of a particular component is added to a system at constant temperature and pressure.

The concept is used especially in the study of solutions. To illustrate in terms of an easily measured quantity, consider the partial molar *volume* of material a in a solution.

$$\overline{V}_a = \left(\frac{\partial V}{\partial n_a}\right)_{T,P,n_b,n_c,\ldots} \quad \textbf{(3.18)}$$

The partial molar volume is the slope of the graph of total volume (V) plotted against the mass of material a (Figure 3.2a). One can think of the partial molar

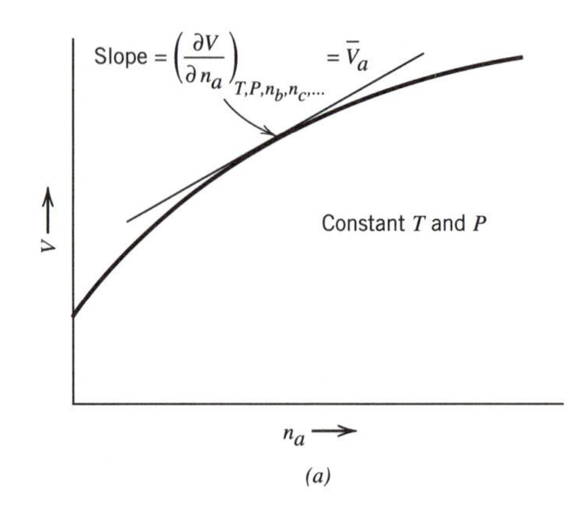

(*a*)

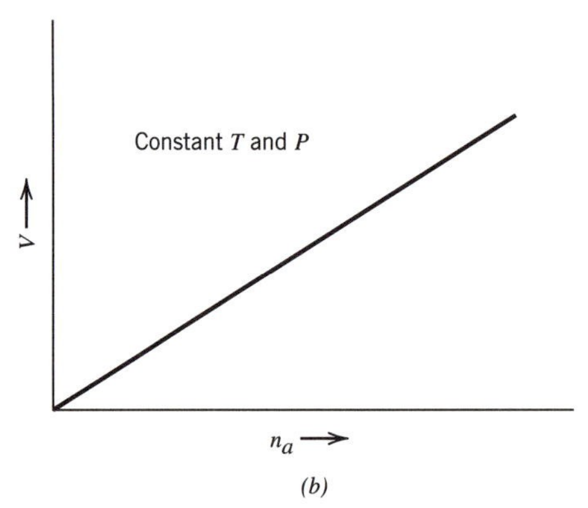

(*b*)

Figure 3.2 (*a*) Volume of a solution ($a + b + c + \cdots$) as a function of moles of a (n_a) added. (*b*) Volume of a as a function of n_a in pure a (slope is $V/n_a = \underline{V}_a$).

volume as the molar volume of material *a* in the solution because it is the rate of change of the volume of the solution with unit additions of material *a*.

If we were adding *a* to *pure a*, the partial molar volume would be equal to the molar volume, $\overline{V}_a = \underline{V}_a$. In this case, the graph of total volume (V) against n_a would be a straight line passing through the origin (Figure 3.2*b*).

In the special case of the G function, the chemical potential is the partial molar Gibbs free energy because it is the change of G with mass of one component *at constant T and P*. If we could measure Gibbs free energy, the slope of the graph of G versus n_a would be $\overline{G}_a$ at constant T, P, n_b, n_c, ... This is not true of other expressions for the chemical potential. For example, the partial molar enthalpy of material a is:

$$\overline{H}_a = \left(\frac{\partial H}{\partial n_a}\right)_{T,P,n_b,n_c,...} \tag{3.19}$$

The chemical potential in terms of enthalpy is:

$$\mu_A = \left(\frac{\partial H}{\partial n_a}\right)_{S,P,n_b,n_c,\ldots} \tag{3.20}$$

Because the constraints on the partial derivatives are not the same, the two are not equal. In fact it can be shown that:

$$\overline{H}_i = \mu_i - T\left(\frac{\partial \mu_i}{\partial T}\right)_{P,n_j \neq n_i} \tag{3.21}$$

3.5 PROPERTY RELATIONS DERIVED FROM *U, H, F,* AND *G*

When applied to the equations for the functions *U, H, F,* and *G,* a mathematical property of exact differentials yields some interesting relationships among other thermodynamic quantities. If *dz* is an exact differential:

$$dz = M\,dx + N\,dy$$

then (see Section A.16)

$$\left(\frac{\partial M}{\partial y}\right)_x = \left(\frac{\partial N}{\partial x}\right)_y$$

This technique may be applied to the equations for *dU, dH, dF,* and *dG,* shown below, to yield what are called the Maxwell relations.

$$dU = T\,dS - P\,dV \tag{3.22}$$

$$dH = T\,dS + V\,dP \tag{3.23}$$

$$dF = -S\,dT - P\,dV \tag{3.24}$$

$$dG = -S\,dT + V\,dP \tag{3.25}$$

Applying this technique to the expression for *dG* yields an expression for the variation of entropy with pressure at constant temperature as follows:

$$dG = -S\,dT + V\,dP$$

$$-\left(\frac{\partial S}{\partial P}\right)_T = \left(\frac{\partial V}{\partial T}\right)_P$$

The relationship could have been derived in terms of molar quantities by simply dividing the equation for *dG* by the total number of moles in the system to yield:

$$-\left(\frac{\partial \underline{S}}{\partial P}\right)_T = \left(\frac{\partial \underline{V}}{\partial T}\right)_P \tag{3.26a}$$

If an equation of state is available to define the relationship among P, $\underline{V}$, and T, then the partial derivative of $\underline{V}$ with respect to T at constant P will be known, and the integration may proceed.

$$\underline{S}_2 - \underline{S}_1 = \int_1^2 d\underline{S} = -\int_{P_1}^{P_2} \left(\frac{\partial \underline{V}}{\partial T}\right)_P dP \tag{3.26b}$$

To compare the properties of a gas with those of condensed phases (i.e., liquid or solids), let us introduce the appropriate P-V-T relationship into the thermodynamic relations. For an ideal gas, $P\underline{V} = RT$, where $\underline{V}$ is the molar volume:

$$\left(\frac{\partial \underline{V}}{\partial T}\right)_P = \frac{R}{P}$$

hence

$$\left(\frac{\partial \underline{S}}{\partial P}\right)_T = -\frac{R}{P}$$

$$d\underline{S} = -\frac{R}{P}\,dP \qquad \text{at constant temperature} \tag{3.27}$$

For a change in pressure from 1 atm to 10 atm at constant temperature:

$$\Delta \underline{S} = \underline{S}_2 - \underline{S}_1 = -\int_{P_1}^{P_2} R\,\frac{dP}{P} = -R \ln P\Big|_1^{10} = -R \ln 10$$

$$\Delta \underline{S} = -19.14 \text{ J/(mol·K)}$$

To repeat the calculation for a solid, we need a description of the temperature dependence of volume at constant pressure, $(\partial \underline{V}/\partial T)_P$. The volumetric thermal expansion coefficient, α_V, is defined as

$$\alpha_V \equiv \frac{1}{\underline{V}}\left(\frac{\partial \underline{V}}{\partial T}\right)_P \tag{3.28a}$$

Hence:

$$\left(\frac{\partial \underline{V}}{\partial T}\right)_P = \underline{V}\alpha_V$$

For a condensed phase, it is reasonable to assume that $\underline{V}$ and α_V are constant over the pressure range of interest, hence:

$$\Delta\underline{S} = \underline{S}_2 - \underline{S}_1 = -\int_1^{10} \underline{V}\,\alpha_V\,dP = -\underline{V}\,\alpha_V(10\text{–}1) = -9\underline{V}\,\alpha_V \qquad \textbf{(3.28b)}$$

To evaluate the expression requires values of $\underline{V}$ and α_V. The molar volume, $\underline{V}$, is simply the molecular weight of the material divided by the density. A typical value for a metal, such as copper, is:

$$\underline{V} = \frac{\text{mol wt}}{\text{density}} = \frac{63.55\ \text{g/mol}^{-1}}{8.96\ \text{g/cm}^3} = 7.09\ \text{cm}^3/\text{mol}$$

$$\underline{V} = 7.09 \times 10^{-6}\ \text{m}^3/\text{mol}$$

Volumetric thermal expansion coefficients of solids are not usually tabulated. Data on thermal expansion are generally available as *linear* thermal expansion coefficients,[2]

$$\alpha_l \equiv \frac{1}{l}\left(\frac{\partial l}{\partial T}\right)_\sigma$$

For isotropic materials (i.e., materials whose properties do not vary with direction), the relationship between the two is:

$$\begin{aligned} \underline{V} &= l^3 \\ \ln \underline{V} &= 3\ \ln/l \\ \frac{dV}{V} &= 3\,\frac{dl}{l} \qquad \text{or} \qquad \frac{1}{V}\left(\frac{dV}{dT}\right)_P = \frac{3}{l}\left(\frac{\partial l}{\partial T}\right)_\sigma \\ \alpha_V &= 3\alpha_l \end{aligned} \qquad \textbf{(3.29)}$$

The value of α_l for copper is $16.7 \times 10^{-6}\ \text{K}^{-1}$.
Substituting these values in Eq. 3.28b yields:

$$\Delta\underline{S} = -3\alpha_l\underline{V}\,\Delta P = -3\underset{(\text{K}^{-1})}{(16.7 \times 10^{-6})}\underset{(\text{m}^3/\text{mol})}{(7.09 \times 10^{-6})} \times \underset{(\text{atm})}{9} \times \underset{(\text{Nm}^{-2}\ \text{atm}^{-1})}{1.013 \times 10^5}$$

$$\Delta\underline{S} = \underline{S}_2 - \underline{S}_1 = -3.24 \times 10^{-4}\ \text{J/(mol·K)}$$

Thus for a change in pressure from 1 to 10 atm, the specific entropy of an ideal gas changes by -19.14 J/(mol·K) compared to a change of -3.24×10^{-4} J/(mol·K)

---

[2]Pressure is uniform triaxial stress. The linear (unidimensional) analogy for constant pressure is constant stress, σ.

for a solid, such as copper. Because the specific volumes and thermal expansion coefficients are very much lower than the same quantities for gases, the specific entropy of solids, and liquids for that matter, is relatively independent of pressure. Knowing this, we can derive a useful simplification in entropy calculations involving solids at pressures other than one atmosphere. For example, the standard entropy of copper at 298 K is 33.14 ± 0.21 J/(mol·K). The correction to account for the change in pressure to 10 atm [-3.24×10^{-4} J/(mol·K)] is less than the experimental uncertainty in the value of the standard entropy. Of course, some correction will be required for very large pressure changes, but these situations are unusual.

3.6 IDEAL GAS

In Section 1.15 it was asserted that the enthalpy of an ideal gas is independent of pressure at constant temperature. Using the techniques described in this chapter, it is possible to prove this assertion as follows:

$$d\underline{H} = T\,d\underline{S} + \underline{V}\,dP \tag{3.23}$$

$$\left(\frac{\partial \underline{H}}{\partial P}\right)_T = T\left(\frac{\partial \underline{S}}{\partial P}\right)_T + \underline{V}$$

From Eq. 3.26a:

$$\left(\frac{\partial \underline{S}}{\partial P}\right)_T = -\left(\frac{\partial \underline{V}}{\partial T}\right)_P \quad \text{and} \quad \left(\frac{\partial \underline{H}}{\partial P}\right)_T = \underline{V} - T\left(\frac{\partial \underline{V}}{\partial T}\right)_P$$

For an ideal gas, $P\underline{V} = RT$

$$\left(\frac{\partial \underline{H}}{\partial P}\right)_T = \underline{V} - T\frac{R}{P} = \underline{V} - \underline{V}$$

$$\left(\frac{\partial \underline{H}}{\partial P}\right)_T = 0 \tag{3.30a}$$

Using a similar technique, but operating on Eq. 3.22, it can be shown that the internal energy of an ideal gas is independent of volume at constant temperature,

$$\left(\frac{\partial \underline{U}}{\partial \underline{V}}\right)_T = 0 \tag{3.30b}$$

3.7 ENTROPY OF MIXING

The relationships derived thus far enable us to calculate the entropy of mixing of two *ideal* gases. To begin, let us calculate the change in entropy of one gas, labeled *a*, as it expands isothermally from V_a to V_T in the apparatus shown schematically in Figure 3.3*a*.

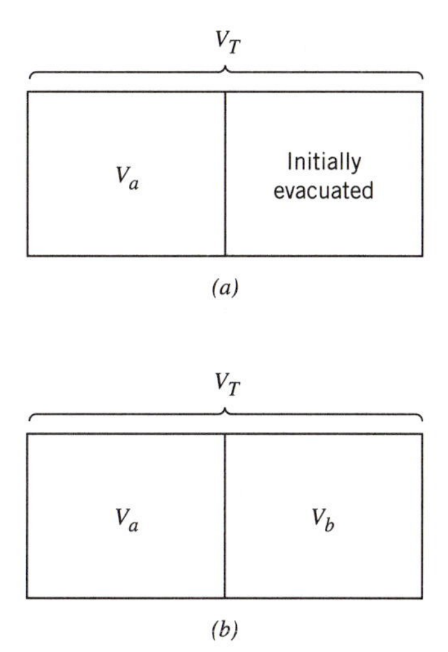

Figure 3.3 Entropy changes for (*a*) one gas and (*b*) two gases.

Applying the Euler method (Section A.18) to the expression for changes in the molar Helmholtz free energy, dF:

$$dF = -S\,dT - P\,dV$$

$$\left(\frac{\partial S}{\partial V}\right)_P = \left(\frac{\partial P}{\partial T}\right)_V = \frac{nR}{V} \qquad \text{ideal gas}$$

$$\int_{S_1}^{S_2} dS = \int_{V_a}^{V_T} \frac{nR}{V}\,dV = n \int_{V_a}^{V_T} R\,d\ln V$$

$$S_2 - S_1 = nR \ln \frac{V_T}{V_a}$$

Now consider two gases on either side of a partition, a on the left, and b on the right, each at the same temperature and pressure (Figure 3.3*b*). Upon removal of the partition, the two gases will mix. Because ideal gases do not interact, we can calculate the total entropy change as the sum of the two entropy changes of the individual gases:

$$S_2 - S_1 = n_a(\underline{S}_2 - \underline{S}_1)_a + n_b(\underline{S}_2 - \underline{S}_1)_b$$

$$S_2 - S_1 = Rn_a \ln\left(\frac{V_a + V_b}{V_a}\right) + Rn_b \ln\left(\frac{V_a + V_b}{V_b}\right)$$

Per mole of mixture, this becomes

$$\frac{S_2 - S_1}{n_a + n_b} = \frac{n_a}{n_a + n_b} R \ln\left(\frac{V_a + V_b}{V_a}\right) + \frac{n_b}{n_a + n_b} R \ln\left(\frac{V_a + V_b}{V_b}\right)$$

Noting that $n_a/(n_a + n_b) = x_a$, the mole fraction of a, and that the term $(S_2 - S_1)/(n_a + n_b)$ is the entropy of mixing per mole of the mixture,

$$\Delta \underline{S}_{\text{mix}} = x_a R \ln\left(\frac{V_a + V_b}{V_a}\right) + x_b R \ln\left(\frac{V_a + V_b}{V_b}\right)$$

For an ideal gas, the volumes V_a and V_b are proportional to the number of moles, n_a and n_b, respectively, at constant temperature and pressure.

$$\frac{V_a + V_b}{V_a} = \frac{n_a + n_b}{n_a} = \frac{1}{x_a}$$

Therefore:

$$\Delta \underline{S}_{\text{mix}} = -R[x_a \ln x_a + x_b \ln x_b] \qquad \textbf{(3.31)}$$

In general, the entropy of mixing of n ideal gases is:

$$\Delta \underline{S}_{\text{mix}} = -R \sum_{i=1}^{i=n} x_i \ln x_i \qquad \textbf{(3.32)}$$

where x_i = mole fraction of component i.

3.8 HEAT CAPACITY

In Section 1.13 the constant pressure heat capacity was defined as

$$C_P \equiv \left(\frac{\partial \underline{H}}{\partial T}\right)_P$$

A more general way of defining heat capacities operates through the entropy function. Noting that $dS = \delta Q_{\text{rev}}/T$, and that the term δQ is the product of a heat capacity and the temperature change ($\delta Q = C\, dT$), the expression for dS can be written as

$$d\underline{S}_{\text{I}} = C_{\text{I}} \frac{dT}{T}$$

or

$$C_{\text{I}} \equiv T\left(\frac{\partial \underline{S}}{\partial T}\right)_{\text{I}} \qquad \textbf{(3.33)}$$

where C_{I} represents the heat capacity under the condition I. For example, C_P is the heat capacity at constant pressure.

To see that Eq. 3.33 yields familiar results, consider:

$$d\underline{H} = T\,d\underline{S} + \underline{V}\,dP$$

$$\left(\frac{\partial \underline{H}}{\partial T}\right)_P = T\left(\frac{\partial \underline{S}}{\partial T}\right)_P = C_P$$

The usefulness of the more general definition of heat capacity becomes apparent when we ask for the heat capacity of a material under special conditions such as constant magnetic field and pressure. In this case, the heat capacity is:

$$C_{\mathcal{H}} = T\left(\frac{\partial \underline{S}}{\partial T}\right)_{\mathcal{H},P}$$

where $\mathcal{H}$ is the magnetic field intensity.

3.9 VARIATION OF HEAT CAPACITY

The variation of heat capacity C_P with pressure at constant temperature can be derived using a technique similar to the one used to derive Eq. 3.30a. Specific enthalpy is a function of temperature and pressure.

$$\underline{H} = \underline{H}(T, P)$$

$$d\underline{H} = \left(\frac{\partial \underline{H}}{\partial T}\right)_P dT + \left(\frac{\partial \underline{H}}{\partial P}\right)_T dP$$

From the definition of heat capacity, the equation for $d\underline{H}$, and Eq. 3.30a:

$$d\underline{H} = C_P\,dT + \left[\underline{V} - T\left(\frac{\partial \underline{V}}{\partial T}\right)_P\right] dP$$

Applying the Euler relationship (Section A.18) yields:

$$\left(\frac{\partial C_P}{\partial P}\right)_T = -T\left(\frac{\partial^2 \underline{V}}{\partial T^2}\right) dP \qquad \textbf{(3.34)}$$

For an ideal gas, $P\underline{V} = RT$:

$$\left(\frac{\partial C_P}{\partial P}\right)_T = -T\frac{\partial}{\partial T}\left(\frac{R}{P}\right)_P = 0$$

Thus the heat capacity of an ideal gas is independent of pressure. The variation of heat capacity with pressure for nonideal gases can be derived from the equation of state, the P-V-T relation.

For solids and liquids, the relation between volume and temperature at constant pressure is expressed by the thermal expansion coefficient α_v, which is defined as

$$\alpha_V \equiv \frac{1}{\underline{V}}\left(\frac{\partial \underline{V}}{\partial T}\right)_P \quad \text{or} \quad \alpha_V \underline{V} = \left(\frac{\partial \underline{V}}{\partial T}\right)_P \tag{3.35a}$$

The variation of heat capacity with pressure is then:

$$\left(\frac{\partial C_P}{\partial P}\right)_T = -T\frac{\partial}{\partial T}(\alpha_V \underline{V})_P = -T\alpha_V\left(\frac{\partial \underline{V}}{\partial T}\right)_P - T\underline{V}\left(\frac{\partial \alpha_V}{\partial T}\right)_P$$
$$\left(\frac{\partial C_P}{\partial P}\right)_T = -T\underline{V}\alpha_V^2 - T\underline{V}\left(\frac{\partial \alpha_V}{\partial T}\right)_P \tag{3.35b}$$

The variation of heat capacity of a solid or liquid is a function of the square of the thermal expansion coefficient and the temperature dependence of the thermal expansion coefficient. If α_V is small and independent of temperature, then C_P will not vary with pressure. If α_V is known as a function of T, the dependence of C_P on P can be derived from Equation 3.35b.

3.10 ISENTROPIC PRESSURE–TEMPERATURE RELATIONSHIP

The relationship between pressure and temperature for an adiabatic, reversible (isentropic) change in a closed system was derived in Section 1.17.

The same equation can be derived using the methods of this chapter. Based on the discussion in Chapter 2, we know that an adiabatic, reversible process in a closed system is one that can be characterized as isentropic—that is, one in which the entropy of the material does not change. The mathematical expression describing the change of T with P at constant S is $(\partial T/\partial P)_S$.

Applying the rule in Section A.15 of the Appendix, Eq. A14 yields:

$$\left(\frac{\partial T}{\partial P}\right)_S = -\frac{\left(\frac{\partial \underline{S}}{\partial P}\right)_T}{\left(\frac{\partial \underline{S}}{\partial T}\right)_P} = -\frac{T\left(\frac{\partial \underline{S}}{\partial P}\right)_T}{T\left(\frac{\partial \underline{S}}{\partial T}\right)_P} = -\frac{T}{C_P}\left(\frac{\partial \underline{S}}{\partial P}\right)_T \tag{3.36}$$

From Eq. 3.26a, $-(\partial \underline{S}/\partial P)_T = (\partial \underline{V}/\partial T)_P$, we have

$$\left(\frac{\partial T}{\partial P}\right)_S = +\frac{T}{C_P}\left(\frac{\partial \underline{V}}{\partial T}\right)_P \tag{3.37}$$

For an ideal gas ($P\underline{V} = RT$), $(\partial \underline{V}/\partial T)_P = R/P$

$$\left(\frac{\partial T}{\partial P}\right)_S = \frac{T}{C_P}\frac{R}{P}$$

$$\frac{dT}{T} = \frac{R}{C_P}\frac{dP}{P}$$

Integrating between states 1 and 2, assuming a constant C_P:

$$\int_{T_1}^{T_2}\frac{dT}{T} = \frac{R}{C_P}\int_{P_1}^{P_2}\frac{dP}{P}$$

$$\frac{T_2}{T_1} = \left(\frac{P_2}{P_1}\right)^{R/C_P}$$

The result is the same as the one derived as Eq. 1.30 in Section 1.17.

The change of temperature with relatively modest pressure changes is considerable. For example, the change in temperature for a diatomic gas compressed from a pressure of 1 atm to 10 atm is:

$$\left(\frac{T_2}{T_1}\right) = \left(\frac{10}{1}\right)^{R/(7/2R)} = 10^{2/7} = 1.93$$

A gas initially at 273 K and 1 atm would upon isentropic compression to 10 atm achieve a temperature of 527 K. This change is much larger than the corresponding change in a solid, as illustrated in Section 3.11.

3.11 ISENTROPIC COMPRESSION OF SOLIDS

To determine the effect of an adiabatic, reversible change in pressure (or stress) on the temperature of a solid, we may proceed as we did in Section 3.10. Equation 3.37 is valid for all materials—gases, liquids, or solids.

$$\left(\frac{\partial T}{\partial P}\right)_{\underline{S}} = \frac{T}{C_P}\left(\frac{\partial \underline{V}}{\partial T}\right)_P$$

In the case of solids, the term $(\partial \underline{V}/\partial T)_P$ is expressed in terms of the *volumetric* thermal expansion coefficient, α_V. For isotropic solids α_V can be shown to be three times the *linear* thermal expansion coefficient α_l. Then:

$$\left(\frac{\partial T}{\partial P}\right)_{\underline{S}} = \frac{T}{C_P}\frac{\underline{V}}{\underline{V}}\left(\frac{\partial \underline{V}}{\partial T}\right)_P = \frac{T\underline{V}\alpha_V}{C_P} = \frac{3T\underline{V}\alpha_l}{C_P} \quad \textbf{(3.38)}$$

At constant entropy,

$$\frac{dT}{T} = \frac{3\underline{V}\alpha_l}{C_P} dP$$

Integrating, assuming that α_l, $\underline{V}$, and C_P are independent of P:

$$\ln \frac{T_2}{T_1} = \frac{3\underline{V}\alpha_l}{C_P} (P_2 - P_1)$$

Applying values for copper for a change in pressure from 1 atm to 10 atm, we have

$$\ln \frac{T_2}{T_1} = \frac{3(7.09 \times 10^{-6})(16.7 \times 10^{-6})}{30} 9(1.013 \times 10^5)$$

$$\ln \frac{T_2}{T_1} = 10.76 \times 10^{-6} = \ln \left[\frac{T_1 + (T_2 - T_1)}{T_1}\right] = \ln \left[1 + \frac{\Delta T}{T_1}\right]$$

Note that $\ln(1 + x) = x$ when x is small

$$\frac{\Delta T}{T} = 10.76 \times 10^{-6}$$

at 300 K, $\Delta T = 3.23 \times 10^{-3}$ K.

In contrast to gases, the temperature of solids (and liquids) is relatively independent of applied pressure for adiabatic, reversible compressions. This holds true for modest pressure changes. Very large pressure changes, such as those produced by explosives, can result in larger temperature changes in solids.

3.12 THERMOELASTIC EFFECT

The change in temperature of a solid upon *elastic* (reversible) deformation under *adiabatic* conditions is called the thermoelastic effect. In a sense, we dealt with this phenomenon in Section 3.11 when we calculated the temperature change of a metal resulting from an adiabatic, reversible change in pressure from 1 atm to 10 atm.

One can think of pressure as a uniform, *triaxial stress,* the same stress in each of three orthogonal directions. The thermoelastic effect can also be observed under *uniaxial stress,* that is, when a force is applied to a solid in *one* direction. Thermoelasticity in the uniaxial case is interesting because the direction of the temperature change (positive or negative) is not the same for all classes of solids. The direction of the temperature change for elongated elastomers (rubber), for example, is different from the change in metals.

To establish the background for thermoelasticity, consider the basic property relation (Eq. 3.4) with an additional term to account for uniaxial forces and changes in length:

$$dU = T\,dS - P\,dV + f\,dl \qquad \textbf{(3.39)}$$

where f is the uniaxial force acting on the solid[3] and dl is the change in linear dimension. The difference in the signs of the $P\,dV$ term and the $f\,dl$ term arises because pressure is considered to be positive when it is *compressive,* and the uniaxial force on a solid is considered to be positive when it is *tensile.*

The G function is defined as

$$G = U + PV - TS$$

and the corresponding expression for dG is

$$dG = -S\,dT + P\,dV + f\,dl \qquad \textbf{(3.40)}$$

When an axial force f is applied isentropically to a solid cylinder of length l and cross-sectional area A, the temperature change of the solid may be expressed as $(\partial T/\partial f)_{S,V}$. If we assume that deformations take place at constant volume, the P-V terms may be ignored, and the expression becomes $(\partial T/\partial f)_S$. This may be expanded to (using the method of Eq. A.14 in Appendix A):

$$\left(\frac{\partial T}{\partial f}\right)_S = -\frac{\left(\frac{\partial S}{\partial f}\right)_T}{\left(\frac{\partial S}{\partial T}\right)_f} = -\frac{T\left(\frac{\partial S}{\partial f}\right)_T}{T\left(\frac{\partial S}{\partial T}\right)_f}$$

$$\left(\frac{\partial T}{\partial f}\right)_S = -\frac{T\left(\frac{\partial S}{\partial f}\right)_T}{nC_f} \qquad \textbf{(3.41)}$$

where n is the number of moles in the sample being stretched, and C_f is the heat capacity at constant force.

Applying cross differentiation (Eq. A.13) to Eq. 3.40 at constant volume yields

$$\left(\frac{\partial S}{\partial l}\right)_T = -\left(\frac{\partial f}{\partial T}\right)_l \qquad \textbf{(3.42)}$$

[3]The symbol chosen to represent force is a lowercase f to avoid confusion with the upper case F used for the Helmholtz free energy.

Applying the chain rule (Eq. A.11) to the left-hand side and the method of Eq. A.14 to the right-hand side of Eq. 3.42, we have

$$\left(\frac{\partial S}{\partial f}\right)_T \left(\frac{\partial f}{\partial l}\right)_T = -\left[-\frac{\left(\frac{\partial l}{\partial T}\right)_f}{\left(\frac{\partial l}{\partial f}\right)_T}\right] = +\left(\frac{\partial l}{\partial T}\right)_f \left(\frac{\partial f}{\partial l}\right)_T$$

$$\left(\frac{\partial S}{\partial f}\right)_T = \left(\frac{\partial l}{\partial T}\right)_f$$

Substituting in Eq. 3.41 yields

$$\left(\frac{\partial T}{\partial f}\right)_S = -\frac{T\left(\frac{\partial l}{\partial T}\right)_f}{nC_f} = -\frac{T\frac{l}{l}\left(\frac{\partial l}{\partial T}\right)_f}{nC_f} = -\frac{Tl\,\alpha_l}{nC_f} \qquad \textbf{(3.43)}$$

In dealing with forces applied to deform solids, it is more useful to think in terms of force per unit area (i.e., stress), σ.

$$\sigma = \frac{f}{A}$$

$$f = \sigma A \qquad \textbf{(3.44)}$$

$$\left(\frac{\partial f}{\partial \sigma}\right)_s = A$$

Multiplying both sides of Eq. 3.43 by Eq. 3.44 gives

$$\left(\frac{\partial T}{\partial f}\right)_s \left(\frac{\partial f}{\partial \sigma}\right)_A = -\frac{TlA\,\alpha_l}{nC_f} = -\frac{TV\alpha_l}{nC_f}$$

where the product of A and l is the volume, V, and V/n is $\underline{V}$, the molar volume. The heat capacity term C_f is replaced by C_σ, which, in turn, is essentially the constant pressure heat capacity, C_P.

$$\left(\frac{\partial T}{\partial \sigma}\right)_s = -\frac{T\underline{V}\alpha_l}{C_P} \qquad \textbf{(3.45)}$$

Assuming that the changes in temperature (ΔT) are small:

$$\Delta T = -\left(\frac{T\underline{V}\alpha}{C_P}\right)\Delta\sigma \qquad \textbf{(3.46)}$$

Let us evaluate the magnitude of this change for the case of iron stressed to level of 10,000 psi or 6.9×10^7 N/m^2. For iron:

$$
\begin{aligned}
\text{Molecular weight} &= 55.85 \text{ g/mol} \\
\text{Density} &= 7.87 \text{ g/cm}^3 \\
\underline{V} &= 7.10 \text{ cm}^3\text{/mol} = 7.10 \times 10^{-6} \text{ m}^3\text{/mol} \\
\alpha_1 &= 11.8 \times 10^{-6} \text{ K}^{-1} \\
C_\sigma = C_P &= 25.1 \text{ J/(mol·K)} \\
\text{Temperature (initial)} &= 300 \text{ K} \\
\Delta T &= -\frac{(300)(7.1 \times 10^{-6})(11.8 \times 10^{-6})(6.9 \times 10^7)}{25.1} \\
\Delta T &= -0.07 \text{ K}
\end{aligned}
$$

Two conclusions can be reached. First, the temperature change that occurs when elastically, a metal, such as iron, is stressed, adiabatically, and reversibly in *tension,* is negative. Second, the temperature change attendant on a reversible, adiabatic change in stress is small in the case of metals.

The change of temperature with stress for elastomers (rubber) differs from the change in metals. For rubber, the temperature change upon elongation shows an initial drop as in the case of metals, but then rises considerably. This phenomenon does not violate the conclusion drawn from Eq. 3.46. The explanation lies in the fact that the thermal expansion coefficient of rubber becomes *negative* when it is in an elongated condition. Reversible elongations of metals are quite small, well under one percent. Rubber, by contrast, can be elongated reversibly by several hundred percent. When in the elongated state (stretched by more than 20 or 30%), rubber will *contract* when heated.[4] The negative thermal expansion coefficient, when introduced into Eq. 3.46, produces a positive temperature change with increased stress (further elongation). The temperature changes induced in elastomers by large, reversible and adiabatic elongations ($>$100%) can be several degrees.

3.13 COMPRESSIBILITY

A physical property of solids that is often required in thermodynamic calculations, but seldom tabulated as such, is the isothermal compressibility β_T, a measure of the change of volume with pressure at constant temperature. It is defined[5] as follows:

$$\beta_T \equiv -\frac{1}{\underline{V}}\left(\frac{\partial \underline{V}}{\partial P}\right)_T \tag{3.47}$$

[4]The reason for this behavior will be discussed in Chapter 10.

[5]We may also define compressibility at constant entropy, $\beta_S = -1/\underline{V}(\partial \underline{V}/\partial P)_S$. The isentropic compressibility is applicable when the deformation takes place rapidly, that is, when there is not time enough for heat to be transferred to keep the temperature constant (e.g., during acoustic measurement of elastic constants).

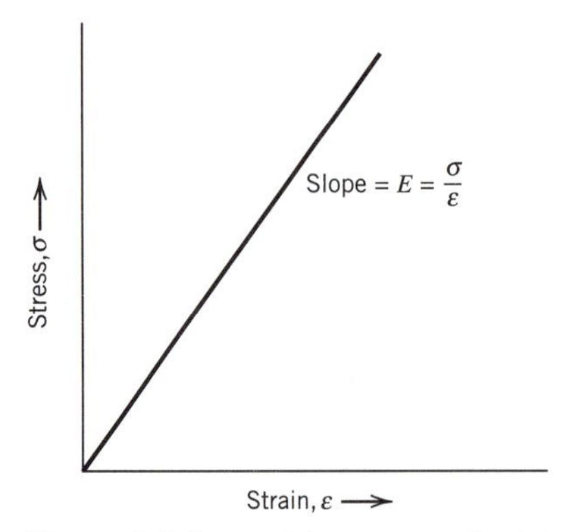

Figure 3.4 Stress (σ) versus strain (ε) for a perfectly elastic substance.

When solids, such as metals and ceramics, are stressed elastically, they usually respond by elongating or contracting linearly, as in Figure 3.4. The constant of proportionality relating the stress σ to the strain ε is the elastic modulus E, sometimes called Young's modulus.

$$E = \frac{\sigma}{\varepsilon} \tag{3.48}$$

In addition to the dimensional change in the direction of the applied stress, a solid will change dimension in directions perpendicular to the applied stress. For example, a solid stressed in tension in the x direction will elongate in the x direction and contract in the y and z directions. The ratio of these changes is known as Poisson's ratio ν. When a stress σ_x is applied in the x direction, the elongation in the x direction will be $\varepsilon_x = \sigma_x/E$. The dimensional change in the y direction (contraction) will be $\varepsilon_y = -\nu\sigma_x/E$.

Assuming that a solid is isotropic,

$$\begin{aligned}
\varepsilon_x &= \frac{\sigma_x}{E} - \frac{\nu\sigma_y}{E} - \frac{\nu\sigma_z}{E} \\
\varepsilon_y &= -\frac{\nu\sigma_x}{E} + \frac{\sigma_y}{E} - \frac{\nu\sigma_z}{E} \\
\varepsilon_z &= -\frac{\nu\sigma_x}{E} - \frac{\nu\sigma_y}{E} + \frac{\sigma_z}{E}
\end{aligned} \tag{3.49}$$

$$\varepsilon_x + \varepsilon_y + \varepsilon_z = \frac{\sigma_x + \sigma_y + \sigma_z}{E} - \frac{2\nu}{E}(\sigma_x + \sigma_y + \sigma_z)$$

$$\varepsilon_x + \varepsilon_y + \varepsilon_z = \frac{1 - 2\nu}{E}(\sigma_x + \sigma_y + \sigma_z)$$

If $\sigma_x = \sigma_y = \sigma_z$, we call the stresses *pressure.* The sign conventions in pressure and stress are different. Pressure is considered to be positive when it is compressive. In contrast, stress is considered to be positive when it is tensile.

$$\Delta P = -\sigma_x$$

Also:

$$\varepsilon_x + \varepsilon_y + \varepsilon_z = 3\,\frac{\Delta l}{l} = \frac{\Delta \underline{V}}{\underline{V}}$$

Therefore:

$$\beta_T = -\frac{1}{\underline{V}}\left(\frac{\Delta \underline{V}}{\Delta P}\right)_T = \frac{3(1 - 2\nu)}{E} \tag{3.50}$$

The isothermal compressibility can thus be calculated from the values of physical properties, E and ν, that are usually tabulated or easily measured. As an example, the compressibility of iron is:

$$\beta_T = \frac{3(1 - 2[0.3])}{196 \times 10^9} = 6.1 \times 10^{-12}\,\frac{1}{\text{N}\cdot\text{m}^2}$$

$$\beta_T = 6.2 \times 10^{-7}\ \text{atm}$$

3.14 MAGNETIC EFFECTS

As an example of a process in which work other than mechanical (P-V or f-l) is done, consider a material in a magnetic field. The work done to change the magnetic flux in the material (Ref. 1) is:

$$\delta W = +\int_0^V \vec{\mathcal{H}}\, d\vec{B}\, dV \tag{3.51}$$

where $\vec{B}$ = magnetic induction
$\vec{\mathcal{H}}$ = magnetic field intensity
V = volume

For the purpose of this analysis, assume that the sample of the material in the magnetic field is isotropic. This eliminates the need to use vector products and vector notation. Assuming that there is no heat flow,

$$dU = +\delta W + V\mathcal{H}\, dB \tag{3.52}$$

By definition the magnetic induction B is

$$B \equiv \mu_0(\mathcal{H} + M) \tag{3.53}$$

where μ_0 = permeability of free space ($4\pi \times 10^{-7}$ volt seconds/amp. meter)
M = magnetization or magnetic moment per unit volume

Substituting Eq. 3.53 in Eq. 3.52 yields:

$$dU = \mu_0 V\mathcal{H}\, d\mathcal{H} + \mu_0 V\mathcal{H}\, dM \tag{3.54a}$$

On a molar basis:

$$d\underline{U} = \mu_0 \underline{V}\mathcal{H}\, d\mathcal{H} + \mu_0 \underline{V}\mathcal{H}\, dM \tag{3.54b}$$

The first term in the equation for dU (Eq. 3.54b) is the energy required to change the magnetic field intensity for the volume in question in vacuum—that is, without the material present. The second term in the equation gives the energy required to change the magnetization in the material. Because we are interested in the thermodynamics of the material, we define dU' as

$$d\underline{U}' = \mu_0 \underline{V}\mathcal{H}\, dM \tag{3.55}$$

When we include the terms for thermal and elastic energy, the relationship becomes (neglecting volume change):

$$d\underline{U}' = T\, d\underline{S} + \mu_0 \underline{V}\mathcal{H}\, dM + \underline{V}\sigma\, d\varepsilon \tag{3.56}$$

The expressions for enthalpy can be modified to take into account magnetic and mechanical effects as follows:

$$\underline{H}' = \underline{U}' - \mu_0 \underline{V}\mathcal{H}M - \underline{V}\sigma\varepsilon \tag{3.57}$$

This yields

$$d\underline{H}' = T\, d\underline{S} - \mu_0 \underline{V}M\, d\mathcal{H} - \underline{V}\varepsilon\, d\sigma \tag{3.58}$$

Equation 3.58 can be used to demonstrate the application of Maxwell's relations to variables other than temperature, pressure, and volume. For example, magnetostriction is the effect of a changing magnetic field on the dimensions of a specimen.[6]

[6]Recognize that we have assumed an isotropic material. Linear magnetostriction is strongly dependent on crystallographic orientation. This dependence must be taken into account in the analysis of single crystals or specimens showing a strong preferred orientation (anisotropy).

Applying the Maxwell relations to the last two terms in Eq. 3.58, we have:

$$\left(\frac{\partial(\underline{V}\varepsilon)}{\partial\mathcal{H}}\right)_{\underline{S},\sigma} = \left(\frac{\partial(\mu_0\underline{V}M)}{\partial\sigma}\right)_{\underline{S},\mathcal{H}}$$

Assuming that $\underline{V}$ is unchanged in the process,

$$\left(\frac{\partial\varepsilon}{\partial\mathcal{H}}\right)_{\underline{S},\sigma} = \mu_0\left(\frac{\partial M}{\partial\sigma}\right)_{\underline{S},\mathcal{H}} \tag{3.59}$$

The magnetostrictive strain ε is

$$\varepsilon = \int_0^{\mathcal{H}}\left(\frac{\partial\varepsilon}{\partial\mathcal{H}}\right)_{\underline{S},\sigma} d\mathcal{H} = \mu_0\int_0^{\mathcal{H}}\left(\frac{\partial M}{\partial\sigma}\right)_{\underline{S},\mathcal{H}} d\mathcal{H}$$

The susceptibility χ is defined as the magnetization divided by the magnetic field.

$$\chi = \frac{M}{\mathcal{H}} \tag{3.60}$$

Substituting in the Eq. 3.59, we have

$$\left(\frac{\partial\varepsilon}{\partial\mathcal{H}}\right)_{\underline{S},\sigma} = \mu_0\left(\frac{\partial(\chi\mathcal{H})}{\partial\sigma}\right)_{\underline{S},\mathcal{H}} = \mu_0\mathcal{H}\left(\frac{\partial\chi}{\partial\sigma}\right)_{\underline{S},\mathcal{H}}$$

$$d\varepsilon = \mu_0\left(\frac{\partial\chi}{\partial\sigma}\right)_{\mathcal{H},\underline{S}}\mathcal{H}\,d\mathcal{H}$$

$$\int_0^{\varepsilon} d\varepsilon = \mu_0\int_{\mathcal{H}=0}^{\mathcal{H}}\left(\frac{\partial\chi}{\partial\sigma}\right)_{\mathcal{H},\underline{S}}\mathcal{H}\,d\mathcal{H}$$

Assuming that $\partial\chi/\partial\sigma$ is constant,

$$\varepsilon = \mu_0\left(\frac{\partial\chi}{\partial\sigma}\right)_{\mathcal{H},\underline{S}}\frac{\mathcal{H}^2}{2} \tag{3.61}$$

This magnetostrictive effect describes the coupling between magnetic and elastic properties of materials, in particular, the change in length of a sample (ε) with a change in the magnetic field ($\mathcal{H}$). The relationship described by Eq. 3.61 holds true for ferromagnetic materials, at small values of $\mathcal{H}$ below their saturation magnetization. When a material reaches saturation magnetization, all its magnetic domains are aligned. Above this level, the effect of changing magnetic field is much lower.

Another effect of changing magnetic field involves the change in temperature of a material with changing magnetic field. This is especially observable at very low temperatures such as those approaching absolute zero. In fact the phenomenon, called adiabatic demagnetization, is used to approach absolute zero using paramagnetic salts. To use the adiabatic demagnetization technique, a sample is cooled to a very low temperature by immersing it in a bath of liquid helium. The sample is then subjected to high magnetic field, and the thermal energy that is evolved in the sample is removed as heat to keep the sample temperature constant. The magnetic field is then removed under adiabatic conditions, and the sample temperature falls to a level lower than the liquid helium bath. To follow this process in thermodynamic terms, consider the specific entropy of a sample as a function of temperature and magnetic field.

$$\underline{S} = \underline{S}\,(T, \mathcal{H})$$

$$d\underline{S} = \left(\frac{\partial \underline{S}}{\partial T}\right)_{\mathcal{H}} dT + \left(\frac{\partial \underline{S}}{\partial \mathcal{H}}\right)_{T} d\mathcal{H} \tag{3.62}$$

The first term in Eq. 3.62 is related to the heat capacity in a constant magnetic field.

$$\left(\frac{\partial \underline{S}}{\partial T}\right)_{\mathcal{H}} = \frac{C_{\mathcal{H}}}{T}$$

To evaluate the second term, consider the expression for dG for a material in a magnetic field without elastic or volume effects.

$$d\underline{G} = -\underline{S}\, dT - \mu_0 \underline{V} M\, d\mathcal{H} \tag{3.63}$$

Applying cross differentiation,

$$\left(\frac{\partial \underline{S}}{\partial \mathcal{H}}\right)_{T} = \mu_0 \underline{V} \left(\frac{\partial M}{\partial T}\right)_{\mathcal{H}}$$

Equation 3.62 then becomes:

$$d\underline{S} = \frac{C_{\mathcal{H}}}{T}\, dT + \mu_0 \underline{V} \left(\frac{\partial M}{\partial T}\right)_{\mathcal{H}} \tag{3.64}$$

For a paramagnetic salt that follows Curie's law, $\chi = \kappa/T$, where κ is a constant:

$$M = \frac{\kappa \mathcal{H}}{T} \tag{3.65}$$

$$\left(\frac{\partial M}{\partial T}\right)_{\mathcal{H}} = -\frac{\kappa \mathcal{H}}{T^2} \tag{3.66}$$

Equation 3.64 then becomes:

$$d\underline{S} = \frac{C_{\mathcal{H}}}{T}\, dT - \frac{\mu_0 \underline{V}\, \kappa}{T^2}\, \mathcal{H} d\mathcal{H} \qquad \textbf{(3.67))}$$

Recognizing that heat added to or subtracted from the sample, dQ, will be TdS for a reversible process, it is useful to express Eq. 3.67 as follows:

$$Td\underline{S} = C_{\mathcal{H}} dT - \mu_0 \underline{V} \frac{\kappa}{T}\, \mathcal{H} d\mathcal{H} \qquad \textbf{(3.68))}$$

The first step in adiabatic demagnetization involves increasing the magnetic field from zero to a value $\mathcal{H}_1$ at a constant temperature, T_1. The termal energy removed as heat is the temperature multiplied by the change in entropy.

$$Q = T_1 \int_{\underline{S}_0}^{\underline{S}_1} d\underline{S} = -\mu_0 \underline{V} \frac{\kappa}{T_1} \int_0^{\mathcal{H}_1} \mathcal{H} d\mathcal{H}$$

$$Q = T_1 (\underline{S}_1 - \underline{S}_0) = -\mu_0 \underline{V} \frac{\kappa}{T_1} \frac{\mathcal{H}_1^2}{2}$$

In the second step in adiabatic demagnetization, the sample is insulated thermally and the magnetic field reduced from $\mathcal{H}_1$ to zero. The entropy change in the sample is zero (dS = 0), because no heat is transferred and the change is reversible.

$$C_{\mathcal{H}} dT = \mu_0 \underline{V} \frac{\kappa}{T}\, \mathcal{H} d\mathcal{H}$$

$$C_{\mathcal{H}} T dT = \mu_0 \underline{V}\, \kappa\, \mathcal{H} d\mathcal{H}$$

$$\int_{T_1}^{T_2} C_{\mathcal{H}} T dT = \mu_0 \underline{V}\, \kappa \int_{\mathcal{H}_1}^{0} \mathcal{H} d\mathcal{H}$$

$$\int_{T_1}^{T_2} C_{\mathcal{H}} T dT = -\mu_0 \underline{V}\, \kappa\, \frac{\mathcal{H}_1^2}{2}$$

The left-hand side of Eq. 3.68 is not integrated because the heat capacity, $C_{\mathcal{H}}$, is a strong function of temperature. Nevertheless, it shows that the temperature after demagnetization T_2, will be less than the temperature before, T_1. Using this technique, temperatures of less than 10^{-5} kelvin have been reached.

REFERENCES

1. Lewis, G. N., and Randall, M., *Thermodynamics,* 2nd rev. ed., K. S. Pitzer and L. Brewer, McGraw-Hill, New York, 1961.
2. Darken, L. S., and Gurry, R. W., *Physical Chemistry of Metals,* McGraw-Hill, New York, 1953.

3. Stull, D. R., and Prophet, H., *JANAF Thermochemical Tables,* 2nd ed., National Standards Research Data Service–National Bureau of Standards, Washington, DC, 1971.
4. Hultgren, R., Orr, R. L., Anderson, P. D., and Kelley K. K., *Selected Values of Thermodynamic Properties of Metals and Alloys,* Wiley, New York, 1963.

PROBLEMS

3.1 Prove that $C_P - C_V = \alpha^2 \underline{V} T/\beta$. Calculate $\gamma = C_P/C_V$ for solid copper at 500 K, given $\underline{V} = 7.115$ cm³/mol, $\alpha = 5.42 \times 10^{-5}$ K^{-1}, and $\beta = 8.37 \times 10^{-12}$ m²/N.

3.2 At -5°C, the vapor pressure of ice is 3.012 mmHg and that of supercooled liquid water is 3.163 mmHg. The latent heat of fusion of ice is 5.85 kJ/mol at -5°C. Calculate ΔG and ΔS per mole for the transition from water to ice at -5°C.

3.3 A strong (nondeforming) capillary tube filled with mercury and closed at 0°C is heated to 10°C. Assuming that the volume of the capillary remains constant, calculate the resulting pressure. For mercury near room temperature the volumetric coefficient of expansion α_V is 18×10^{-5} per degree and the compressibility β is 3.7×10^{-6} per atmosphere.

3.4 **(a)** What can be said about the relative magnitudes of the heat capacities at constant pressure C_P and at constant volume C_V of an alloy such as Invar, which has a vanishingly small expansion coefficient?

(b) Can the difference between the heat capacities at constant pressure and at constant volume $(C_P - C_V)$ assume a negative value?

3.5 A container of liquid lead is to be used as a calorimeter to determine the heat of mixing of two metals, A and B. It has been determined by experiment that the "heat capacity" of the bath is 100 cal/°C at 300°C. With the bath originally at 300°C, the following experiments are performed.

1. A mechanical mixture of 1 g of A and 1 g of B is dropped into the calorimeter. A and B were originally at 25°C. When the two have dissolved, the temperature of the bath is found to have increased 0.20°C.
2. Two grams of a 50:50 (wt %) A-B alloy at 25°C is dropped similarly into the calorimeter. The temperature decreases 0.40°C.

(a) What is the heat of mixing of the 50:50 A-B alloy (per gram of alloy)?

(b) To what temperature does it apply?

3.6 The equilibrium freezing point of water is 0°C. At that temperature the latent heat of fusion of ice (the heat required to melt the ice) is 6030 J/mol.

(a) What is the entropy of fusion of ice at 0°C?

(b) What is the change of Gibbs free energy for ice → water at 0°C (J/mol)?

(c) What is the heat of fusion of ice at -5°C?

$C_{P(\text{ice})} = 0.5$ cal/(g·°C)

$C_{P(\text{water})} = 1.0$ cal/(g·°C)

(d) Repeat parts a and b at -5°C.

3.7 **(a)** What is the specific volume of iron at 298 K, in cubic meters per mole?

(b) Derive an equation for the change of entropy with pressure at constant temperature for a solid, expressed in terms of physical quantities usually available, such as the ones listed as data.

(c) The specific entropy of iron (entropy per mole) at 298 K and a pressure of 100 atm is needed for a thermodynamic calculation. The tabulated "standard entropy" (at 298 K and a pressure of 1 atm) is

$$S^{\circ}_{298} = 27.28 \text{ J/K·mol}$$

What percentage error would result if one assumed that the specific entropy at 298 K and 100 atm were equal to the value of S°_{298} given above?

DATA (for Iron)

$$C_P = 24 \text{ J K}^{-1} \text{ mol}^{-1}$$

$$\text{Compressibility} = 6 \times 10^{-7} \text{ atm}^{-1}$$

$$\text{Linear coefficient of thermal expansion} = 15 \times 10^{-6°} \text{ C}^{-1}$$

$$\text{Density} = 7.87 \text{ g/cm}^3$$

$$\text{Molecular weight} = 55.85 \text{ g/mol}$$

Note: It may be possible to solve this problem without using all the data given.

3.8 Derive an expression for the isentropic compressibility of a solid (β_S) in terms of isothermal compressibility (β_T) and other properties normally tabulated.

$$\beta_S \equiv -\frac{1}{V}\left(\frac{\partial V}{\partial P}\right)_S$$

$$\beta_T \equiv -\frac{1}{V}\left(\frac{\partial V}{\partial P}\right)_T$$

3.9 **(a)** Derive an expression for the change of temperature of a solid material that is compressed adiabaticlly and reversibly in terms of physical quantities usually available, such as the ones listed here.

(b) The pressure on a block of iron is increased by 1000 atm adiabatically and reversibly. What is the temperature change? The initial temperature of the iron is 298 K.

DATA (for Iron)

$$C_P = 24 \text{ J/mol·K}$$

$$\text{Compressibility} = 6 \times 10^{-7} \text{ atm}^{-1}$$

$$\text{Linear coefficient of thermal expansion} = 15 \times 10^{-6} \text{ K}^{-1}$$

$$\text{Density} = 7.87 \text{ g/cm}^3$$

$$\text{Molecular weight} = 55.85 \text{ g/mol}$$

Note: It may be possible to solve this problem without using all the data given.

3.10 A block of rubber weighing 100 g is to be compressed from a pressure of 1 atm to a pressure of 1001 atm (a change in pressure of 1000 atm).

(a) What is the volume of the block of rubber?

(b) The block of rubber is compressed isothermally at 298 K. What is its change of entropy?

(c) How much heat must be added (or removed) from the rubber to maintain it at a constant temperature of 298 K?
Specify whether heat is added or removed.

DATA

For rubber: Volumetric thermal expansion coefficient is $100 \times 10^{-6}\ \mathrm{K}^{-1}$.
Density is about 1.0 g/cm^3.

Note: For the purposes of this problem, assume that the volume of the rubber does not change much during the compression.

3.11 The First Law tells us that energy is a conserved quantity. Is Gibbs free energy also a conserved quantity?

Chapter 4

Equilibrium

The concept of equilibrium is fundamental to the study of thermodynamics. To say that a system is at equilibrium implies that the system is stable, that is, unchanging with time,[1] and that certain properties of the system are uniform throughout. Intuitively we can say that the temperature of the system should be uniform throughout. That does not mean, however, that the system must be homogeneous in form. From our common experience we know that pure ice can exist in equilibrium with pure, liquid water at 0°C and a pressure of one atmosphere. In terms of thermodynamics we would say that the two *phases,* liquid water and ice, can exist in equilibrium. A phase is a portion of matter that is uniform throughout, not only in chemical composition, but also in physical state.

When two or more phases are in equilibrium, there is no driving force for change, and the phases are stable. It is important to understand the criteria for equilibrium because the usefulness of many metallic, polymeric, and ceramic systems depends on the presence, at equilibrium, of various different phases in the material.

---

[1]Stability or equilibrium in the macroscopic sense does not mean that the individual particles or molecules of a given system are stationary. If we were in a position to follow the paths of individual molecules in a system at equilibrium, we would observe very rapid motion. However, for the purpose of classical thermodynamics, the system is at equilibrium, because there is no change in its *macroscopic* state.

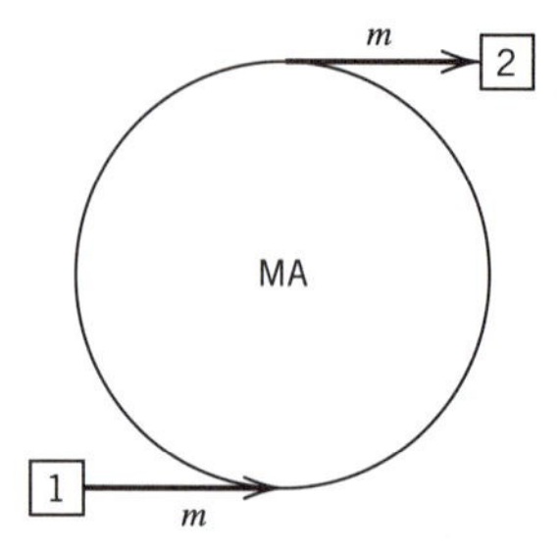

Figure 4.1 Machine MA to transform mass m from state 1 to state 2.

4.1 CONDITION OF EQUILIBRIUM

As a principle, we can assert that two states are in equilibrium when no reversible work can be done by having the system change between those two states. States 1 and 2 are in equilibrium if:

$$\delta W_{\text{rev. } 1\to 2} = 0 \tag{4.1}$$

Consider the diagram in Figure 4.1. The device MA is a steady state machine that converts a material from state 1 to state 2. Material enters the machine in state 1 and leaves in state 2. The criterion for equilibrium between states 1 and 2 is that no reversible work can be done by the machine, MA, based on the exchange of the material or energy between 1 and 2.

The first conclusion one can draw from the foregoing statement is that the temperatures in the two states must be equal, $T_2 = T_1$. This is a consequence of the second law of thermodynamics. If the temperatures were unequal, one could conceivably devise a heat engine to operate between those two temperatures, taking thermal energy as heat from the material at the higher temperature, and rejecting heat to the material at the lower temperature. Since this engine could produce reversible work, the states would not be in equilibrium. Two states that are in equilibrium, therefore, are at the same temperature.

We might be tempted to speak of the equality of pressures in stages 1 and 2, but such a statement is not necessarily true. The two states, 1 and 2, could be at equilibrium with unequal pressures if the two were at different altitudes, for example. To explore this and other variations, let us write the first and second laws of thermodynamics for the passage of an infinitesimal amount of a material i, dn_i, through the steady state machine, MA.

Because the material, i, may be one of several components in states 1 and 2, we must deal with its partial molar enthalpy ($\overline{H}_i$) and partial molar entropy ($\overline{S}_i$). The terms $\underline{PE}_i$ and $\underline{KE}_i$ signify the potential and kinetic energies per mole of i.

The First and Second laws are:

$$(\overline{H} + \underline{PE} + \underline{KE})_{i,1} dn_{i,1} - (\overline{H} + \underline{PE} + \underline{KE})_{i,2} dn_{i,2} + \delta Q + \delta W = dU \quad \textbf{(4.2)}$$

$$\overline{S}_{i,1} dn_{i,1} - \overline{S}_{i,2} dn_{i,2} + \frac{\delta Q}{T} - \frac{\delta \text{lw}}{T} = dS \quad \textbf{(4.3)}$$

Rearranging Eq. 4.3, we have

$$-T\overline{S}_{i,1} dn_{i,1} + T\overline{S}_{i,2} dn_{i,2} - \delta Q + \delta\ \text{lw} = -T\ dS \quad \textbf{(4.4)}$$

Add Eqs. 4.2 and 4.4, to eliminate the δQ term, and note that at steady state the properties of the machine do not change ($dU = 0$, and $dS = 0$). Furthermore, $\delta W + \delta\ \text{lw} = \delta W_{\text{rev}}$ and $\overline{H} - T\overline{S} = \overline{G}$. Also $dn_{i,1} = dn_{i,2} = dn_i$. Therefore:

$$\delta W_{\text{rev}} = (\overline{G} + \underline{PE} + \underline{KE})_{i,2} dn_i - (\overline{G} + \underline{PE} + \underline{KE})_{i,1} dn_i \quad \textbf{(4.5a)}$$

Considering the definition of equilibrium (Eq. 4.1), the condition for equilibrium between states 1 and 2 is:

$$\delta W_{\text{rev}} = 0 = (\overline{G} + \underline{PE} + \underline{KE})_{i,2} dn_i - (\overline{G} + \underline{PE} + \underline{KE})_{i,1} dn_i \quad \textbf{(4.5b)}$$

In most cases we will consider states 1 and 2 will exist at the same potential energy level and kinetic energy level, and Eq. (4.5a) becomes:

$$\delta W_{\text{rev}} = (\overline{G}_{i,2} - \overline{G}_{i,1}) dn_i = \Delta \overline{G}_i = 0 \qquad \text{or} \qquad \overline{G}_{i,1} = \overline{G}_{i,2} \quad \textbf{(4.6a)}$$

This can also be stated as in terms of the chemical potential of i,

$$\mu_{i,1} = \mu_{i,2} \quad \textbf{(4.6b)}$$

That is, states 1 and 2 are in equilibrium with respect to material i if the partial molar Gibbs free energy (or chemical potential) of i is the same in both states.

In single-component systems the partial molar Gibbs free energy of a material, $\overline{G}$, is simply the molar Gibbs free energy $\underline{G}$. The criterion for equilibrium between states 1 and 2 can be written as:

$$\delta W_{\text{rev}} = (\underline{G}_2 - \underline{G}_1) dn = \Delta \underline{G}\ dn \quad \textbf{(4.6c)}$$

Note that if the difference in Gibbs free energy $\Delta G = G_2 - G_1$, between the two states 1 and 2 is negative, then the reversible work term is negative. That means that the material may change spontaneously from state 1 to state 2 because no work needs to be done to force the change; in fact, work can be generated by the change. Actual work need not be done. The potential to do so might be dissipated as lost work, but the potential to do reversible work exists.

If the change in Gibbs free energy in moving the system between states 1 and 2 (ΔG) is positive, then the reversible work term is positive. Work would have to be done *on* the system to force it to change from state 1 to state 2, and the change would not occur spontaneously.

4.2 BAROMETRIC EQUATION

To examine a special case of a single material that can exist at two different altitudes, z_1 and z_2, consider an ideal gas of molecular weight M which fills an isothermal, vertical tube (Figure 4.2). The points 1 and 2 are at different altitudes. If the two are at equilibrium, then one may write for one mole of the material:

$$\underline{G}_2 - \underline{G}_1 + Mg(z_2 - z_1) = 0 \tag{4.7a}$$

where M is the molecular weight expressed in kilograms per mole, $\underline{G}_1$ and $\underline{G}_2$ are the molar Gibbs free energies of the gas, and g is the gravitational constant.

At constant temperature ($dT = 0$), the equation for $d\underline{G}$ (Eq. 3.25 for one mole) becomes, assuming ideal gas behavior:

$$d\underline{G} = \underline{V}\, dP = \frac{RT}{P}\, dP = RT\, d(\ln P)$$

Integrating between states 1 and 2:

$$\underline{G}_2 - \underline{G}_1 = RT \ln \frac{P_2}{P_1}$$

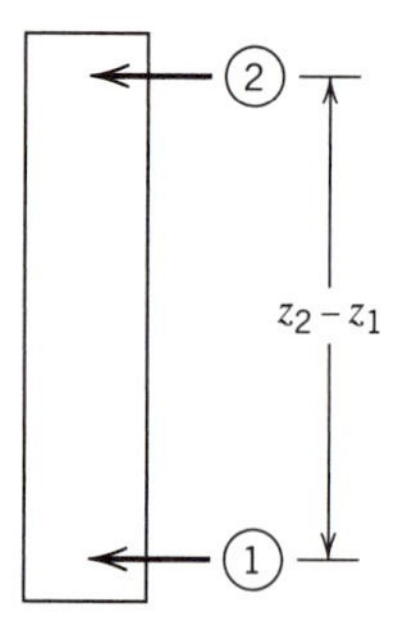

Figure 4.2 Equilibrium between top and bottom of a column of gas at constant temperature.

Hence,

$$RT \ln \frac{P_2}{P_1} + Mg(z_2 - z_1) = 0$$

$$\ln \frac{P_2}{P_1} = -\frac{Mg}{RT}(z_2 - z_1) \quad \textbf{(4.7b)}$$

$$P_2 = P_1 \exp\left[-\frac{Mg}{RT}(z_2 - z_1)\right]$$

Equation 4.7b, called the barometric equation, represents the variation of pressure with height for an ideal gas. Notice that the ratio of P_2 to P_1 is a function of both M, the molecular weight of the gas, and the acceleration of gravity. If there were two gases in the container, with different molecular weights, one would expect the composition at the top of the container to be different from the composition at the bottom. If the gravitational potential were to be enhanced by, for example, using a centrifuge, one could intensify these compositional differences. In fact, one could imagine separation of gases consisting of various isotopes of the same element using such a technique.

4.3 PHASE EQUILIBRIA

If two phases exist at equilibrium (at the same potential and kinetic energy levels), then the *partial molar* Gibbs free energies or chemical potentials (μ) of each of the components in the two phases must be equal. When a material composed of a *single* component exists in different physical states, the two states will be at equilibrium when the *molar* Gibbs free energies of the two states are equal. With only a single component, the partial molar Gibbs free energy is equal to the molar Gibbs free energy.

To illustrate, consider the equilibrium between diamond and graphite, two crystallographic forms of carbon. At room temperature, the Gibbs free energy of diamond is greater than the Gibbs free energy of graphite. Thus, graphite is the state that exists at equilibrium. But, as the pressure is increased, the difference in Gibbs free energy between the two states changes (Figure 4.3). At the point where the two lines on Figure 4.3 cross, diamond and graphite are in equilibrium. At pressures greater than this equilibrium pressure, diamond is the equilibrium state.

The variation of $\underline{G}$ with P at constant temperature ($dT = 0$) can be determined from Eq. 3.25:

$$d\underline{G} = \underline{V}\, dP \quad \textbf{(4.8)}$$

The relationship between $\underline{G}$ and P can take several forms, depending on the P-V-T behavior of the material involved.

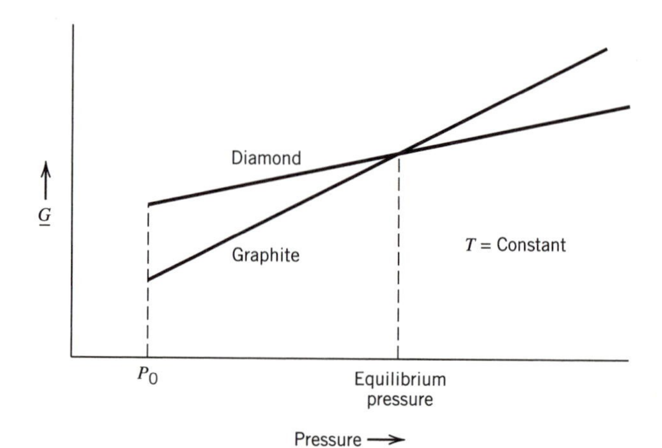

Figure 4.3 Specific Gibbs free energy versus pressure at constant temperature for graphite and diamond.

For an ideal gas:

$$d\underline{G} = \frac{RT}{P}\,dP = RT\,d(\ln P)$$
$$\Delta\underline{G} = RT \ln \frac{P_2}{P_1} \tag{4.9}$$

For an incompressible solid, one in which the specific volume ($\underline{V}_0$) does not change appreciably with pressure:

$$d\underline{G} = \underline{V}_0 dP$$
$$\Delta\underline{G} = \underline{V}_0(P_2 - P_1) \tag{4.10}$$

For a compressible solid:

$$\beta_T = -\frac{1}{\underline{V}}\left(\frac{\partial \underline{V}}{\partial P}\right)_T$$

If β_T is constant:

$$\int_{V_0}^{V} \frac{d\underline{V}}{\underline{V}} = -\beta_T \int_{P_0}^{P} dP$$

$$\ln \frac{\underline{V}}{\underline{V}_0} = -\beta_T(P - P_0)$$

$$\underline{V} = \underline{V}_0 \exp[-\beta_T(P - P_0)]$$

$$\Delta \underline{G} = \int_{P_0}^{P} \underline{V}_0 \exp[-\beta_T(P - P_0)]$$

If P_0 is taken as zero:

$$\Delta \underline{G} = \frac{\underline{V}_0}{\beta_T}[1 - \exp(-\beta_T P)] \quad \textbf{(4.11)}$$

4.4 SPECIAL CASE: CLOSED SYSTEM, CONSTANT VOLUME

The study of materials, especially the solid state physics of electronic materials, sometimes involves the analysis of a *constant volume,* closed system in which kinetic or potential energy differences are unimportant. The conditions for equilibrium in this special case can be derived following the same techniques used in Section 4.1: that is, by combining the First and Second laws to find the condition under which the reversible work between the two states (1 and 2) is zero when an infinitesimal amount of material i is transferred from 1 to 2. The temperature of the two states must be the same, of course, using the argument presented as in Section 4.1.

In the following equations a tilde used in conjunction with a thermodynamic property denotes the partial derivative of that property with respect to the mass of material i at constant temperature, *volume,* and the mass of other materials. This form of notation is adopted to avoid confusion with partial molar quantities, which involve the partial derivative of thermodynamic quantities at constant temperature and *pressure.*

From the First Law:

$$\delta Q + \delta W = (\tilde{U}_2 - \tilde{U}_1)dn_i$$

From the Second Law:

$$\frac{\delta Q}{T} - \frac{\delta \text{lw}}{T} = (\tilde{S}_2 - \tilde{S}_l)dn_i$$

Combining, to eliminate the δQ term, we have

$$\delta W_{\text{rev}} = (\tilde{U}_{i,2} - \tilde{U}_{i,1})dn_i - T(\tilde{S}_{i,2} - \tilde{S}_{i,1})dn_i \quad \textbf{(4.12)}$$

where

$$\tilde{U} = \left(\frac{\partial U}{\partial n_i}\right)_{T,V,n_j \neq n_i}$$

and

$$\tilde{S} = \left(\frac{\partial S}{\partial n_i}\right)_{T,V,n_j \neq n_i}$$

At equilibrium, $\delta W_{\text{rev}} = 0$, hence:

$$\tilde{F}_{i,1} = \tilde{F}_{i,2} \qquad \text{or} \qquad \mu_{i,1} = \mu_{i,2} \tag{4.13}$$

where

$$\tilde{F}_i = \left(\frac{\partial F}{\partial n_i}\right)_{T,V,n_j \neq n_i} = \mu_i$$

This results in the familiar expression $\mu_{i,1} = \mu_{i,2}$, as the condition of equilibrium. However, the Helmholtz free energy is used to derive the chemical potential in this constant volume case. The usefulness of this approach will become apparent in Chapter 10 because in statistical thermodynamics Helmholtz free energy is more easily calculated than Gibbs free energy. Since the two are not appreciably different in the case of solids, the work of theorists is thus made easier.

4.5 FIRST-ORDER TRANSITIONS: VARIATION OF EQUILIBRIUM PRESSURE WITH TEMPERATURE

In Section 4.1 we concluded that when a one-component material exists at equilibrium in two physical states, say, A and B (called states 1 and 2 in Section 4.1), the molar Gibbs free energy of the two states must be equal, $\underline{G}_A = \underline{G}_B$. If the two states are to remain in equilibrium as temperature and pressure change, the changes in each of the molar Gibbs free energies must be the same:

$$d\underline{G}_A = d\underline{G}_B$$

The first derivatives of $\underline{G}$ with respect to T and P are:

$$\left(\frac{\partial \underline{G}}{\partial T}\right)_P = -\underline{S}, \qquad \text{and} \qquad \left(\frac{\partial \underline{G}}{\partial P}\right)_T = \underline{V}$$

If the *first derivatives* are *discontinuous* at the temperature–pressure boundary between A and B (see Figure 4.4), the transition between them is, by definition, a *first-order transition.* Melting and boiling are examples of first-order transitions.

The first derivatives of $\underline{G}$ may be continuous at the temperature–pressure boundary ($\underline{S}_A = \underline{S}_B$ and $\underline{V}_A = \underline{V}_B$). If they are, and the *second derivatives are discontinuous,* then the transition between A and B is a *second-order transition.* Glass transitions in polymers and some magnetic transitions are examples of second-order transitions. The thermodynamics of second-order transitions is considered in Section 4.9.

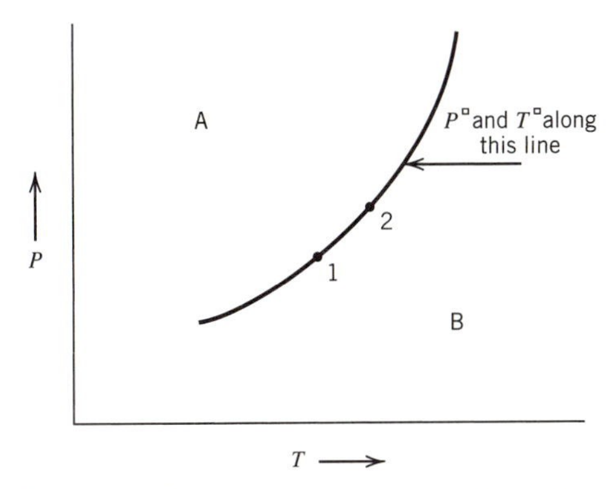

Figure 4.4 Pressure–temperature relationship for equilibrium between phases A and B.

Let us now examine the thermodynamics of first-order transitions. Figure 4.4 represents a *phase diagram* for a single material, one whose composition does not change. In the region labeled A, phase A is the equilibrium phase. To the right in the field labeled B, phase B is the equilibrium phase. Along the line joining the two, these two phases, A and B, can exist in equilibrium. A common example of phases coexisting in equilibrium is seen in the phase diagram for water and steam. In this case, phase A would be water, phase B, steam. The line between the two is called the vapor–pressure curve. The values on this curve represent the pressure of water vapor in equilibrium with pure liquid water at the same pressure[2] at the specified temperature. If the two phases are solid and liquid, the line in the phase diagram represents the melting temperature as a function of pressure.

To describe the shape of the curve in thermodynamic terms, consider the points 1 and 2 on the phase equilibrium line (Figure 4.4). Assume them to be infinitesimally apart in pressure and temperature. We may use the superscript □ to indicate equilibrium values and write

$$d\underline{G}_A = -\underline{S}_A dT^{\square} + \underline{V}_A dP^{\square}$$

$$d\underline{G}_B = -\underline{S}_B dT^{\square} + \underline{V}_B dP^{\square}$$

Because phases A and B are in equilibrium on the line, the molar Gibbs free energies of the two phases are equal at each temperature. Therefore changes in $\underline{G}_A$ and $\underline{G}_B$ are equal.

$$d\underline{G}_A = d\underline{G}_B = -\underline{S}_A dT^{\square} + \underline{V}_A dP^{\square} = -\underline{S}_B dT^{\square} + \underline{V}_B dP^{\square}$$

[2]Note that both the vapor and the liquid are at the *same* pressure. For example, at 293 K water vapor at 2.339×10^3 Pa is in equilibrium with liquid water at the same pressure. If the total pressure on the system is increased (by adding a nonreacting gas), the vapor pressure of the water will change, as discussed in Section 4.7.

Rearranging

$$(\underline{S}_B - \underline{S}_A)dT^{\square} = (\underline{V}_B - \underline{V}_A)dP^{\square} \quad \textbf{(4.14a)}$$

or

$$\Delta\underline{S}\, dT^{\square} = \Delta\underline{V}\, dP^{\square} \quad \textbf{(4.14b)}$$

Because phases A and B are in equilibrium:

$$\Delta\underline{G} = \Delta\underline{H} - T^{\square}\,\Delta\underline{S} = 0 \quad \textbf{(4.14c)}$$

$$\Delta\underline{S} = \frac{\Delta\underline{H}}{T^{\square}} \quad \textbf{(4.14d)}$$

Combining Eqs. 4.14b–4.14d, we have

$$\begin{aligned} \Delta\underline{H}\,\frac{dT^{\square}}{T^{\square}} &= \Delta\underline{V}\, dP^{\square} \\ \frac{dP^{\square}}{dT^{\square}} &= +\frac{1}{\Delta\underline{V}}\frac{\Delta\underline{H}}{T^{\square}} \end{aligned} \quad \textbf{(4.15)}$$

This is called the Clapeyron equation.

If the transition is from solid to liquid, then the change in enthalpy is the change in specific enthalpy upon melting or $\Delta\underline{H}$ of fusion ($\Delta\underline{H}_{fus}$). The change in volume, $\Delta\underline{V}$, is the specific volume of the liquid minus the specific volume of the solid.

As an example, consider the change of the melting point of tin resulting from a pressure change of 500 atm (about 5×10^7 Pa) For tin:

Melting temperature = 505 K at 1 atm pressure

Heat of fusion = 7196 J/mol

Molecular weight = 118.7 g/mol

Density of solid tin = 7.30 g/cm^3

Change of volume upon melting = +2.7%

To calculate $\Delta\underline{V}$ for use in Eq. 4.15:

$$\underline{V} = \frac{\text{mol wt}}{\text{density}} = \frac{118.7}{7.3} = 16.26\,\frac{\text{cm}^3}{\text{mol}} = 16.26 \times 10^{-6}\,\frac{\text{m}^3}{\text{mol}}$$

$$\Delta V = (0.027)(16.26 \times 10^{-6}) = 4.39 \times 10^{-7}\,\frac{\text{m}^3}{\text{mol}}$$

Assuming that $\Delta \underline{V}$ and $\Delta \underline{H}$ do not change with temperature, and that $\Delta T \ll T$,

$$\frac{\Delta P^{\square}}{\Delta T^{\square}} = \frac{1}{4.39 \times 10^{-7}} \frac{7196}{505} = 3.25 \times 10^{7} \left(\frac{N}{\text{m}^2} \cdot \frac{1}{\text{K}} \right)$$

$$\text{For } \Delta P^{\square} = (500)(1.013 \times 10^{5}) \text{ Pa}$$

$$\Delta T^{\square} = \frac{(500)(1.013 \times 10^{5})}{3.25 \times 10^{7}} = +1.58 \text{ K}$$

Note that the melting temperature rises as the pressure rises because all the terms in Eq. 4.15 are positive. In the case of water, the opposite is true because the specific volume of water decreases upon melting.

4.6 CLAPEYRON EQUATION IN VAPOR EQUILIBRIA

Suppose the two phases in equilibrium are a vapor phase (as phase B) and a condensed phase (liquid or solid) as phase A. If we assume that the specific volume of the vapor is very much larger than the specific volume of the condensed phase, then:

$$\frac{dP^{\square}}{dT^{\square}} = \frac{1}{\underline{V}_{(\text{vapor})}} \frac{\Delta \underline{H}_{\text{vap}}}{T^{\square}}$$

where $\Delta \underline{H}_{\text{vap}}$ is the heat of vaporization.

If we assume the vapor to be an ideal gas:

$$\frac{dP^{\square}}{dT^{\square}} = \frac{P^{\square}}{RT^{\square}} \frac{\Delta \underline{H}_{\text{vap}}}{T^{\square}}$$

$$\frac{dP^{\square}}{P^{\square}} = \frac{\Delta \underline{H}_{\text{vap}}}{R} \frac{dT^{\square}}{(T^{\square})^2} = -\frac{\Delta \underline{H}_{\text{vap}}}{R} d\left(\frac{1}{T^{\square}}\right) \quad \textbf{(4.16)}$$

$$d(\ln P^{\square}) = -\frac{\Delta \underline{H}_{\text{vap}}}{R} d\left(\frac{1}{T^{\square}}\right)$$

From Eq. 4.16 it is apparent that if the natural logarithm of the vapor pressure is plotted against the inverse temperature (absolute), the slope of the line will be equivalent to the negative enthalpy of evaporation divided by the gas constant as in Figure 4.5. This relationship can serve as the basis of a convenient determination of the enthalpy of evaporation of a condensed phase. As an example, let us calculate the enthalpy of evaporation of water from vapor pressure data taken at 293 and 298 K. For water:

T **(K)**	***P*** **(bars) [1 bar = 10^5 P]**
293	0.02339
298	0.03169

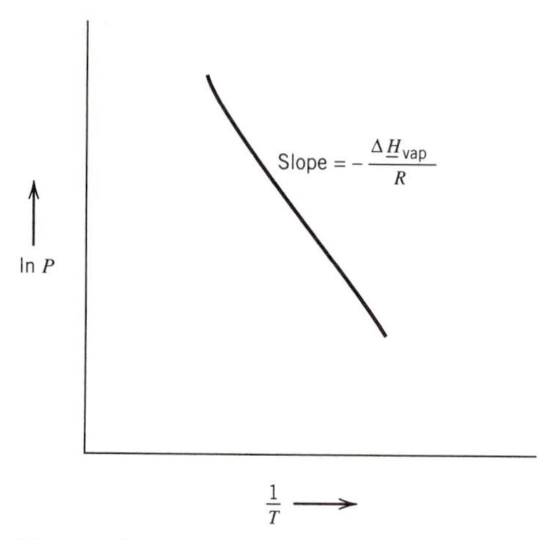

Figure 4.5 Natural logarithms of saturation pressure versus inverse of absolute temperature.

Assuming that the value of $\Delta \underline{H}_{\text{vap}}$ is constant over the small (5 K) temperature range, we can integrate Eq. 4.16 between the limits of pressure and temperature in the preceding tabulation:

$$\int_{P_1}^{P_2} d(\ln P^{\square}) = -\frac{\Delta \underline{H}_{\text{vap}}}{R} \int_{T_1}^{T_2} d\left(\frac{1}{T^{\square}}\right)$$

$$\ln \frac{P_2^{\square}}{P_1^{\square}} = -\frac{\Delta \underline{H}_{\text{vap}}}{R} \left[\frac{1}{T_2^{\square}} - \frac{1}{T_1^{\square}}\right]$$

$$\ln \frac{P_2^{\square}}{P_1^{\square}} = +\frac{\Delta \underline{H}_{\text{vap}}}{R} \left[\frac{T_2^{\square} - T_1^{\square}}{T_1^{\square} T_2^{\square}}\right]$$

$$\ln \left(\frac{3169}{2339}\right) = \frac{5}{(293)(298)} \frac{\Delta \underline{H}_{\text{vap}}}{8.314}$$

$$\Delta \underline{H}_{\text{vap}} = 44{,}092 \text{ J/mol} = 2450 \text{ J/g}$$

In the preceding example the enthalpy of evaporation was calculated from two vapor pressure data points based on Eq. 4.16. Of course, if an algebraic expression for the vapor pressure of a material is available, Eq. 4.16 may be used directly to determine its enthalpy of evaporation (or enthalpy of sublimation if the material is a solid). For example, the vapor pressure of liquid gallium is (Ref. 1):

$$\ln P \text{ (atm)} = 8.346 - \frac{31{,}848}{T} + 0.7579 \ln T - 0.7234 \times 10^{-3} T$$

To apply the relationship in Eq. 4.16:

$$d(\ln P) = +\frac{31{,}848\ dT}{T^2} + \frac{0.7579\ dT}{T} - 0.7234 \times 10^{-3} dT$$

Rearranging the terms:

$$d(\ln P) = (31{,}848 + 0.7579T - 0.7234 \times 10^{-3}T^2)\frac{dT}{T^2}$$

or

$$d(\ln P) = -(31{,}848 + 0.7579T - 0.7234 \times 10^{-3}T^2)d\left(\frac{1}{T}\right)$$

Comparing with Eq. 4.16,

$$\Delta \underline{H}_{\text{vap}} = R(31{,}848 + 0.7579T - 0.7234 \times 10^{-3}T^2)$$

Based on the given vapor pressure equation (for Gallium), the enthalpy of evaporation is a function of temperature. If the vapor pressure equation is of the simple form $\ln P = A/T + B$, then the enthalpy of evaporation is not a function of temperature. In reality, this usually means that the vapor pressure may be a function of temperature, but the vapor pressure data is not known with sufficient precision to justify the use of a more complex equation.

4.7 VARIATION OF VAPOR PRESSURE OF A CONDENSED PHASE WITH TOTAL APPLIED PRESSURE

Figure 4.6 is the phase diagram for a one-component system in which the vapor and the liquid are at equilibrium. When we speak of the equilibrium vapor pressure of a material, we assume that the total pressure in the system is the vapor pressure, that is, just the pressure generated by the vapor. For example, if the substance is water, the total pressure on the system would be one atmosphere at 100°C. At temperatures lower than 100°C, the pressure on the system would be less.

We now ask how the vapor pressure of the liquid changes when the total pressure on the system exceeds the vapor pressure. The total pressure of the system can be changed arbitrarily by introducing an inert gas, such as argon, into the system. In that case, the pressure of water vapor is called the partial pressure of water. The total pressure is the sum of all the partial pressures: water and argon in this case.

Based on Eq. 4.8 ($d\underline{G} = -\underline{S}\ dT + \underline{V}\ dP$, which at constant temperature is, $d\underline{G} = \underline{V}\ dP$), as the pressure on a liquid is increased, its specific Gibbs free energy increases. If the Gibbs free energy of the liquid increases, then for equilibrium to be maintained, the Gibbs free energy of that same material in the vapor state must also

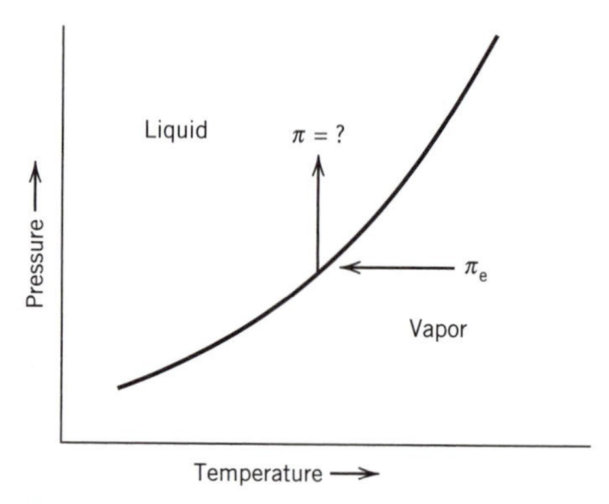

Figure 4.6 Effect of total pressure on vapor pressure.

increase. This means that increasing the pressure on the liquid increases the vapor pressure.

To explore this relationship, let π_e represent the equilibrium vapor pressure, and π represents the vapor pressure of the liquid under a total pressure P_T. As the liquid is pressurized from π_e to P_T, its specific Gibbs free energy changes as follows, assuming, for the purpose of the calculation, that the liquid is incompressible (Eq. 4.10):

$$d\underline{G}_l = \underline{V}_l dP$$
$$\Delta\underline{G}_l = \underline{V}_l(P_T - \pi_e) \tag{4.17}$$

The Gibbs free energy of the gas must also change. Assuming it is an ideal gas,

$$\Delta\underline{G}_v = RT \ln \frac{\pi}{\pi_e} \tag{4.18}$$

These changes in Gibbs free energy between Eqs. 4.15 and 4.18 must be equal for the two to be in equilibrium. Hence,

$$\underline{V}_l(P_T - \pi_e) = RT \ln \frac{\pi}{\pi_e}$$

If $P_T >> \pi_e$,

$$\underline{V}_l P_T = RT \ln \frac{\pi}{\pi_e} \tag{4.19}$$

In general this effect is small, as can be illustrated by putting some typical values into the equation. Let us calculate the fractional change of the vapor pressure of water at 298 K when the total pressure of the system is increased to 10 atm. At this temperature, the equilibrium vapor pressure of water, π_e, is 3170 Pa (0.031 atm).

For water:

$$\underline{V} = \frac{\text{mol wt}}{\text{density}} = \frac{18}{1} = 18\ \frac{\text{cm}^3}{\text{mol}} = 18 \times 10^{-6}\ \frac{\text{m}^3}{\text{mol}}$$

$$\underline{V}_l P_T = RT \ln \frac{\pi}{\pi_e}$$

$$(18 \times 10^{-6})(10 \times 1.013 \times 10^5) = 8.314(298) \ln \frac{\pi}{\pi_e}$$

$$\ln \frac{\pi}{\pi_e} = 0.0074 = \ln\left(\frac{\pi_e + \Delta\pi}{\pi_e}\right) = \ln\left(1 + \frac{\Delta\pi}{\pi_e}\right)$$

Noting that $\ln(1 + x) = x$ when x is small

$$\frac{\Delta\pi}{\pi_e} = 0.0074$$

Thus when the total pressure on the liquid–vapor equilibrium system is increased to 10 atm, the vapor pressure of water changes by 0.74%.

4.8 VARIATION OF VAPOR PRESSURE WITH PARTICLE SIZE

One important manifestation of the effect discussed in Section 4.7 is the variation of vapor pressure of a material with particle size. The pressure on a condensed phase that exists in the form of a small droplet is the sum of the imposed gas pressure and the internal pressure generated by the surface tension of the material. If the surface tension of a material is represented by γ, and the radius by r, then from Figure 4.7:

$$2\pi r\gamma = \pi r^2\, \Delta P \tag{4.20}$$

$$\Delta P = \frac{2\gamma}{r}$$

The total pressure then exerted on the liquid (or solid) is the vapor pressure plus the pressure imposed by the surface tension. The vapor pressure of the droplet, π, is based on Eq. 4.19:

$$\underline{V}_l\left(\frac{2\gamma}{r}\right) = RT \ln \frac{\pi}{\pi_e} \tag{4.21}$$

$$\ln \frac{\pi}{\pi_e} = \frac{2\underline{V}_l\gamma}{RT}\,\frac{1}{r}$$

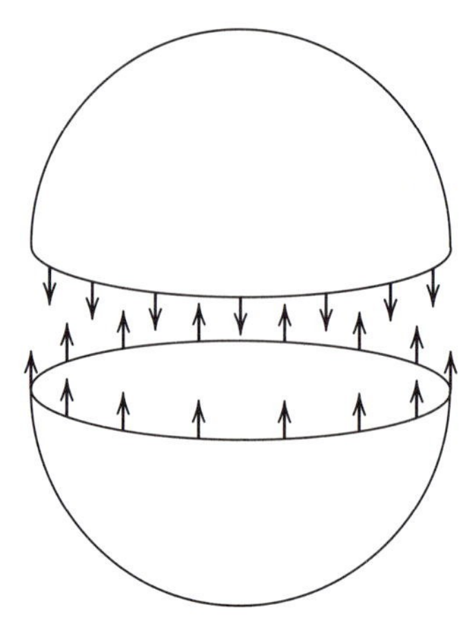

Figure 4.7 The force holding two hemispheres together is $2\pi r\gamma$; the force pushing them apart is $\pi r^2\, \Delta P$, where $\Delta P = 2\gamma/r$.

Thus π, the vapor pressure of a material, is a function of the particle size: the smaller the size, the greater the vapor pressure. This means that for a substance with a variety of particle sizes, the small particles will tend to shrink and the larger ones to grow because the vapor pressure above the small particles is greater than the vapor pressure above the large particles. Material would then move through the gas phase from small particles to larger ones. This process is called coarsening.

4.9 SECOND-ORDER TRANSITIONS

As indicated in Section 4.5, a second-order transition between two states (A and B) is one in which the *first derivatives* of $\underline{G}$ with respect to T and P ($\underline{S}$ and $\underline{V}$) are *continuous* at the boundary between the two, and the *second derivatives* of $\underline{G}$ with respect to T and P are *discontinuous.* Examples of such transitions are some order–disorder phenomena in solid solutions, ferromagnetic transitions, and some transitions between superconducting and nonsuperconducting states.

Let us derive the thermodynamic relationships involved in second-order transitions. If the first derivatives are continuous, then at the boundary between the two states A and B:

$$\underline{S}_{\mathrm{A}} = \underline{S}_{\mathrm{B}}$$

Specific entropy is a function of T and P:

$$\underline{S} = \underline{S}(T,P)$$

Taking the total differential,

$$d\underline{S} = \left(\frac{\partial \underline{S}}{\partial T}\right)_P dT + \left(\frac{\partial \underline{S}}{\partial P}\right)_T dP$$

Applying the methods illustrated in Chapter 3 (Section 3.4), and assuming that both A and B are condensed phases,

$$d\underline{S} = \frac{C_P}{T}\,dT - \left(\frac{\partial \underline{V}}{\partial T}\right)_P dP$$
$$d\underline{S} = \frac{C_P}{T}\,dT - \underline{V}\alpha_V\,dP \qquad \textbf{(4.22)}$$

Because $\underline{S}_A = \underline{S}_B$, $d\underline{S}_A = d\underline{S}_B$,

$$d\underline{S}_A - d\underline{S}_B = 0 = \frac{C_{P,B} - C_{P,A}}{T^\square}\,dT^\square - (\underline{V}_B\alpha_B - \underline{V}_A\alpha_A)dP^\square$$

But

$$\underline{V}_B = \underline{V}_A = \underline{V}$$

Hence:

$$\frac{dP^\square}{dT^\square} = \frac{\Delta C_P}{\underline{V}T^\square\,\Delta\alpha} \qquad \textbf{(4.23)}$$

Note that when the specific volume $\underline{V}$ does not change at the temperature–pressure boundary, the slope of the curve of specific volume as a function of temperature $(\partial\underline{V}/\partial T)_P$, the thermal expansion coefficient, does change.

Another thermodynamic condition may be derived using a similar technique, but operating on the specific volume.

$$\underline{V} = \underline{V}(T,P)$$
$$d\underline{V} = \left(\frac{\partial \underline{V}}{\partial T}\right)_P dT + \left(\frac{\partial \underline{V}}{\partial P}\right)_T dP$$
$$d\underline{V} = \underline{V}\alpha_V dT - \underline{V}\beta_T dP$$

at the boundary between A and B, $d\underline{V}_A = d\underline{V}_B$, and $\underline{V}_A = \underline{V}_B = \underline{V}$

Hence,

$$(\alpha_B - \alpha_A)dT^\square = (\beta_B - \beta_A)dP^\square \tag{4.24}$$

$$\frac{dP^\square}{dT^\square} = \frac{\Delta\alpha}{\Delta\beta}$$

4.10 SUPERCONDUCTIVITY: AN EXAMPLE

One of the second-order transitions described in Section 4.9 can be demonstrated in the case of superconductivity. This phenomenon was discovered in 1911 when Onnes observed that the electrical resistivity of mercury vanished completely below 4.2 K, the superconducting critical temperature for mercury. Since then many elements and compounds have been shown to undergo the superconducting transition.

For a particular material, the critical temperature T_c has been shown experimentally to be a function of the magnetic field present, just as the temperature of transition between liquid and vapor is a function of the pressure. The regimes of superconducting and nonsuperconducting (normal) states for type I superconductors can be displayed on a diagram with two axes: magnetic field $\mathcal{H}$ and temperature T (Figure 4.8). At temperatures below the zero-field critical temperature, there is a critical magnetic field ($\mathcal{H}_c(T)$) at which the superconducting transition takes place. The critical temperature, which is a function of the magnetic field, is a maximum at $\mathcal{H} = 0$. Experiments with many superconductors have shown that the dependence of critical magnetic field on temperature is given approximately by:

$$\mathcal{H}_c = \mathcal{H}_0 \left[1 - \left(\frac{T}{T_c}\right)^2\right] \tag{4.25}$$

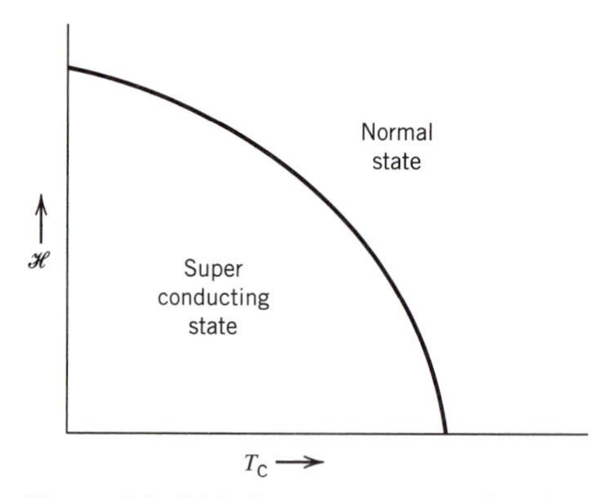

Figure 4.8 Critical temperature as a function of magnetic field for a superconducting material.

We can treat the equilibrium line between superconducting and normal phases in a manner similar to the way we treated the *P-T* relationship for phase equilibrium (Sections 4.5 and 4.9). In this case, the intensive variables of interest are T and $\mathcal{H}$. The equation describing the change of Gibbs free energy with T and $\mathcal{H}$ is (Eq. 3.62):

$$d\underline{G} = -\underline{S}\, dT - \mu_0 \underline{V} M\, d\mathcal{H}$$

In an ideal superconductor (type I), the magnetic induction B is zero when the material is in the superconducting state (Meissner effect). Because $B = \mu_0 (\mathcal{H} + M)$, $M = -\mathcal{H}$ in the superconducting state. Let us assume that the magnetization in the normal state (M_n) is very low, and can be neglected. Therefore:

$$-\underline{S}_s dT_c + \mu_0 \underline{V} \mathcal{H}_c d\mathcal{H}_c = -\underline{S}_n dT_c \tag{4.26}$$

$$(\underline{S}_n - \underline{S}_c) dT_c = -\mu_0 \underline{V} \mathcal{H}_c d\mathcal{H}_c$$

where the subscript s denotes the superconducting phase and the subscript n denotes the normal phase.

The latent heat of the transition, L, is

$$L = T(\underline{S}_n - \underline{S}_c) \tag{4.27}$$

$$L = -\mu_0 T \underline{V} \mathcal{H}_c \frac{d\mathcal{H}_c}{dT_c} \tag{4.28}$$

Note that this latent heat is zero both at $T = 0$ and at $\mathcal{H}_c = 0$. If this is true, then $\underline{S}_n = \underline{S}_s$ (i.e., the entropy term, S), is continuous at zero field. The transition at this zero field is thus not a first-order transition. Rather, it can be shown to be a second-order transition in which there is a discontinuity in the second derivative of the Gibbs free energy, namely, in the heat capacities. From the definition of heat capacity,

$$C_n = T\left(\frac{\partial \underline{S}}{\partial T}\right)_n$$

$$C_n - C_s = T\left(\frac{d\underline{S}_n}{dT} - \frac{d\underline{S}_c}{dT}\right)$$

Differentiating Eq. 4.26 with respect to T, and assuming that $\underline{V}$ is not a function of temperature:

$$C_n - C_s = -\underline{V} T \mu_0 \left[\mathcal{H}_c \frac{\partial^2 \mathcal{H}}{\partial T^2} + \left(\frac{d\mathcal{H}}{dT}\right)^2\right]$$

At $T = T_c$ where $\mathcal{H}_c = 0$,

$$C_n - C_s = -VT_c\mu_0 \left(\frac{\partial \mathcal{H}}{dT}\right)^2_{\text{at } T=T_c}$$

Thus the heat capacities of the superconducting and normal states are different.

REFERENCE

1. Alcock, C. B., Itkin, V. P., and Horrigan, M. K., *Can. Metall. Q. 23* (3), 309–314, 1984.

PROBLEMS

4.1 **(a)** At 298 K, what is the Gibbs free energy change (ΔG) for the following reaction?

$$C_{\text{graphite}} = C_{\text{diamond}}$$

(b) Is the diamond thermodynamically stable relative to graphite at 298 K?

(c) What is the change of Gibbs free energy of diamond when it is compressed isothermally from 1 atm to 1000 atm?

(d) Assuming that graphite and diamond are incompressible ($\beta = 0$), calculate the pressure at which the two exist at equilibrium at 298 K.

(e) What is the Gibbs free energy of diamond relative to graphite at 900 K? To simplify the calculation, assume that the heat capacities of the two materials are equivalent.

(f) Diamond is synthesized from graphite at high pressures and high temperatures. The need for high pressure should be obvious from your calculations, but why is the process carried out at high temperatures?

DATA

Density of graphite is 2.25 g/cm^3

Density of diamond is 3.51 g/cm^3

	$\Delta H^\circ_{f(298)}$ (kJ/mol)	S°_{298} (J K^{-1} mol^{-1})
Diamond	1.897	2.38
Graphite	0	5.74

4.2 At 1 atm pressure, ice melts at 273.15 K. In addition, $\Delta \underline{H}_{\text{fus}} = 6.009$ kJ/mol, density of ice = 0.92 g/cm^3, and density of liquid water = 1.00 g/cm^3.

(a) What is the melting point of ice under 50 atm pressure?

(b) The blade of an ice skate is ground to a knife edge on each side of the skate. If the width of the knife edge is 0.001 in., and the length of the skate in contact with the ice is 3 in., calculate the pressure exerted on the ice by a 150 lb man. (Assume that the skater uses one edge.)

(c) What is the melting point of ice under this pressure?

4.3 At 400°C, liquid zinc has a vapor pressure of 10^{-4} atm. Estimate the boiling temperature of zinc, knowing that its heat of vaporation is approximately 28 kcal/mol.

4.4 Trouton's rule is expressed as follows: $\Delta \underline{H}_{vap} = 90T_b$ in joules per mole, where T_b is the boiling point (K)

The boiling temperature of mercury is 630 K. Estimate the partial pressure of liquid Hg at 298 K. Use Trouton's rule to estimate the heat of vaporization of mercury.

4.5 Liquid water under an air pressure of 1 atm at 25°C has a larger vapor pressure than it would have in the absence of air pressure. Calculate the increase in vapor pressure produced by the pressure of the atmosphere on the water. Water has a density of 1 g/cm^3; the vapor pressure (in the absence of the air pressure) is 3167.2 Pa.

4.6 Zinc is to be purified by distillation (evaporation followed by condensation).

(a) What is the level of vacuum (pressure) needed to distill zinc at 600°C (873 K)?

(b) How much energy in the form of heat (in joules per mole of zinc) is required for the distillation? Assume that the heat of condensation cannot be recovered because the zinc is condensed in a separate apparatus.

DATA

For zinc:

$$\ln P \text{ (atm)} = -15{,}250 + 1.255 \ln T \text{ (K)} - 21.79T\ (K)$$

4.7 The boiling point of silver ($P = 1$ atm) is 2450 K. The enthalpy of evaporation of liquid silver is 255,000 J/mol at its boiling point. Assume, for the purpose of this problem, that the heat capacities of liquid and vapor are the same.

Write an equation for the vapor pressure of silver, in atmospheres, as a function of kelvin temperature.

The equation should be suitable for use in a tabulation, NOT in differential form. Put numerical values in the equation based on the data given.

4.8 Zinc may exist as a solid, a liquid, or a vapor. The equilibrium pressure-temperature relationship between solid zinc and zinc vapor is given by the vapor pressure equation for the solid. A similar relation exists for liquid zinc.

At the triple point all three phases, solid, liquid, and vapor exist in equilibrium. That means that the vapor pressures of the liquid and the solid are the same. The vapor pressure of *solid* Zn varies with T as:

$$\ln P \text{ (atm)} = \frac{-15{,}755}{T} - 0.755 \ln(T) + 19.25$$

and the vapor pressure of *liquid* Zn varies with T as:

$$\ln P\ (\text{atm}) = \frac{-15{,}246}{T} - 1.255\ \ln(T) + 21.79$$

Calculate:

(a) The boiling point of Zn under 1 atm.
(b) The triple-point temperature.
(c) The heat of evaporation of Zn at the normal (1 atm) boiling point.
(d) The heat of fusion of Zn at the triple-point temperature.
(e) The differences between the heat capacities of solid and liquid Zn.

4.9 A particular material has a latent heat of vaporization of 5000 J/mol. This heat of vaporization does not change with temperature or pressure. One mole of the material exists in a two-phase equilibrium (liquid–vapor) in a container of volume $V = 1$ L, a temperature of 300 K, and pressure of 1 atm. The container (constant volume) is heated until the pressure reaches 2 atm. (Note that this is NOT a small ΔP.) The vapor phase can be treated as an ideal monatomic gas and the molar volume of the liquid can be neglected relative to that of the gas.

Find the fraction of material in the vapor phase in the initial and final states.

4.10 The melting point of gold is 1336 K, and the vapor pressure of *liquid* gold is given by:

$$\ln P\ (\text{atm}) = 23.716 - \frac{43{,}522}{T} - 1.222\ \ln T\ (\text{K})$$

(a) Calculate the heat of vaporization of gold at its melting point.

Answer parts b, c, and d numerically *only* if the data given in this problem statement are sufficient to support the calculation. If there are not enough data, write ''solution not possible.''

(b) What is the vapor pressure of *solid* gold at its melting point?
(c) What is the vapor pressure of *solid* gold at 1200 K?
(d) What is the heat of fusion of solid gold?

Chapter 5

Chemical Equilibrium

In Chapter 4 we discussed the criterion for thermodynamic equilibrium and derived several thermodynamic relationships for one-component systems. That was an exploration of *physical* equilibrium. In this chapter we consider equilibria among different chemical species, the study of *chemical* equilibrium.

5.1 THERMODYNAMIC ACTIVITY

In the study of chemical equilibrium we need to calculate values of the partial molar Gibbs free energy for materials under different conditions. These calculations are generally made using the concept of thermodynamic activity. The thermodynamic activity of a component varies with its physical form and concentration. It is evaluated relative to a *standard state* at the same temperature. The activity is set equal to one at the standard state.

Activity is defined in terms of a quantity called fugacity. The activity is the ratio of the fugacity of a material to its fugacity in the standard state. Fugacity is most easily understood as a property of a gas. One can consider it to be the pressure of the gas corrected for nonideality. To illustrate, the change of the molar Gibbs free energy of a *single* gas with pressure at constant temperature is:

$$d\underline{G} = \underline{V}\, dP \quad (d\underline{G} = -\underline{S}\, dT + \underline{V}\, dP, \text{ with } dT = 0)$$

If the gas is ideal ($\underline{V} = RT/P$), then:

$$d\underline{G} = RT\frac{dP}{P} = RT\, d\ \ln P \qquad \textbf{(5.1)}$$

For a real gas, we can define a function, fugacity (f), by analogy to pressure:

$$d\underline{G} = RT\, d\ \ln f \qquad \textbf{(5.2a)}$$

$$\text{Lim}\ f/p = 1 \text{ as } P \rightarrow 0$$

At low pressures, as pressure approaches zero, the gas behaves ideally and its fugacity is equal to its pressure. Equation 5.2a was derived for a single, ideal gas. It also applies to mixtures of gases, in which case the molar Gibbs free energy ($\underline{G}$) is replaced by the partial molar Gibbs free energy ($\overline{G}$). For the component i,

$$d\overline{G}_i = RT\, d\ \ln f_i \qquad \textbf{(5.2b)}$$

The fugacity of a condensed phase (solid or liquid) is equal to the fugacity of the vapor in equilibrium with it, because, at equilibrium, the partial molar Gibbs free energies of the two are equal (Eq. 4.6a).

By integrating Eq. 5.2a, we have an expression for the difference in partial molar Gibbs free energy between two states at constant temperature in terms of fugacity:

$$\Delta\overline{G} = \overline{G}_2 - \overline{G}_1 = RT\int_{f_1}^{f_2} d\ \ln f = RT\ \ln\frac{f_2}{f_1} \qquad \textbf{(5.3a)}$$

This is sometimes written as

$$\mu_2 - \mu_1 = RT\ \ln\frac{f_2}{f_1} \qquad \textbf{(5.3b)}$$

As indicated earlier, the thermodynamic *activity* of a material is defined as the ratio of the fugacity of that material to the fugacity in its standard state:

$$a_i \equiv \frac{f_i}{f_i^\circ} \qquad \textbf{(5.4)}$$

where f_i° is the fugacity of material i in its standard state.

Although the standard state of a material can be arbitrarily defined, certain conventions generally are followed. The standard state of a gas is usually taken to be the pure gas at one atmosphere pressure. The usual standard states for liquids and solids are the pure liquid or solid under one atmosphere pressure. Other standard states used in special cases (e.g., when dealing with dilute solutions) are covered in Chapter 7.

It is important to recognize the difference between a *standard* state and a *reference* state. A reference state completely defines the condition of the material in terms of temperature, pressure, and physical form: gas, liquid, or crystal structure in the case of a solid. Reference states were discussed when we considered enthalpies of formation in Chapter 1 (Section 1.18). We defined the reference state to be a pure element in its equilibrium condition at 298 K, and we set its enthalpy to zero. The enthalpy of the element at a temperature other than 298 K is not zero but can be calculated from a knowledge of its heat capacity.

The standard state is defined only in terms of pressure and physical form. Temperature is not mentioned because thermodynamic activity is an isothermal concept. Activity is the ratio of the fugacity of a material to the fugacity of the standard state *at the same temperature.*

The difference in specific Gibbs free energy between a material under an arbitrary set of conditions and a standard state (at the same temperature, of course) is:

$$\int_{\underline{G}^\circ}^{\overline{G}} dG_i = \overline{G}_i - \underline{G}_i^\circ = RT \ln \frac{f_i}{f_i^\circ} = RT \ln a_i \qquad \textbf{(5.5a)}$$

where f° is the fugacity in the standard state and $\underline{G}^\circ$ is the molar Gibbs free energy in the standard state.

If the material in question is an ideal gas, then the difference in Gibbs free energy between any state and the standard state can be related to pressure difference between the two because the fugacity of an ideal gas is equal to its pressure.

$$\int_{\underline{G}^\circ}^{\overline{G}} dG_i = \overline{G}_i - \underline{G}_i^\circ = RT \ln \frac{p_i}{p_i^\circ} = RT \ln a_i \qquad \textbf{(5.5b)}$$

For example, the activity of an *ideal* gas at a pressure of 2 atm is 2 because the fugacity of the standard state is one atmosphere. Note that thermodynamic activity has no units. It is the ratio of two quantities, each having the same units. If the pressure on a pure gas is less than one atmosphere, and if one atmosphere is chosen as the standard state, the activity of the pure gas will be less than one. Pure nitrogen at a pressure of 0.1 atm has an activity of one-tenth, if pure nitrogen at one atmosphere is taken as the standard state.

The fugacity of a condensed phase, a liquid or a solid, is equal to the fugacity of the vapor in equilibrium with it. For example, if the equilibrium vapor pressure of a liquid is 0.01 atm at a specified temperature, then its fugacity is 0.01 atm, assuming that the vapor is ideal.

The value of thermodynamic activity changes not only with pressure but also with composition. To look ahead a bit to the study of solutions (Chapter 7), the fugacity of a material in an *ideal solution* varies linearly with mole fraction of that substance over the entire composition range (see Figure 5.1). The fugacity of material B is the mole fraction of B multiplied by the fugacity of pure B. If pure material B is taken as the standard state, then the activity of B is simply the mole fraction of B in this solution. For example, in a liquid A–B ideal solution consisting of 80 mol % B, the

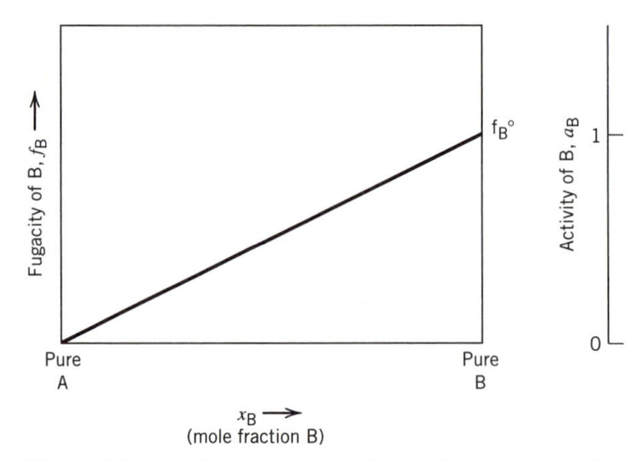

Figure 5.1 Fugacity as a function of mole fraction x in an ideal solution.

activity of B is 0.8, if pure B is taken as the standard state. The activity of A is 0.2, if pure A is taken as the standard state.

$$f_B = x_B f_B^P \tag{5.6}$$

where f_B^P is the fugacity of pure B
Then

$$a_B = \frac{f_B}{f_B^\circ} = \frac{x_B f_B^P}{f_B^\circ}$$

If

$$f_B^P = f_B^\circ \qquad \text{then} \quad a_B = x_B \qquad \text{(ideal solution)} \tag{5.7}$$

5.2 CHEMICAL EQUILIBRIUM

The expression for a chemical reaction in which b moles of material B react with c moles of C to form d moles of D and e moles of material E can be written as:

$$bB + cC = dD + eE$$

From Equation 4.5:

$$\delta W_{rev} = \Delta G = d\overline{G}_D + e\overline{G}_E - b\overline{G}_B - c\overline{G}_C$$

From Equation 5.5:

$$\overline{G}_B = \underline{G_B^\circ} + RT \ln a_B$$

The same holds true for reactant C, and products D and E, thus ΔG can be written as:

$$\Delta G = d(\underline{G}^{\circ}_{D} + RT \ln a_{D}) + e(\underline{G}^{\circ}_{E} + RT \ln a_{E}) - b(\underline{G}^{\circ}_{B} + RT \ln a_{B}) - c(\underline{G}^{\circ}_{C} + RT \ln a_{C}) \qquad \textbf{(5.8)}$$

$$\Delta G = \mathrm{d}\underline{G}^{\circ}_{D} + \mathrm{e}\underline{G}^{\circ}_{E} - \mathrm{b}\underline{G}^{\circ}_{B} - \mathrm{c}\underline{G}^{\circ}_{C} + RT \ln \frac{a^{d}_{D} a^{e}_{E}}{a^{b}_{B} a^{c}_{C}}$$

$$\Delta G = \Delta G^{\circ} + RT \ln J_{a}$$

where $\Delta G^{0} = \mathrm{d}\underline{G}^{\circ}_{D} + \mathrm{e}\underline{G}^{\circ}_{E} - \mathrm{b}\underline{G}^{\circ}_{B} - \mathrm{c}\underline{G}^{\circ}_{C}$, and J_a is the product of the activities of the reactants and products raised to the appropriate power (their stoichiometric coefficients):

$$J_{a} = \frac{a^{d}_{D} a^{e}_{E}}{a^{b}_{B} a^{c}_{C}} \qquad \textbf{(5.9)}$$

Equilibrium is achieved when no reversible work can be done between two states: that is, where ΔG is zero. There will be situations of interest to us in which the ΔG term is not zero. As discussed in Chapter 6, reversible work can be done on chemical compounds to change them to others by subjecting them to electrical potentials. This is the subject of electrochemistry. For the time being, however, we will study the case of chemical equilibrium in which $\Delta G = 0$. This means that:

$$\Delta G^{\circ} = -RT \ln J_{a(\text{equilibrium})} = -RT \ln K_{a} \qquad \textbf{(5.10)}$$

where K_a is the special value of the product J_a when the system is at chemical equilibrium. The quantity K_a is known as the equilibrium constant for a reaction, expressed in terms of activities.

5.3 GASEOUS EQUILIBRIA

To explore the applications of Eq. 5.10, let us consider two examples. In the first, we will deal with a tank that contains pure oxygen at a total pressure of one atmosphere. It is well established that under most normal conditions, oxygen exists primarily in the diatomic form, O_2. We know, however, that it can also exist in the monatomic state. Let us calculate the equilibrium concentration of monatomic oxygen in the tank at 1000 K. It is known that the standard Gibbs free energy of formation of monatomic oxygen is 187,000 J per mole of monatomic oxygen at 1000 K. Thus for the reaction:

$$\tfrac{1}{2}O_{2} \rightarrow O \qquad \textbf{(5.11)}$$

the standard Gibbs free energy change is:

$$\Delta G° = \Delta G°_{f,O} - \tfrac{1}{2}\Delta G°_{f,O_2} = 187{,}800 - 0 = 187{,}800 \text{ J}$$

From Eq. 5.10:

$$\Delta G° = -RT \ln \frac{a_O}{a_{O_2}^{1/2}} = -RT \ln \frac{P_O}{P_O^{1/2}} \qquad \textbf{(5.12)}$$

assuming that both monatomic and diatomic oxygen behave as ideal gases.[1] Solving:

$$\ln \frac{P_O}{P_{O_2}^{1/2}} = -\frac{\Delta G°}{RT} = -\frac{187{,}800}{(8.314)(1000)} = -22.58$$

$$\frac{P_O}{P_{O_2}^{1/2}} = 1.5 \times 10^{-10}$$

Since

$$P_O + P_{O_2} = 1 \text{ atm}, P_{O_2} = 1 - P_O$$

$$P_O = (1.5 \times 10^{-10})(1 - P_O)^{1/2}$$

$$P_O = 1.5 \times 10^{-10}$$

Thus the activity of diatomic oxygen in the tank is very nearly one, and the activity of monatomic oxygen, measured by its pressure, is approximately 1.5×10^{-10} at 298 K, a low pressure. This tells us that the predominant species of oxygen is O_2 at 1000 K.

As a second example, consider the equilibrium in the water–hydrogen–oxygen system. At 2000 K we would like to operate a system so that the partial pressure of oxygen will be 1×10^{-10} atm. The question is, ''What should be the ratio of hydrogen to water vapor to achieve this condition?'' We know that the standard Gibbs free energy of formation of water is given, in joules, by the equation:

$$H_2 + \tfrac{1}{2}O_2 \rightarrow H_2O \qquad \textbf{(5.13)}$$

$$\Delta G° = -246{,}000 + 54.84T$$

At 2000 K

[1]Each activity term is the ratio of the pressure of the species to the pressure in its standard state, one atmosphere. The activity has a numerical value of P, the pressure in atmospheres, but not dimensions. Activities are dimensionless ratios.

Applying Eq. 5.10:

$$\ln K_a = -\frac{\Delta G^\circ}{RT} = 8.20$$

$$K_a = 3652 = \frac{a_{H_2O}}{a_{H_2}a_{O_2}^{1/2}} \tag{5.14a}$$

Assuming that the three gases (hydrogen, oxygen, and water vapor) are ideal, and that in each case the standard state is the gas at one atmosphere pressure, the activity of each gas is simply equal numerically to its partial pressure. Thus:

$$K_a = 3652 = \frac{P_{H_2O}}{P_{H_2}P_{O_2}^{1/2}} \tag{5.14b}$$

Because we want to set the partial pressure of oxygen to 10^{-10} atm,

$$\frac{P_{H_2O}}{P_{H_2}} = 3652P_{O_2}^{1/2} = 3652 \times 10^{-5}$$

The total pressure on the system is one atmosphere ($P_{H_2O} + P_{H_2} + P_{O_2} = 1$ atm). Then let $P_{H_2O} = x$ and $P_{H_2} = 1 - x$, because P_{O_2} is negligible in terms of total pressure.

$$\frac{x}{1-x} = 3.65 \times 10^{-2}$$

$$x = 0.0352$$

At one atmosphere, a hydrogen–water mixture in which the water content is 3.52% will, at equilibrium, have an oxygen partial pressure of 10^{-10} atm at 2000 K. This illustrates that it is possible to obtain low oxygen pressures by chemical means. In many cases, this technique is less expensive than using a vacuum apparatus to achieve the same oxygen activity.

5.4 SOLID–VAPOR EQUILIBRIA

Equation 5.10 can also be used to calculate the equilibria between solids and gases. Consider the equilibrium among Cu_2O, Cu, and O_2:

$$4Cu(s) + O_2(g) \rightarrow 2Cu_2O(s) \tag{5.15}$$

where (s) indicates a solid phase and (g) a gas phase.

If we know the $\Delta G°$ for the chemical reaction, we should be able to calculate the pressure of oxygen in equilibrium with solid copper and solid cuprous oxide. If the pressure of oxygen exceeds this equilibrium pressure, we would expect the reaction to proceed to the right (i.e., Cu_2O will be formed). If the pressure of oxygen is less than the equilibrium pressure, then the reaction should proceed to the left (i.e., Cu_2O should decompose into Cu and O_2).

As an example, let us calculate the pressure of oxygen below which cuprous oxide (Cu_2O) dissociates into its elements at a temperature of 1000 K. In a practical sense, we are asking whether cuprous oxide can be reduced to metallic copper at 1000 K by lowering the pressure. Conversely, we are asking what level of vacuum is required to prevent the oxidation of copper upon heating to 1000 K in pure oxygen.

From Eq. 5.10:

$$\Delta G° = -RT \ln K_a = -RT \ln \frac{a^2 Cu_2O}{a_{Cu}^4 P_{O_2}} \qquad \textbf{(5.16)}$$

We know that the standard Gibbs free energy for the reaction represented by Eq. 5.16 (twice the Gibbs free energy of formation of cuprous oxide) is given by

$$\Delta G° = -339{,}000 - 14.24T \ln T + 247T \qquad \textbf{(5.17)}$$

where $\Delta G°$ is in joules.

The activity of cuprous oxide is 1 because during the reaction the material exists as the pure solid, and its standard state is pure Cu_2O. We are trying to calculate the oxygen pressure at which pure copper will coexist with it, that is, the point at which the activity of copper will become 1. The activity of oxygen is just equal to its pressure, in atmospheres, because the standard state for gases is the pure gas at one atmosphere pressure. At oxygen pressures greater than this equilibrium pressure, copper will have an activity less than 1 and will not exist as metallic copper. Proceeding with the calculation at 1000 K:

$$\Delta G° = -190{,}360 = 8.314 \times 1000 \ln P_{O_2}$$

$$\ln P_{O_2} = -22.90, \; P_{O_2} = 1.14 \times 10^{-10} \text{ atm}$$

The oxygen pressure calculated is 1.14×10^{-10} atm. At this pressure, metallic copper and cuprous oxide can exist in equilibrium. At higher oxygen pressures, only cuprous oxide will exist in equilibrium; at lower pressures, metallic copper will exist.[2]

By making a series of these calculations at different temperatures, one could

---

[2]The reaction of metals to their oxides can be used to remove oxygen from inert gases (see Problem 5.4). For example, if in argon containing oxygen as an impurity, pressure is equilibrated with copper at 1000 K, the oxygen pressure in the gas will be reduced to 1.14×10^{-10} atm. At a total pressure of one atmosphere, the gas would contain only 1.14×10^{-8}% oxygen. When this technique is used, the reacting metals are called ''getters.'' Other elements, such as titanium, are even more effective in the process of gettering.

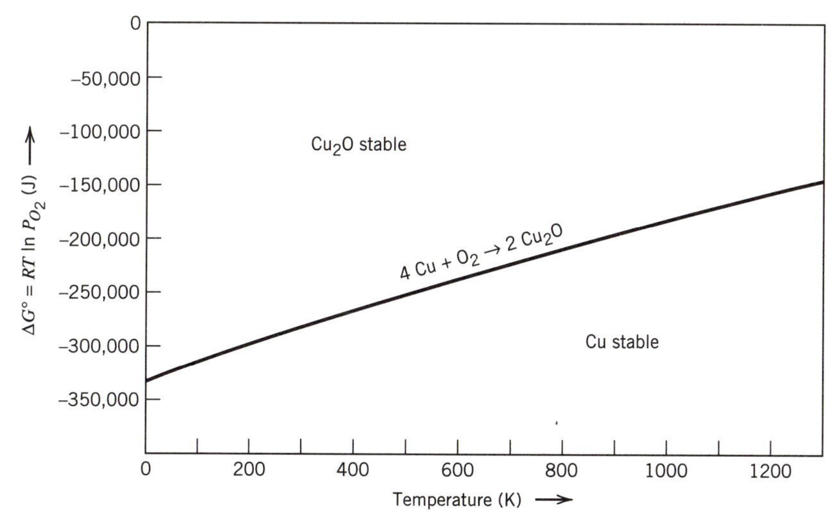

Figure 5.2 ΔG° versus T for the reaction $4Cu + O_2 \rightarrow 2Cu_2O$, showing regions of stability of Cu_2O and Cu.

evolve a plot (Figure 5.2) in which the quantity, ΔG° or $RT \ln P_{O_2}$, is plotted as the ordinate, and absolute temperature serves as the abscissa. The line represents the equilibrium between the two phases, copper and cuprous oxide (Cu_2O), in the presence of oxygen. At pressures above the line, cuprous oxide exists as the equilibrium phase; below the line, we have copper as the equilibrium phase.

The phase diagram can be made more complete by recognizing that another oxide of copper exists, cupric oxide (CuO). We can write the reaction for the formation of cupric oxide from Cu_2O and O_2:

$$2Cu_2O + O_2 \rightarrow 4CuO$$

From a knowledge of the standard Gibbs free energy change (ΔG°) for this reaction as a function of temperature, we calculate the oxygen pressures for the equilibrium between the two copper compounds. When these are plotted in Figure 5.3 on the same axes as in Figure 5.2, we have a diagram of the stability ranges for the oxides of copper and metallic copper as a function of temperature and oxygen pressure.

5.5 SOURCES OF INFORMATION ON ΔG°

For a chemical reaction, the Gibbs free energy change at a temperature T, ΔG°_T, is the sum of the Gibbs free energies of formation of the products, less the sum of the Gibbs free energies of formation of the reactants at the same temperature (similar to calculation of ΔH° in Section 1.18).

$$\Delta G^\circ_T = \sum_{\text{products}} n_p \, \Delta \underline{G}^\circ_{f,T} - \sum_{\text{reactants}} n_r \, \Delta \underline{G}^\circ_{f,T} \tag{5.18}$$

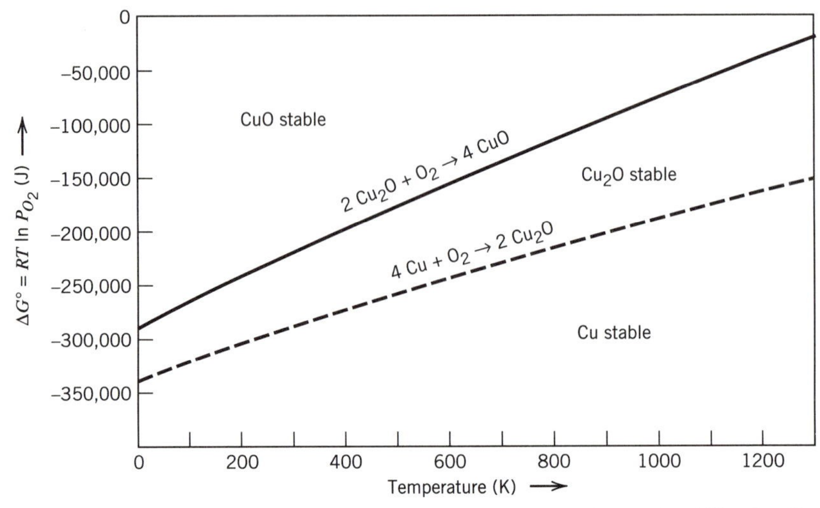

Figure 5.3 $\Delta G°$ versus T for oxides of copper, showing regions of stability for Cu, Cu_2O, and CuO.

where n is the number of moles of each of the species involved in the reaction, n_p for products and n_r for reactants.

Values of ΔG_T° can also be calculated from values of ΔH_T° and ΔS_T° as follows (assuming that there are no phase transformations between 298 K and T):

$$\Delta G_T^\circ = \Delta H_T^\circ - T\,\Delta S_T^\circ \tag{5.19}$$

where

$$\Delta H_T^\circ = \Delta H_{298,T}^\circ + \int_{298}^{T} \Delta C_p^\circ dT \tag{5.20}$$

and

$$\Delta S_T^\circ = \Delta S_{298,T}^\circ + \int_{298}^{T} \left(\frac{\Delta C_p^\circ}{T}\right) dT \tag{5.21}$$

In addition, actual values of the standard Gibbs free energy of formation for chemical compounds are available in many references. In some cases, values of ΔG_f° for compounds are tabulated at 100 K intervals from zero to the highest temperature for which they are known, and at 298 K (Ref. 1). Other works report data as algebraic equations. A sample of these values is given in Table 5.1. The form of these equations is often (Ref. 2):

$$\Delta G_{f,T}^\circ = A + BT + CT \ln T \tag{5.22}$$

Table 5.1 Standard Gibbs Free Energy Changes

Reaction	A	B	C	Range (°K)
$2Al(s) + \frac{3}{2}O_2(g) = Al_2O_3(s)$	−1,676,990	−7.23	+366.7	298–923
$2Al(l) + \frac{3}{2}O_2(g) = Al_2O_3(s)$	−1,697,700	−15.69	+385.9	923–1800
$C(s) + 2H_2(g) = CH_4(g)$	−69,120	+22.26	−65.35	298–1200
$C(s) + \frac{1}{2}O_2(g) = CO(g)$	−111,710		−87.65	298–2500
$C(s) + O_2(g) = CO_2(g)$	−394,130		−0.84	298–2000
$2CaO(s) + SiO_2(s) = Ca_2SiO_4(s)$	−120,360		−5.02	298–1700
$2Co(s) + O_2(g) = 2CoO(s)$	−476,560		+155.23	298–1300
$2Cr(s) + \frac{3}{2}O_2(g) = Cr_2O_3(s)$	−1,120,270		+259.83	298–2100
$4Cu(s) + O_2(g) = 2Cu_2O(s)$	−339,000	−14.24	+247.0	298–1356
$2Cu_2O(s) + \frac{1}{2}O_2(g) = 2CuO(s)$	−146,230	−11.08	+185.35	298–1300
$H_2(g) + \frac{1}{2}O_2(g) = H_2O(g)$	−246,740		+54.81	298–2500
$Mg(s) + \frac{1}{2}O_2(g) = MgO(s)$	−603,960	−5.36	+142.05	298–923
$Mg(l) + \frac{1}{2}O_2(g) = MgO(s)$	−608,140	−0.44	+112.76	923–1380
$Mn(s) + \frac{1}{2}O_2(g) = MnO(s)$	−384,720		+72.80	298–1500
$Ni(s) + \frac{1}{2}O_2(g) = NiO(s)$	−244,560		+98.53	298–1725
$Pb(s) + \frac{1}{2}O_2 = PbO(s)$	−221,120	−6.27	+141.6	298–600
$S_2 + 2O_2 = 2SO_2$	−724,840		+144.85	298–2000
$Si(s) + 2Cl_2 \rightarrow SiCl_4(g)$	−616,300	−6.61	+178.24	298–600
$Si(s) + SiO_2(s) = 2SiO(g)$	+666,930	25.07	−508.6	298–1700
$Si(s) + O_2(g) = SiO_2(s)$	−881,150	−5.45	+218.49	298–1700
$Zr(s) + O_2(g) = ZrO_2(s)$	−1,087,800	−18.44	+322.17	298–2000

(s) = solid
(l) = liquid
(g) = gas
Source: Values from O. Kubaschewski and E. L. L. Evans.

This form arises because the ΔC_p° terms in Eqs. 5.20 and 5.21 usually may be treated as independent of temperature for the purpose of calculating ΔG°.[3] To illustrate:

If the ΔC_p° term is constant, then:

$$\Delta H_T^\circ = \Delta H_{298}^\circ + \Delta C_p^\circ (T - 298) \tag{5.23}$$

Similarly, the entropy of formation can be written as

$$\Delta S_T^\circ = \Delta S_{298}^\circ + \Delta C_p^\circ \ln\left(\frac{T}{298}\right) \tag{5.24}$$

[3]The experimental information used to calculate values of ΔG° usually is not precise enough to justify the introduction of the temperature variation of ΔC_p° into the calculation.

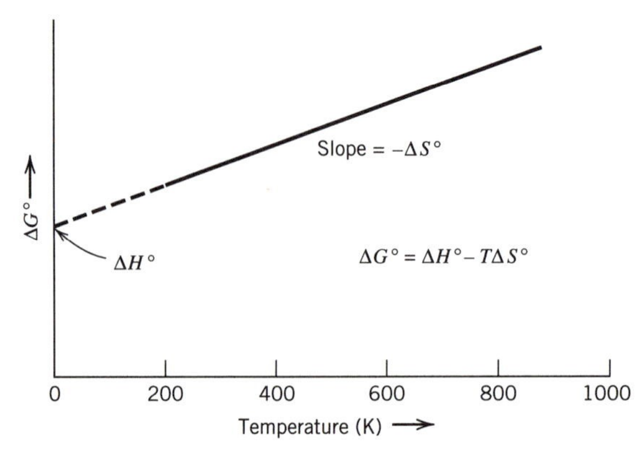

Figure 5.4 $\Delta G^\circ = \Delta H^\circ - T\Delta S^\circ$ versus T, showing that the intercept of 0 K is ΔH° and the slope is $-\Delta S^\circ$ if $\Delta C_P^\circ = 0$.

Combining all these, we have:

$$\Delta G_T^\circ = \Delta H_{298}^\circ + \Delta C_p^\circ(T - 298) - T\,\Delta S_{298}^\circ - T\,\Delta C_p^\circ(\ln T - \ln 298)$$

$$\Delta G_T^\circ = \Delta H_{298}^\circ - 298\,\Delta C_p^\circ - T[\Delta S_{298}^\circ - \Delta C_p^\circ(298)] - \Delta C_p^\circ T \ln T \tag{5.22}$$

$$\Delta G_T^\circ = A + BT + CT \ln T$$

where

$$A = \Delta H_{298}^\circ - 298\Delta C_p^\circ$$

$$B = -(\Delta S_{298}^\circ - \Delta C_p^\circ \ln 298)$$

$$C = -\Delta C_p^\circ$$

If ΔC_p°, the difference in heat capacities between reactants and products, is small, or if the data upon which the standard free values are not sufficiently precise to justify the inclusion of differences in heat capacity, then the ΔG_T° equation assumes the form $A + BT$, where A is the standard enthalpy change of the reaction and B is the standard entropy change of the reaction, namely:

$$\Delta G_T^\circ = \Delta H^\circ - T\,\Delta S^\circ \tag{5.25}$$

When data in the form of Eq. 5.25 are presented graphically, the result is a straight line on a plot of ΔG_T° versus temperature (Figure 5.4), with ΔH° as its intercept at 0 K, and $-\Delta S^\circ$ as its slope.

5.6 ELLINGHAM DIAGRAMS

Values of Gibbs free energy changes for chemical reactions are often presented in graphical form, with ΔG° as the ordinate and T, the absolute temperature, as the

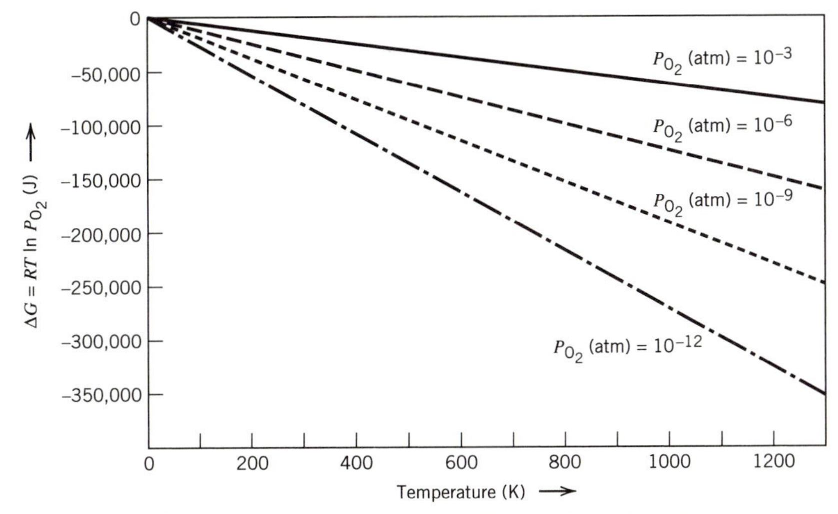

Figure 5.5 Plot of $\Delta G°$ versus T, where $\Delta G° = RT \ln P_{O_2}$.

abscissa. When values of ΔG_f° for oxides are displayed, the Ellingham diagram offers a simple and useful way to estimate equilibrium oxygen pressures as a function of temperature.[4] To establish the foundation for these graphs, let us use the axes in Figure 5.2 to plot lines of constant oxygen pressure. The relationship among the variables is:

$$\Delta \underline{G}° = +RT \ln P_{O_2}$$

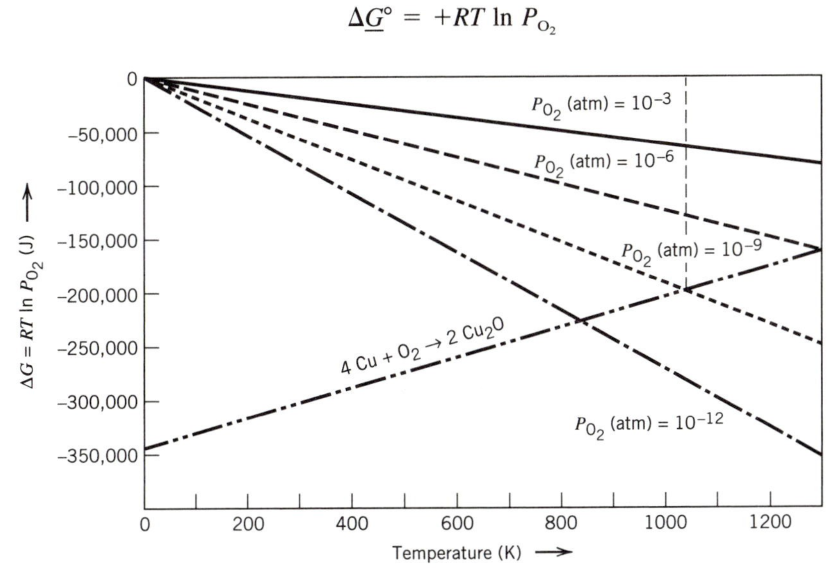

Figure 5.6 Superposition of Figures 5.2 and 5.5. At 1050 K, equilibrium oxygen pressure in the reaction $4Cu + O_2 \rightarrow 2Cu_2O$ is 10^{-9} atm.

[4]The same may be done for chlorides, fluorides, and other compounds (Ref. 4).

Thus for every value of oxygen pressure, there is a simple, linear relationship between $\Delta G°$ and T. The intercept of the line on the $\Delta \underline{G}°$ axis is zero, and the slope is $+R \ln P_{O_2}$ (Figure 5.5).

If the two diagrams (Figures 5.2 and 5.5) are superimposed, as in Figure 5.6, the intersections of the constant pressure lines and the copper–cuprous oxide equilibrium line give the equilibrium oxygen pressures for this reaction at various temperatures. For example, at 1050 K, the oxygen pressure in equilibrium with Cu_2O and Cu is 10^{-9} atm.

Figure 5.7 is an Ellingham diagram, a graph of lines representing the values of $\Delta G_f°$ for several important oxides (Ref. 3). A series of scales runs along the right-hand side of the graph. The first relates to oxygen pressures and is the intersection of the constant oxygen pressure lines (see Figure 5.5) with the scale. The other end of the line passes through the upper left-hand corner of the graph, as in Figure 5.5.

The other two scales on the right-hand side of Figure 5.7 relate to other measures of the oxygen pressure. Based on the calculations in Section 5.3 (Eq. 5.14b), we know that the equilibrium pressure of oxygen in a system at any temperature may be established by controlling the ratio of H_2 to H_2O in the vapor.

$$2H_2 + O_2 \rightarrow 2H_2O$$

$$K_a = \frac{P_{H_2O}^2}{P_{H_2}^2 P_{O_2}}$$

$$P_{O_2} = \frac{1}{K_a}\left(\frac{P_{H_2O}}{P_{H_2}}\right)^2$$

The Ellingham diagrams take advantage of this relationship to give the P_{H_2}/P_{H_2O} ratio (or simply the H_2/H_2O ratio) in equilibrium with oxides and metals at various temperatures. The diagram is used as a nomograph. A straightedge is placed on the marker labeled ''H'' on the 0 K line and on the intersection of the line representing the reaction of interest and the temperature. The H_2/H_2O ratio is read on the scale indicated in the diagram (Figure 5.7).

Oxygen pressure in a system can also be controlled using the CO-CO_2-O_2 equilibrium. The reaction among these three species is:

$$2CO + O_2 \rightarrow 2CO_2$$

The equilibrium constant for the reaction is given by:

$$K_a = \frac{P_{CO_2}^2}{P_{CO}^2 P_{O_2}}$$

Solving for the oxygen pressure:

$$P_{O_2} = \frac{1}{K_a}\left(\frac{P_{CO_2}}{P_{CO}}\right)^2$$

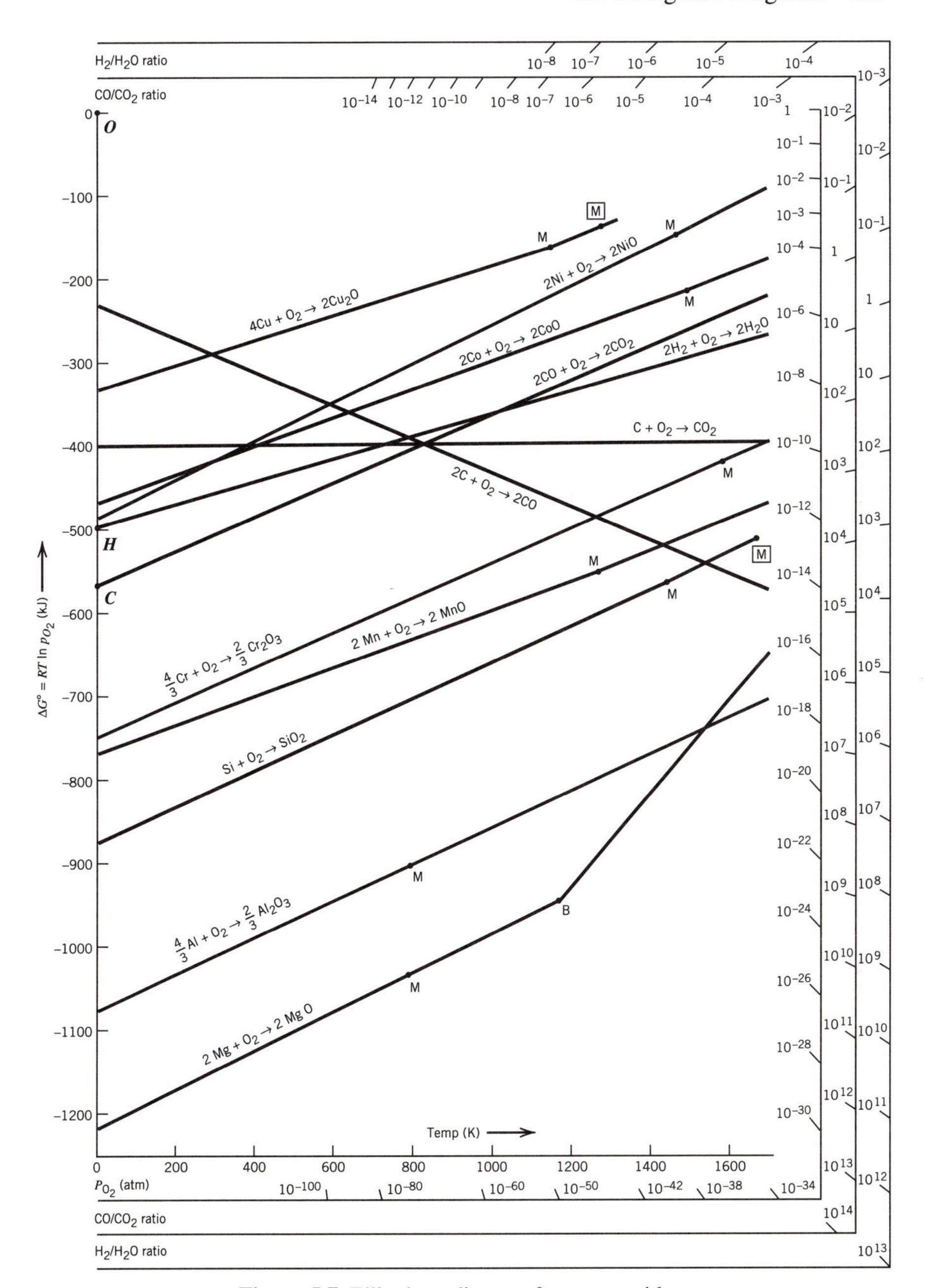

Figure 5.7 Ellingham diagram for some oxides.

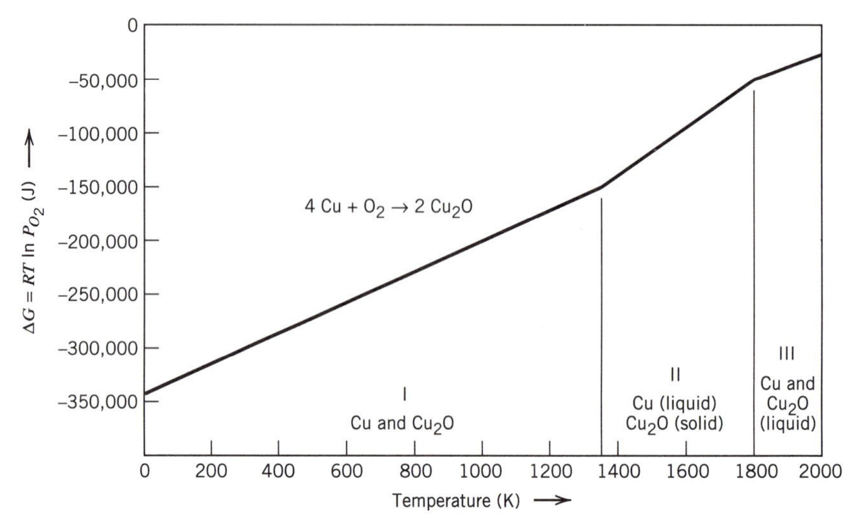

Figure 5.8 Plot of $\Delta G^\circ = RT \ln P_{O_2}$ versus for $4Cu + O_2 \rightarrow 2Cu_2O$, showing change in the slope of the curve at the melting points of condensed phases.

Values of the CO/CO_2 ratios can be read from Ellingham diagrams using the marker labeled "C," using the same technique described for H_2/H_2O ratios.

In Section 5.4 the form of the ΔG° versus T curve was discussed, and it was shown that the curve representing equilibria on the Ellingham diagram is a straight line if the difference in heat capacity between the reactants and products (ΔC_p°) is zero, or if the precision of the data do not warrant the use of a more complex equation such as Eq. 5.22. In this case the intersection of the line with the ordinate at zero of temperature represents the change in standard enthalpy for the reaction. The slope of the line represents the negative of the standard change in entropy for the reaction. In fact, even if the more complex form of the ΔG° equation is used, the curves on Ellingham diagrams are hard to distinguish from straight lines.

Lines on Ellingham diagrams often have sharp breaks in them (discontinuities in slope). These sharp breaks are caused by phase transformations. To illustrate, consider the reaction between copper, oxygen, and cuprous oxide. The line representing this equilibrium from the Ellingham diagram is shown as Figure 5.8. In region I the line represents the equilibrium between gaseous oxygen and *solid* copper and *solid* cuprous oxide.

$$4Cu(s) + O_2 \rightarrow 2Cu_2O(s) \qquad \Delta S^\circ = -129.2 \text{ J/(mol·K)}$$

At a temperature of 1356 K copper melts. The entropy of melting is roughly 9.6 J/(mol·K) (per mole of copper).

$$Cu(s) \rightarrow Cu(l) \qquad \Delta S^\circ = 9.7 \text{ J/(mol·K)}$$

For the reaction *liquid* copper plus oxygen gas to form solid cuprous oxide:

$$4Cu(l) + O_2 \rightarrow 2Cu_2O(s) \qquad \Delta S^\circ = -168 \text{ J/(mol·K)}$$

Thus the slopes of the lines in regions II and I will be different. The additional sharp bend in the line at the border between zones II and III accounts for the melting of cuprous oxide.

5.7 VARIATION OF EQUILIBRIUM CONSTANT WITH TEMPERATURE

The variation of equilibrium constant with temperature can be determined by knowing the Gibbs energy change for a reaction as a function of temperature. From Eq. 3.25, we can write

$$d(\Delta G^\circ) = -\Delta S^\circ \, dT \tag{5.26}$$

We can also write:

$$\Delta G^\circ = \Delta H^\circ - T\,\Delta S^\circ$$

$$\Delta S^\circ = \frac{\Delta H^\circ - \Delta G^\circ}{T}$$

Substituting for ΔS° in Eq. 5.26:

$$\begin{aligned} d(\Delta G^\circ) &= \frac{\Delta G^\circ}{T}\,dT - \frac{\Delta H^\circ}{T}\,dT \\ d(\Delta G^\circ) - \frac{\Delta G^\circ}{T}\,dT &= -\frac{\Delta H^\circ}{T}\,dT \end{aligned} \tag{5.27}$$

Multiplying Eq. 5.27 by $1/T$, we obtain:

$$\frac{d(\Delta G^\circ)}{T} - \frac{\Delta G^\circ}{T^2}\,dT = -\frac{\Delta H^\circ}{T^2}\,dT$$

Combining terms yields:

$$d\left(\frac{\Delta G^\circ}{T}\right) = \Delta H^\circ \, d\left(\frac{1}{T}\right) \tag{5.28}$$

But $\Delta G^\circ = -RT \ln K_a$.

$$d(-R \ln K_a) = \Delta H^\circ \, d\left(\frac{1}{T}\right)$$

Finally,

$$d(\ln K_a) = -\frac{\Delta H^\circ}{R} \, d\left(\frac{1}{T}\right) \tag{5.29}$$

From Eq. 5.29 it is apparent that a graph with the natural logarithm of K_a as the ordinate and $1/T$ (absolute temperature) as the abscissa (Figure 5.9) will yield a line whose slope is $-\Delta H^\circ$ divided by R.

This especially useful equation is known as the van't Hoff relationship. Knowledge of equilibrium constants (K_1 and K_2) at two temperatures (T_1 and T_2) can yield information on ΔH° of the reaction, assuming that it is constant between these two temperatures:

$$\int_{K_1}^{K_2} d(\ln K_a) = -\frac{\Delta H^\circ}{R} \int_{T_1}^{T_2} d\left(\frac{1}{T}\right)$$
$$\ln \frac{K_2}{K_1} = -\frac{\Delta H^\circ}{R}\left[\frac{1}{T_2} - \frac{1}{T_1}\right] \tag{5.30}$$

Conversely, knowledge of an equilibrium constant at one temperature and the ΔH° can yield information on the equilibrium constant at other temperatures.

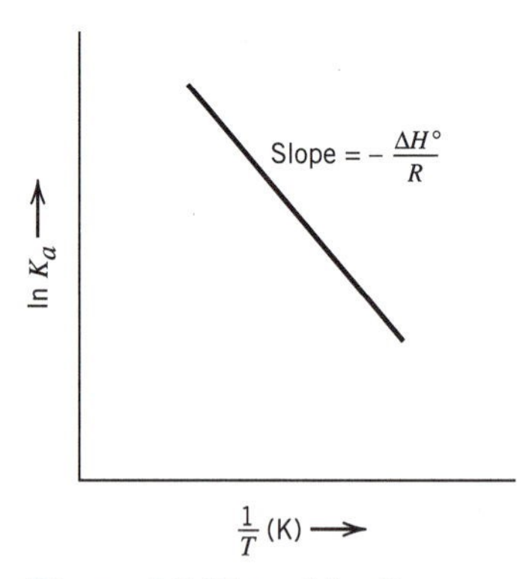

Figure 5.9 Plot of ln K_a versus $1/T$. For the reaction represented, the slope of the line is $-\Delta H^\circ/R$.

As an example, consider the decomposition of calcium carbonate (limestone) into calcium oxide and carbon dioxide:

$$CaCO_3 \rightarrow CaO + CO_2$$

$$K_a = \frac{a_{CaO} P_{CO_2}}{a_{CaCO_3}}$$

The following data are available for the pressure of carbon dioxide in equilibrium with CaO and $CaCO_3$:

Temperature (K)	Pressure (atm)
1030	0.10
921	0.01

Based on these data, let us calculate the heat effect of the reaction (change in enthalpy), and let us estimate the temperature at which the equilibrium partial pressure of carbon dioxide will be one atmosphere. If we assume that the activities of CaO and $CaCO_3$ are each 1, and $\Delta H°$ for the reaction is constant in this range, then from Eq. 5.30:

$$\ln \frac{K_2}{K_1} = -\frac{\Delta H°}{R}\left[\frac{1}{T_2} - \frac{1}{T_1}\right] = \frac{\Delta H_0}{R}\left[\frac{T_2 - T_1}{T_2 T_1}\right]$$

$$\ln\left(\frac{0.10}{0.01}\right) = \ln 10 = \frac{\Delta H_0(109)}{(8.314)(921)(1030)}$$

$$\Delta H° = +166{,}000 \text{ J}$$

Thus, the reaction is endothermic, requiring 166,600 J per mole of calcium carbonate decomposed. With this value in hand, and mindful of the dangers of extrapolation of data beyond the range of measurement, we can proceed to estimate the temperature at which the pressure of carbon dioxide is one atmosphere—that is, the temperature (T_3) at which the pressure of carbon dioxide is one atmosphere ($K_3 = 1$):

$$\ln \frac{K_3}{K_2} = \ln 10 = -\frac{166{,}000}{R}\left[\frac{1}{T_3} - \frac{1}{1030}\right]$$

$$T_3 = 1168 \text{ K}$$

Of course, if an algebraic expression for $\Delta G°$ of a reaction as a function of T is available, the value of $\Delta S°$ for the reaction is derived simply by using Eq. 5.26. The value of $\Delta H°$ can be evaluated using Eq. 5.28. For example, for the oxidation of copper to cuprous oxide, $4Cu(s) + O_2(g) \rightarrow 2Cu_2O(s)$, we know $\Delta G°$ in joules:

$$\Delta G° = -339{,}000 - 14.24T \ln T + 247T$$

The entropy change for the reaction (ΔS°) is:

$$-\Delta S^\circ = \frac{d(\Delta G^\circ)}{dT} = -14.24 \ln T - 14.24 + 247$$

$$\Delta S^\circ = -232.76 + 14.24 \ln T$$

The enthalpy change for the reaction (ΔH°) is:

$$\Delta H^\circ \, d\left(\frac{1}{T}\right) = d\left(\frac{\Delta G^\circ}{T}\right) = d\left(-\frac{339{,}000}{T} - 14.24 \ln T + 247\right)$$

$$\Delta H^\circ \, d\left(\frac{1}{T}\right) = +\frac{339{,}000}{T^2}\,dT - 14.24\frac{dT}{T} = 339{,}000\frac{dT}{T^2} - 14.24T\frac{dT}{T^2}$$

$$\Delta H^\circ \, d\left(\frac{1}{T}\right) = -(339{,}000 + 14.24T)d\left(\frac{1}{T}\right)$$

$$\Delta H^\circ = -339{,}000 + 14.24T$$

5.8 GASES DISSOLVED IN METALS (SIEVERT'S LAW)

Gases, such as hydrogen and oxygen, dissolve in liquid and solid metals. In the gaseous state, hydrogen and oxygen exist in diatomic form, H_2 and O_2. When dissolved, however, they are found in monatomic form, *H* and *O*. In a chemical reaction a gas that is in solution is indicated by an underscore.

For hydrogen dissolving in copper:

$$H_2(g) = 2H \text{ (in copper solution)} \tag{5.31}$$

The equilibrium constant for this reaction is:

$$K_a = \frac{a_H^2}{a_{H_2(g)}} \tag{5.32}$$

Since hydrogen gas at moderate pressures behaves as an ideal gas, its activity is just its pressure in atmospheres because the standard state is hydrogen at one atmosphere pressure.

The activity of materials dissolved as dilute solutions (such as gases in metals) is usually defined in a different way. Let us assume that hydrogen activity is linearly related to its concentration in the metal (Figure 5.10), a reasonable assumption based on experimental information.

The units of concentration of hydrogen in the metal can be the volume of gaseous hydrogen dissolved in a given mass of copper—for example, cubic centimeters of

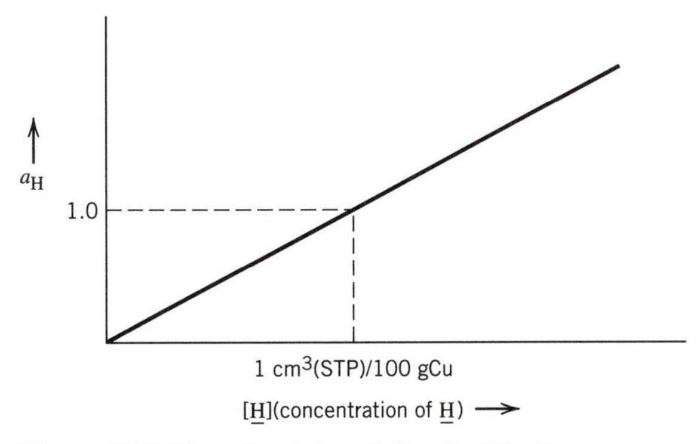

Figure 5.10 Plot of activity of dissolved hydrogen versus concentration.

gas at standard temperature and pressure (STP = 298 K and 1 atm) per 100 g of copper.

We can arbitrarily pick a concentration of dissolved hydrogen equivalent to 1 cm^3 (STP) per 100 g of copper as its standard state. With that standard state, and assuming linearity of a_H versus H

$$a_H = [H] \tag{5.33}$$

where [H] is in cubic centimeters (STP) per 100 g of copper.
The relationship in Eq. 5.32 then becomes:

$$K_a = \frac{[H]^2}{P_{H_2}} \tag{5.34}$$

or

$$[H]^2 = K_a P_{H_2}$$
$$[H] = K_a^{1/2} P_{H_2}^{1/2} \tag{5.35}$$

Thus, the amount of hydrogen dissolved in copper at a given temperature is proportional to the *square root* of the hydrogen gas pressure. This relationship, which is true of most diatomic gases dissolved in metals, is known as Sievert's law.

5.9 CHEMICAL EQUILIBRIUM AND ADIABATIC FLAME TEMPERATURES

In the discussion of adiabatic flame temperatures in Chapter 1 (Section 1.20) we noted that the method used in the calculation assumed that the chemical reaction involved in the combustion proceeds to completion. This condition is not necessarily met at very high temperatures. In the example cited, the combustion of methane in pure oxygen would produce an adiabatic flame temperature exceeding 5100 K *if the reaction to carbon dioxide and water were to proceed to completion.* Based on chemical equilibrium principles discussed in this chapter, the reaction would not proceed to completion, and a lower adiabatic flame temperature would be achieved. To calculate the correct adiabatic flame temperature, First Law energy balances and chemical equilibrium must be taken into account simultaneously.

To illustrate the procedure needed, consider the case of a stream of monatomic hydrogen at 298 K forming diatomic hydrogen in a burner (or a torch). The reaction is:

$$2H \rightarrow H_2$$

The data needed for the calculation are as follows:
For H:

$$\Delta \underline{H}^{\circ}_{f,298} = +218{,}000 \text{ J/mol}$$

$$\Delta \underline{G}^{\circ}_{f,298} = +203{,}290 \text{ J/mol}$$

$$\underline{S}^{\circ}_{298} = +114.72 \text{ J/(mol·K)}$$

For H_2:

$$\Delta \underline{H}^{\circ}_{298} = 0$$

$$\underline{S}^{\circ}_{298} = +130.68 \text{ J/(mol·K)}$$

For the purposes of keeping the calculation simple, assume that the heat capacity (C_p) is $5R/2$ and $7R/2$ for H and H_2, respectively.

The basis for the calculation is 2 mols of H, and the system is the burner. The First Law yields:

$$\Sigma \underline{H}_i n_i = \Sigma \underline{H}_o n_o$$

$$2(218{,}000) = \int_{298}^{T} C_{p,H_2} dT = \tfrac{7}{2}R(T - 298)$$

$$T = 15{,}300 \text{ K! (assuming the reaction proceeds to completion)}$$

Let us now calculate the extent of the reaction at $T = 15{,}300$ K

$$2H \rightarrow H_2$$

$$\Delta G^{\circ}_T = \Delta H^{\circ}_T - T\,\Delta S^{\circ}_T$$

The enthalpy change (ΔH°) is:

$$\Delta H^\circ_T = -2(218{,}000) + [\tfrac{7}{2} R - 2\ (\tfrac{5}{2} R)]\ (T - 298)$$

The entropy change (ΔS°) is:

$$\Delta S^\circ_T = \Delta S^\circ_{298} + \int_{298}^{T} \frac{\Delta C^\circ_p}{T}\, dT$$

$$\Delta S^\circ_T = -98.76 - \tfrac{3}{2} R \ln \left(\frac{T}{298}\right)$$

Combining:

$$\Delta G^\circ_T = -432{,}000 + 86.29T + 12.47T \ln \left(\frac{T}{298}\right) \qquad \textbf{(5.36a)}$$

$$\Delta G^\circ_{15{,}300\ \mathrm{K}} = 1.64 \times 10^6 = -RT \ln K_a$$

$$K_{a(15{,}300\ \mathrm{K})} = 2.5 \times 10^{-6}$$

To solve for x, the equilibrium number of moles of H_2 (or the extent of the reaction), we set up the following table of mole fractions:

	Initial Moles	Final Moles	Final Mole Fraction
H	2	$2 - 2x$	$\dfrac{2 - 2x}{2 - x}$
H_2	0	x	$\dfrac{x}{2 - x}$
Total moles (final)		$2 - x$	

If the total pressure is 1 atm:

$$P_{H_2} = \frac{x}{2 - x} \qquad P_H = \frac{2 - 2x}{2 - x}$$

$$K_a = \frac{x(2 - x)}{4(1 - x)^2} = 2.5 \times 10^{-6} \quad \text{at} \quad 15{,}300\ \mathrm{K} \qquad \textbf{(5.36b)}$$

$$x = 10^{-5}$$

The mole fraction of H_2 at equilibrium at 15,300 K is then $10^{-5}/(2 - 10^{-5})$, or about 5×10^{-6}. Clearly, the reaction does not proceed to completion at 15,300 K.

To calculate the correct temperature, we must write the First Law in terms that allow for an incomplete reaction, assuming that only x moles of H_2 are formed for every two moles of H fed to the burner. There would be $(2 - 2x)$ moles of H left unreacted.

$$\begin{aligned} 436{,}000 &= \tfrac{7}{2}R(x)(T - 298) + (2 - 2x)(218{,}000) \\ &\quad + (2 - 2x)(\tfrac{5}{2}R)(T - 298) \\ x[436{,}000 + \tfrac{3}{2}R(T - 298)] &= 5R[T - 298] \end{aligned} \tag{5.37}$$

At this point we must consider both equations 5.36a and 5.37 and solve for the value of x that satisfies both of them. This can be visualized as a graph with axes x and T. Both equations (5.36a and 5.37) are plotted on the axes (Figure 5.11). The intersection of the two is the adiabatic flame temperature that will be achieved. In this case it is about 4000 K.

In Chapter 1 (Section 1.20) we calculated the adiabatic temperature rise for the complete polymerization of methyl methacrylate, and we noted that at the calculated final temperature (578 K or 305°C), the polymerization reaction might not take place because of chemical equilibrium conditions. For the polymerization reaction M(l) = $1/nP_n$, the enthalpy of polymerization is expressed as:

$$\Delta \underline{H}^\circ_p = \frac{1}{n}\,\Delta \underline{H}^\circ_f(P_n) - \Delta \underline{H}^\circ_f(\text{M}) \tag{5.38}$$

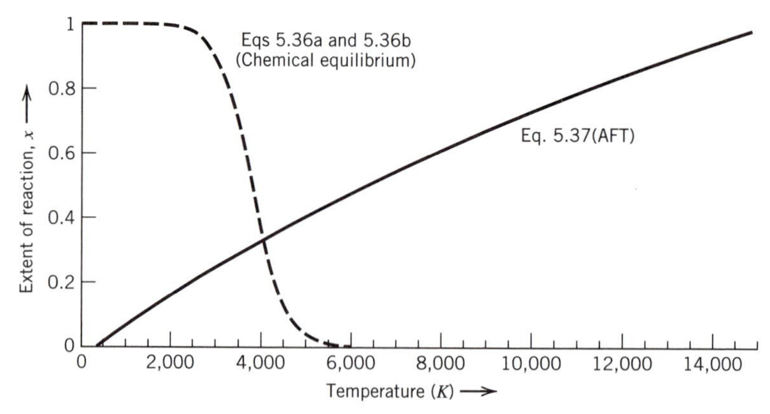

Figure 5.11 The extent of the reaction $2H \rightarrow H_2$ as a function of temperature for the AFT and chemical equilibrium calculations (Eqs. 5.37 and 5.36a, respectively).

A similar expression for the entropy of polymerization is:

$$\Delta \underline{S}^{\circ}{}_{p} = \frac{1}{n} \Delta \underline{S}^{\circ}_{f}(P_{n}) - \Delta \underline{S}^{\circ}_{f}(M) \tag{5.39}$$

The standard Gibbs free energy of polymerization is $\Delta G^{\circ}_{p} = \Delta \underline{H}^{\circ}_{p} - T\,\Delta \underline{S}^{\circ}_{p}$. The entropies of polymerization are negative quantities. Thus as the temperature rises, the standard Gibbs free energy of polymerization increases. At the temperature at which ΔG°_{p} becomes positive, the driving force for polymerization no longer exists, as demonstrated in Eqs. 5.40–5.42. In the polymer literature, this point is called the ''ceiling temperature.'' Consider one step in a polymerization reaction, the addition of one molecule of the monomer (M) to a polymer (P_i) with i segments (mers) in a chain.

$$P_i + M \rightarrow P_{i+1} \tag{5.40}$$

At equilibrium $K = [P_{i+1}]/[P_i][M]$. If we assume that the concentrations of P_i and P_{i+1} are about equal, then $K = [M]^{-1}$. From the equation $\Delta G^{\circ} = -RT \ln K$, the concentration of M at equilibrium is:

$$\ln[M]_{eq} = \frac{\Delta \underline{H}^{\circ}_{p}}{RT} - \frac{\Delta \underline{S}^{\circ}}{R} \tag{5.41}$$

At the ceiling temperature T_{ceil}, the equilibrium concentration of M will be just the concentration of the pure monomer, because no polymerization will take place. Rearranging Eq. 5.41:

$$T_{ceil} = \frac{\Delta \underline{H}^{\circ}_{p}}{\Delta \underline{S}^{\circ}_{p}} + R \ln [M]_{pure} \tag{5.42}$$

Let us estimate the ceiling temperature when the concentration of pure M is taken as the standard state, and ln[M] is zero for a polymer. In the case of methyl methacrylate $\Delta \underline{S}^{\circ}_{p} = -118$ J/(mol·K) and $\Delta \underline{H}^{\circ}_{p} = -56{,}100$ J/mol.

$$T_{ceil} = \frac{\Delta \underline{H}^{\circ}_{p}}{\Delta \underline{S}^{\circ}{}_{p}} = \frac{56{,}100}{118} = 475 \text{ K } (202^{\circ}\text{C})$$

The ceiling temperature of methyl methacrylate is thus 475 K or 202°C, well below the 578 K or 503°C calculated as the adiabatic temperature rise assuming complete reaction (polymerization).

From the two examples presented in this section, we can conclude that the adiabatic temperature changes calculated using the method presented in Section 1.20 are valid only if the reaction in question *proceeds to completion.* If this does not happen, the calculation must satisfy both the enthalpy criterion of Section 1.20 and chemical equilibrium conditions illustrated in this chapter.

REFERENCES

1. Stull, D. R., and Prophet, H., *JANAF Thermochemical Tables,* 2nd ed., National Standards Research Data Service–National Bureau of Standards, Washington, DC, 1971.
2. Kubaschewski, O., and Evans, E. L. L., *Metallurgical Thermochemistry,* Wiley, New York, 1956.
3. Ellingham, H. J. T., Reducibility of Oxides and Sulfides in Metallurgical Processes, *J. Soc. Chem. Ind., 63,* 125, 1944.
4. Glassner, A., *The Thermochemical Properties of the Oxides, Fluorides, and Chlorides to 2500°K,* ANL-5750, U.S. Government Printing Office, Washington, DC, 1957.

PROBLEMS

5.1 Calculate the partial pressure of monatomic hydrogen in hydrogen gas at 2000 K and 1 atm pressure.

$$\text{For } \tfrac{1}{2}H_2(g) \rightarrow H(g)$$

$$\Delta H^\circ_{298} = 217{,}990 \text{ J}$$

$$\Delta S^\circ_{298} = 49.35 \text{ J/K}$$

Assume that the heat capacity of H(g) $= \tfrac{3}{2}R$.
The heat capacity of H_2 can be assumed to be 31 J/(mol·K)

5.2 For the reaction:

$$Co(s) + \tfrac{1}{2}O_2(g) = CoO(s)$$

$$\Delta G^\circ = -59{,}850 + 19.6T$$

where ΔG° is in calories and T is in kelvin.

(a) Calculate the oxygen equilibrium pressure (atm) over Co and CoO at 1000°C.

(b) What is the uncertainty in the value calculated in part a if the error in the ΔH° term is estimated to be ± 500 cal?

5.3 Calculate the temperature at which silver oxide (Ag_2O) begins to decompose into silver and oxygen upon heating:

(a) in pure oxygen at $P = 1$ atm

(b) in air at $P_{\text{Total}} = 1$ atm

DATA

$$\Delta H_f \text{ for } Ag_2O = -7300 \text{ cal/mol}$$

Standard Entropies at 298° K
[cal/(mol·K)]

Ag_2O	29.1
O_2	49.0
Ag	10.2

Assume that $\Delta C_p = 0$ for the decomposition reaction.

5.4 One step in the manufacture of specially purified nitrogen is the removal of small amounts of residual oxygen by passing the gas over copper gauze at approximately 500°C. The following reaction takes place:

$$2Cu(s) + \tfrac{1}{2}O_2(g) \rightarrow Cu_2O(s)$$

(a) Assuming that equilibrium is reached in this process, calculate the amount of oxygen present in the purified nitrogen.
(b) What would be the effect of raising the temperature to 800°C? Or lowering it to 300°C? What is the reason for using 500°C?
(c) What would be the effect of increasing the gas pressure?
For $2Cu + \frac{1}{2}O_2 \rightarrow Cu_2O(s)$, $\Delta G°$ (in calories) is $-39{,}850 + 15.06T$.

5.5 The solubility of hydrogen ($P_{H_2} = 1$ atm) in liquid copper at 1200°C is 7.34 cm^3 (STP) per 100 g of copper. Hydrogen in copper exists in monatomic form.

(a) Write the chemical equation for the dissolution of H_2 in copper.
(b) What level of vacuum (atm) must be drawn over a copper melt at 1200°C to reduce its hydrogen content to 0.1 cm^3 (STP) per 100 g?
(c) A 100 g melt of copper at 1200°C contains 0.5 cm^3 (STP) of H_2. Argon is bubbled through the melt slowly so that each bubble equilibrates with the melt. How much argon must be bubbled through the melt to reduce the H_2 content to 0.1 cm^3 (STP) per 100 g?

Note: STP means standard temperature and pressure (298 K and 1 atm).

5.6 The following equilibrium data have been determined for the reaction:

$$NiO(s) + CO(g) \rightarrow Ni(s) + CO_2(g)$$

***T* (°C)**	***K* × 10⁻³**
663	4.535
716	3.323
754	2.554
793	2.037
852	1.577

(a) Plot the data using appropriate axes and find $\Delta H°$, K, and $\Delta G°$ at 1000 K.
(b) Will an atmosphere of 15% CO_2, 5% CO, and 80% N_2 oxidize nickel at 1000 K?

5.7 At 1 atm pressure and 1750°C, 100 g of iron dissolves 35 cm^3 (STP) of nitrogen. Under the same conditions, 100 g of iron dissolves 35 cm^3 of hydrogen.

Argon is insoluble in molten iron. How much gas will 100 g of iron dissolve at 1750°C and 760 mm pressure under an atmosphere that consists of:

(a) 50% nitrogen and 50% hydrogen?
(b) 50% argon and 50% hydrogen?
(c) 33% nitrogen, 33% hydrogen, and 34% argon?

5.8 Solid silicon in contact with solid silicon dioxide is to be heated to a temperature of 1100 K in a vacuum furnace. The two solid phases are not soluble in each other, but it is known that silicon and silicon dioxide can react to form gaseous silicon monoxide. For the reaction:

$$Si(s) + SiO_2(s) \rightarrow 2SiO(g)$$

the Gibbs free energy change (J) is:

$$\Delta G^\circ = 667{,}000 + 25.0T \ln T - 510\,T$$

(a) Calculate the equilibrium pressure of SiO gas at 1100 K.
(b) For the reaction above, calculate ΔH° and ΔS° at 1100 K.
(c) Using the Ellingham chart (Figure 5.7), estimate the pressure of oxygen (O_2) in equilibrium with the materials in the furnace.

5.9 What is the pressure of uranium (gas) in equilibrium with uranium dicarbide (UC_2) and pure carbon at 2263 K?

DATA

At 2263 K, ΔG_f° for UC_2 is −82,000 cal/mol

Vapor pressure of pure uranium is:

$$\ln P \text{ (uranium)} = 25.33 - \frac{100{,}000}{T} \; T \ln K$$

5.10 The direct reduction of iron oxide by hydrogen may be represented by the following equation:

$$Fe_2O_3 + 3H_2 \rightarrow 2Fe + 3H_2O$$

What is the enthalpy change, in joules, for the reaction? Is it exothermic or endothermic?

$$2Fe + \tfrac{3}{2}O_2 \rightarrow Fe_2O_3 \qquad \Delta G^\circ = -810{,}250 + 254.0T$$
$$H_2 + \tfrac{1}{2}O_2 \rightarrow H_2O \qquad \Delta G^\circ = -246{,}000 + 54.8T$$

5.11 Calcium carbonate decomposes into calcium oxide and carbon dioxide according to the reaction:

$$CaCO_3 \rightarrow CaO + CO_2$$

DATA

for the pressure of carbon dioxide in equilibrium with CaO and $CaCO_3$:

Temperature (K)	Pressure (atm)
1030	0.10
921	0.01

(a) What is the heat effect (ΔH) of the decomposition of one mole of $CaCO_3$? Is the reaction endothermic or exothermic?
(b) At what temperature will the equilibrium pressure of CO_2 equal one atmosphere?

5.12 In the carbothermic reduction of magnesium oxide, briquettes of MgO and carbon are heated at high temperature in a vacuum furnace to form magnesium (gas) and carbon monoxide (gas).

(a) Write the chemical reaction for the process.
(b) What can you say about the relationship between the pressure of magnesium gas and the pressure of carbon monoxide?
(c) Calculate the temperature at which the sum of the pressures of Mg (gas) and CO reaches one atmosphere.
With T in kelvin, the free energies of formation, in calories, of the relevant compounds are:

$$\text{MgO} \qquad \Delta G_f^\circ = -174{,}000 + 48.7T$$

$$\text{CO} \qquad \Delta G_f^\circ = -28{,}000 - 20.2T$$

5.13 Metallic silicon is to be heated to 1000°C. To prevent the formation of silicon dioxide (SiO_2), it is proposed that a hydrogen atmosphere be used. Water vapor, which is present as an impurity in the hydrogen, can oxidize the silicon.

(a) Write the chemical equation for the oxidation of silicon to silicon dioxide by water vapor.
(b) Using the accompanying data, where ΔG° is in joules, determine the equilibrium constant for the reaction at 1000°C (1273 K).
(b) What is the maximum content of water in the hydrogen (ppm) that is permitted if the oxidation at 1000°C is to be prevented?
(d) Check the answer to part c on the Ellingham diagram (Figure 5.7).

DATA

$$H_2(g) + \tfrac{1}{2}O_2(g) \rightarrow H_2O(g) \qquad \Delta G^\circ = -246{,}000 + 54.8T$$
$$Si(s) + \tfrac{1}{2}O_2(g) \rightarrow SiO_2(s) \qquad \Delta G^\circ = -902{,}000 + 174T$$

5.14 Solid barium oxide (BaO) is to be prepared by the decomposition of the mineral witherite ($BaCO_3$) in a furnace open to the atmosphere (p = 1 atm).

(a) Write the equation of the decomposition (witherite and BaO are immiscible).
(b) Based on the accompanying data, what is the heat effect of the decomposition of the witherite (J/mol). Specify whether heat is to be added (endothermic) or evolved (exothermic).
(c) How high must the temperature be raised to raise the carbon dioxide pressure above the mineral to one atmosphere?

DATA

	Thermodynamic Properties [KCAL/(g·mol)][1]	
	$\Delta G^\circ_{f(298)}$	$\Delta H^\circ_{f(298)}$
CO_2	−94	−94
BaO	−126	−133
$BaCO_3$	−272	−291

[1]Assume that $C_{P_{CO_2}} + C_{P_{BaO}} = C_{P_{BaCO_3}}$.

5.15 As the Ellingham diagram indicates, Mg has a very stable oxide. Therefore Mg metal can be obtained from the oxide ore by a two-step process. First the oxide is converted to a chloride. In the second step the chloride is converted to metal Mg by passing H_2 gas over liquid $MgCl_2$ at 1200°C. The reaction in this last step is:

$$MgCl_2(l) + H_2(g) \rightarrow Mg(g) + 2HCl(g)$$

(a) Calculate the equilibrium pressure of $H_2(g)$, Mg(g), and HCl(g) if the total pressure is maintained constant at 1 atm.

(b) Calculate the maximum vapor pressure of H_2O that can be tolerated in the hydrogen without causing the oxidation of the Mg vapor.

5.16 A common reaction for the gasification of coal is:

$$H_2O(g) + C(s) \rightarrow H_2(g) + CO(g)$$

(a) Write the equilibrium constant for this reaction and compute its value at 1100 K.

(b) If the total gas pressure is kept constant at 10 atm, calculate the fraction of H_2O that reacts.

(c) If the reaction temperature is increased, will the fraction of water reacted increase or decrease? Explain your answer.

Use the data in Table 5.1.

Chapter 6

Electrochemistry

Chapter 5 discussed chemical reactions and chemical equilibrium. The approach to the subject was based on the condition for equilibrium, that *no reversible work* could be done as the reactants changed to the products ($W_{rev} = \Delta G = 0$). The reactants and products were assumed to be in contact with one another.

In this chapter, we consider the subject of electrochemistry, which differs from the study of chemical equilibrium in that we are not limited to the condition $\Delta G = 0$. In fact, we will deal with situations in which we can force the occurrence of reactions that would not otherwise proceed spontaneously. This can be accomplished by doing work on the system and by separating the chemical species from one another. We will also consider situations in which work can be derived *directly* from chemical reactions that can take place spontaneously, without using heat engines.

To illustrate this last point, consider the reaction between hydrogen and oxygen to yield water.

$$H_2 + \tfrac{1}{2}O_2 \rightarrow H_2O \tag{6.1}$$

Let us suppose that we are trying to extract energy from this reaction in the form of electricity. The hydrogen and oxygen could be combined (burned in a torch) and the energy of the flame used to boil water as part of a heat engine as illustrated in

Chapter 2. The heat engine would produce electrical work, consistent with the second law of thermodynamics. Not all the heat generated, however, could be converted into useful work (electricity). The maximum efficiency of the conversion would depend on the highest temperature reached in the heat engine and the lowest temperature to which heat could be rejected. In addition to this Second Law limitation, there are inefficiencies introduced by friction in various forms. As a point for comparison, most modern power plants operate at an overall conversion efficiency of about 33%.

6.1 ELECTROCHEMICAL CELL

To begin our discussion of alternative means for converting energy, notice that in the reaction of hydrogen and oxygen to produce water, electrons are transferred from the hydrogen atoms to the oxygen atoms. Each hydrogen atom changes from oxidation state zero to oxidation state one; that is, it loses an electron. Oxygen, at the same time, gains two electrons:

$$H_2 + \tfrac{1}{2}O_2 \rightarrow (2H^+ \cdot O^{2-}) \qquad \text{or} \qquad H_2O$$

If by some method we could conduct the reaction to extract work from the electron flow, we could avoid the "Second Law tax": that is, the inefficiency inherent in a heat engine that is imposed by the Second Law. To examine this possibility, consider the electrochemical cell illustrated in Figure 6.1.

In the illustrated cell, the reaction

$$H_2 \rightarrow 2H^+ + 2e^- \tag{6.2}$$

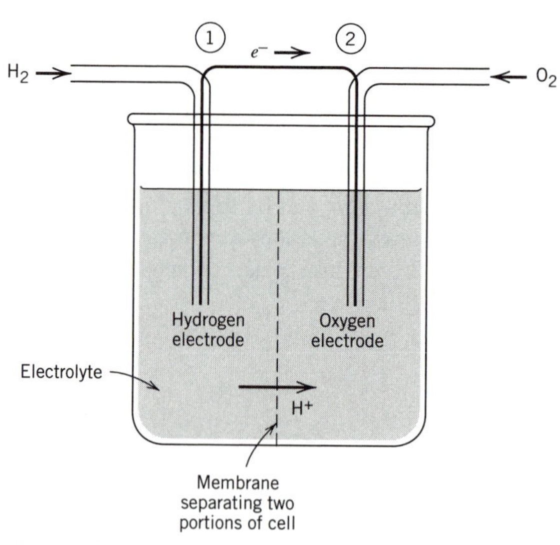

Figure 6.1 Schematic diagram of a hydrogen–oxygen fuel cell.

takes place at the hydrogen electrode. The hydrogen breaks down into hydrogen ions and electrons. The material that separates the hydrogen electrode from the oxygen electrode, a liquid solution in this case, is called the electrolyte. An *electrolyte* must be able to *conduct ions,* but it *should not conduct electrons.* This property is necessary because we want to separate the electron flow from the ion flow. Because of this property, hydrogen ions, but not electrons, may move through the electrolyte to the oxygen electrode. The electrons flow through the wire connecting the two electrodes. As they do, we may extract some work from them.

At the oxygen electrode, the following reaction takes place:

$$\tfrac{1}{2}O_2 + 2H^+ + 2e^- \rightarrow H_2O \tag{6.3}$$

Adding Eqs. 6.2 and 6.3 yields the expression for the overall reaction, which is the same as if the hydrogen were burned in oxygen:

$$H_2 + \tfrac{1}{2}O_2 \rightarrow H_2O \tag{6.1}$$

By means of an electrochemical cell such as the one in Figure 6.1, the reaction of hydrogen and oxygen to form water (or the reverse) can be conducted in such a way as to separate the electronic flow from the movement of ions.

When the reaction is carried out in the direction indicated by Eq. 6.1, this apparatus is called a fuel cell. The open circuit voltage of this fuel cell (the voltage that is measured when no current is allowed to flow) when the hydrogen pressure and the oxygen pressure are each equal to one atmosphere will be 1.229 V. The means for calculating this voltage are presented later in the chapter.

The reaction described in Eq. 6.1 can be conducted in reverse. If an electrical potential greater than the open circuit potential is applied to the cell with polarity opposite to the natural cell potential, then the reaction (Eq. 6.1) will be reversed so that water is dissociated into hydrogen at one electrode and oxygen at the other. In this mode, the apparatus is considered to be a hydrogen–oxygen generator and the process is called the electrolysis of water.

6.2 NOMENCLATURE

The nomenclature adopted in electrochemistry is as follows:

The electrode at which *oxidation* (increase of oxidation number)[1] takes place is called the *anode.*

---

[1]The oxidation state, or oxidation number, or valence of an ion refers to the imbalance of electrical charge of the ion. Positive oxidation states indicate an excess of positive charge. The hydrogen ion (H^+) is in oxidation state +1. In diatomic hydrogen (H_2), the oxidation state of hydrogen is zero. When the apparatus of Figure 6.1 is operated as a fuel cell, the hydrogen electrode is called the anode and the oxygen electrode is called the cathode. The reverse nomenclature applies when the apparatus is used as a hydrogen–oxygen generator.

The electrode at which *reduction* takes place (reduction in oxidation state) is called the *cathode.*

We note in passing that pratical hydrogen–oxygen fuel cells are generally operated with alkaline electrolytes, not with acid electrolytes implied by Eqs. 6.2 and 6.3. In acid solutions, the hydrogen ion is present in much larger concentrations than the hydroxyl ion OH^-. To describe the reactions in a practical hydrogen–oxygen fuel cell, the equations should be written with the hydroxyl ion rather than the hydrogen ion as the transport medium. The acid (hydrogen-ion-rich) electrolyte is used in this example because it makes the illustration easier to understand. The equations should be:

$$H_2 + 2OH^- \rightarrow 2H_2O + 2e^-$$

$$\tfrac{1}{2}O_2 + 2e^- + H_2O \rightarrow 2OH^-$$

$$H_2 + \tfrac{1}{2}O_2 \rightarrow H_2O$$

6.3 CALCULATION OF CELL VOLTAGE

In the apparatus shown in Figure 6.1, consider the cell to be the system. If the apparatus is considered to be isothermal and at steady state, then,

$$W_{rev} = \Delta G \quad \textbf{(6.4)}$$

Suppose the electrodes 1 and 2 are at different electrical potential levels, ϕ_1 and ϕ_2. Then,

$$dW_{1-2} = -(\phi_2 - \phi_1)\, dQ \quad \textbf{(6.5)}$$

where dQ is the differential quantity of negative charge (electrons) transferred between the two states. Integrating the equation,

$$W_{1-2} = -(\phi_2 - \phi_1)Q \quad \textbf{(6.6)}$$

The difference $(\phi_2 - \phi_1)$ is called E, the electromotive force (emf) of the cell (voltage). The charge Q is measured in moles of electrons that flowed in the process. A mole of electrons is just a quantity equal to Avogadro's number of electrons (6.022×10^{23}). The term Q in terms of the number of electrons n_e and the charge on the electron e is:

$$Q = n_e e$$

$$Q = zN_A e$$

where z is the number of moles of electrons and N_A is Avogadro's number. This is usually written as:

$$Q = z\mathcal{F} \tag{6.7}$$

where $\mathcal{F}$ is the Faraday constant, the electrical charge on one mole of electrons. It is the product of the charge of the electron multiplied by N_A, Avogadro's number (1.6×10^{-19} C/electron multiplied by 6.022×10^{23} electrons). The value of $\mathcal{F}$ is 96,480 C per mole of electrons. The work done is:

$$W = -Ez\mathcal{F} \tag{6.8}$$

where E is the voltage of the cell and z is the number of moles of electrons transferred for the chemical equation as written. For example, $z = 2$ for the reaction in Eq. 6.1 (see Eqs. 6.2 and 6.3).

Combining with Eq. 6.4 yields

$$\Delta G = -Ez\mathcal{F} \tag{6.9}$$

This is the relationship between the electrical potential of a cell and the change in Gibbs free energy between the reactants and products.[2] If each of the reactants and products in an electrochemical cell is in its standard state, then the electrochemical potential observed is called the standard potential of the cell.

$$\Delta G^\circ = -E^\circ z\mathcal{F} \tag{6.10}$$

As an example, let us calculate the standard potential of the hydrogen–oxygen fuel cell. At 298 K, the Gibbs free energy[3] of formation of water from hydrogen and oxygen, each in its standard state, is $-237{,}191$ J/mol, and the number of electrons involved in the reaction is 2 (see Eqs. 6.2 and 6.3).

$$\Delta G^\circ = -237{,}191 = -E^\circ z\mathcal{F} = -E^\circ \times 2 \times 96{,}480$$

$$E^\circ = +1.229 \text{ V}$$

6.4 DIRECTION OF REACTION

From Eq. 6.9, if the electrochemical potential of the cell is greater than zero (positive), then the change in Gibbs free energy (ΔG) is negative, and the reaction may proceed spontaneously as written. In this mode, the cell will, when connected to a circuit, operate as a generator of electrical current, as in a battery that is being discharged. The reaction, of course, could be pushed in the opposite direction simply by applying a voltage in the opposite direction greater than the open circuit voltage

---

[2]The sign convention for polarity is different in some European publications. That is, the equation would read $\Delta G = +Ez\mathcal{F}$

[3]Gibbs free energy, usually expressed in joules, is also given in the equivalent units of volt-coulombs (V·C).

of the cell. In this case, the electrical current imposed on the cell would reverse the discharge reaction described above, and the cell (battery) would be charged.

6.5 HALF-CELL REACTIONS

From Eq. 6.10 it can be seen that the standard potential (voltage) of a cell can be calculated from the standard Gibbs free energy change of a reaction as tabulated in Table 5.1. In electrochemistry, another form of tabulation is often used. This involves the use of half-cell reactions. In this method the reactions at the anode and at the cathode are written separately. Tables of voltages for these half-cell reactions are given in Table 6.1. Half-cell reactions are tabulated in the form:

$$X = X^{+z} + ze^-$$

Table 6.1 Standard Oxidation Potentials

Electrode Reaction at 298 K	E°_H (V)
Acid Solutions	
$Li \rightarrow Li^+ + e$	3.045
$K \rightarrow K^+ + e$	2.925
$Cs \rightarrow Cs^+ + e$	2.923
$Ba \rightarrow Ba^{2+} + 2e$	2.90
$Ca \rightarrow Ca^{2+} + 2e$	2.87
$Na \rightarrow Na^+ + e$	2.714
$Mg \rightarrow Mg^{2+} + 2e$	2.37
$Al \rightarrow Al^{3+} + 3e$	1.66
$Zn \rightarrow Zn^{2+} + 2e$	0.763
$Fe \rightarrow Fe^{2+} + 2e$	0.440
$Cr^{2+} \rightarrow Cr^{3+} + e$	0.41
$Cd \rightarrow Cd^{2+} + 2e$	0.403
$Sn \rightarrow Sn^{2+} + 2e$	0.136
$Pb \rightarrow Pb^{2+} + 2e$	0.126
$Fe \rightarrow Fe^{2+} + 3e$	0.036
$H_2 \rightarrow 2H^+ + 2e$	0.000
$Cu^+ \rightarrow Cu^{2+} + e$	−0.153
$Cu \rightarrow Cu^{2+} + 2e$	−0.337
$2I^- \rightarrow I_2 + 2e$	−0.5355
$Fe^{2+} \rightarrow Fe^{3+} + e$	−0.771
$Ag \rightarrow Ag^+ + e$	−0.7991
$Hg \rightarrow Hg^{2+} + 2e$	−0.854
$2\,Br^- \rightarrow Br_2\,(l) + 2e$	−1.0652
$Cl^- \rightarrow \frac{1}{2}Cl_2 + e$	−1.3595

Source: W. M. Latimer, *The Oxidation States of the Elements and Their Potentials in Aqueous Solutions,* (2nd ed., Prentice-Hall, Englewood Cliffs, NJ, 1952).

The reference point for these tables is the standard hydrogen electrode which is considered, by definition, to have a zero voltage when the reactant, hydrogen, and product, hydrogen ions, are in their standard states. The standard state for hydrogen gas is the pure gas at one atmosphere pressure. The standard state for ions in aqueous solutions is generally taken as the ion in a concentration of one mole per liter of solution.[4]

The standard half-cell potential for the reduction of oxygen to water at 298 K is 1.229 V.

For A:

$$O_2 + 4H^+ + 4e^- \rightarrow 2H_2O \qquad E^\circ = 1.229\ \text{V}$$

The standard potential for hydrogen oxidation is zero.

For B:

$$H_2 \rightarrow 2H^+ + 2e^- \qquad E^\circ = 0$$

The sum of the two is +1.229 V.

Another illustration of the calculation of cell potential using half-cell data is provided by the Daniell cell in which zinc and copper are the electrodes (Figure 6.2). The reactions that take place are:

$$\text{I: } Zn \rightarrow Zn^{2+} + 2e^- \qquad E^\circ_{\text{I}} = +0.763\ \text{V}$$

$$\text{II: } Cu \rightarrow Cu^{2+} + 2e^- \qquad E^\circ_{\text{II}} = -0.337\ \text{V}$$

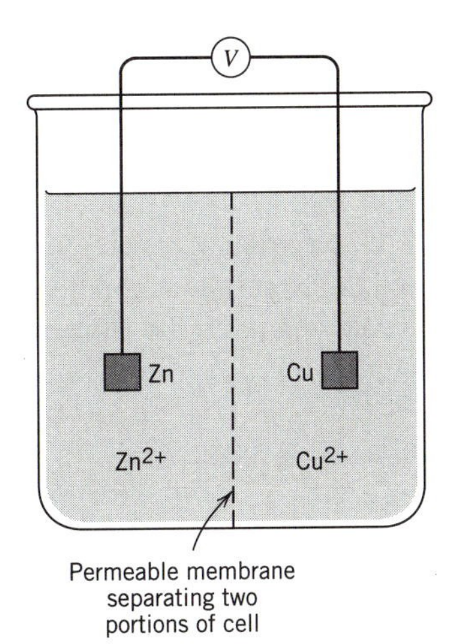

Figure 6.2 Schematic diagram for a Daniell cell.

[4]Table 6.1 lists electrode potentials for acidic solutions, those in which the hydrogen ion concentration $[H^+]$ is one mole per liter.

These can be combined as follows:

$$Zn + Cu^{2+} \rightarrow Cu + Zn^{2+}$$

$$E^\circ = E^\circ_{\mathrm{I}} - E^\circ_{\mathrm{II}} = 0.763 + 0.337$$

$$E^\circ = 1.10 \text{ V}$$

6.6 VARIATION OF VOLTAGE WITH CONCENTRATION: NERNST EQUATION

Consider the chemical equation

$$b\mathrm{B} + c\mathrm{C} = d\mathrm{D} + e\mathrm{E}$$

Based on the principles of Chapter 5, the Gibbs free energy change for this reaction can be written as follows:

$$\Delta G = \Delta G^\circ + RT \ln \frac{a_{\mathrm{D}}^d \, a_{\mathrm{E}}^e}{a_{\mathrm{B}}^b \, a_{\mathrm{C}}^c}$$

and

$$\Delta G = \Delta G^\circ + RT \ln J_a \tag{6.11}$$

Combining Eqs. 6.9, 6.10 and 6.11 yields

$$-Ez\mathcal{F} = -E^\circ z\mathcal{F} + RT \ln J_a$$

$$E = E^\circ - \frac{RT}{z\mathcal{F}} \ln J_a \tag{6.12}$$

This relationship between the electromotive force of a cell and the activities of reactants and products is called the Nernst equation.

In many presentations this equation (with the temperature assumed to be 298 K) is written as

$$E = E^\circ - \frac{0.05916}{z} \log_{10} J_a \tag{6.13}$$

(Note the change in the base of the logarithmic term.)

6.7 POURBAIX DIAGRAMS

The regions of stability of various chemical forms (phases) can be displayed in graphical form on axes appropriate to the problem at hand. In Section 5.4, we displayed the stability of copper compounds as a function of temperature and oxygen pressure. Pourbaix diagrams display the stability of compounds in electrical and chemical terms at a given temperature. The axes in Pourbaix diagrams are electrical potential as the ordinate and pH as the abscissa.

The term pH is defined as the negative logarithm (base 10) of the hydrogen ion activity in an aqueous solution. The standard state for ions in aqueous solutions is at a concentration of one mole per liter. Assuming linearity of activity with concentration, the activity of H^+ is equal to its concentration in moles per liter.

$$\mathrm{pH} = -\log_{10}(a_{\mathrm{H}^+}) = -\log_{10}[\mathrm{H}^+] \tag{6.14}$$

Let us use this definition to construct a very simple Pourbaix diagram, the one for the stability of hydrogen, oxygen, and water at a temperature of 298 K. In a cell containing pure water, consider the reaction at the oxygen electrode:

$$\mathrm{O_2} + 4\mathrm{H}^+ + 4\mathrm{e}^- \rightarrow 2\mathrm{H_2O} \qquad E^\circ = 1.229\ \mathrm{V}$$

Based on Eq. 6.13:

$$E = E^\circ - \frac{RT}{z\mathcal{F}} \ln J_a = E^\circ - \frac{0.05916}{4} \log_{10} \frac{a_{\mathrm{H_2O}}}{a_{\mathrm{H}^+}^4 P_{\mathrm{O_2}}} \tag{6.15}$$

The activity of water is 1, because it is in its standard state, and the number of electrons involved in the reaction is 4, thus:

$$E = E^\circ + \frac{0.05916}{4} \log_{10} [\mathrm{H}^+]^4 P_{\mathrm{O_2}} \tag{6.16}$$

At a pressure of oxygen of one atmosphere,

$$E = 1.229 + 0.05916 \log_{10} [\mathrm{H}^+] \tag{6.17}$$

Equation 6.17 describes the stability of water relative to the evolution of oxygen as a function of pH. At a pH of 0, the range of stability of liquid water is from zero to +1.229 V. Above that voltage, oxygen is evolved at the anode. Below that voltage, hydrogen is evolved at the cathode. The Pourbaix diagram (Figure 6.3) shows the regions of stability of pure water, and hydrogen, and oxygen at one atmosphere pressure at 298 K. The regions of stability, of course, change with gas pressure (Eq. 6.16). Figure 6.4 is the same as Figure 6.3 with the addition of lines for gas pressures of 10^{-4} atm.

The Pourbaix diagram for hydrogen, oxygen, and water is, as stated, a simple one, and really not very useful. Pourbaix diagrams for more complex systems (Ref. 1) can be derived from a knowledge of the equilibrium constants among the species present and the electrode potentials of the reactions involved. Figure 6.5 is an

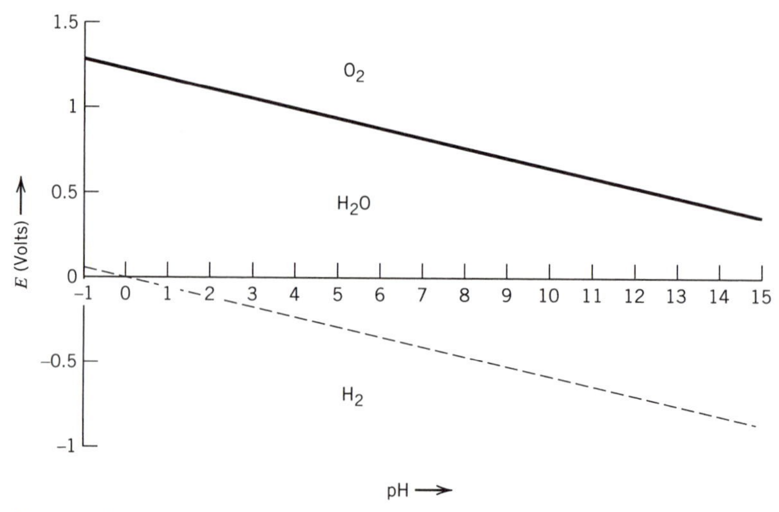

Figure 6.3 Pourbaix diagram for water, hydrogen, and oxygen; P_{H_2} and P_{O_2} = 1 atm.

example showing the stability ranges of various forms of copper. Metallic copper is stable in the region labeled Cu; that is, it is not attacked through corrosion. In the regions labeled Cu_2O and CuO, the respective oxides are stable. If a piece of metallic copper were introduced into an aqueous medium in the pH–electric potential regions of solid oxide stability, the appropriate oxide would be formed on the surface of copper, and the metal would be protected from dissolution (corrosion). This is called passivation. In the regions labeled Cu^{2+} and CuO^{2-}, the ionic forms of copper are stable.

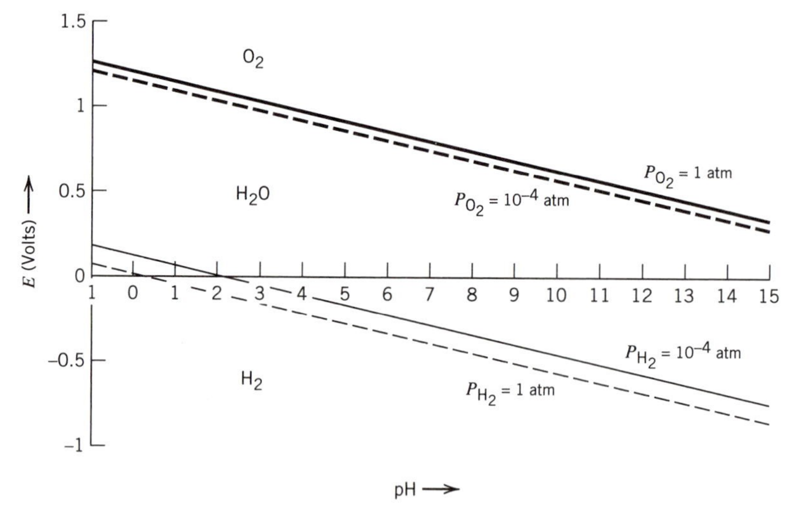

Figure 6.4 Pourbaix diagram for water, hydrogen, and oxygen showing P_{H_2} and P_{O_2} at both 1 and 10^{-4} atm.

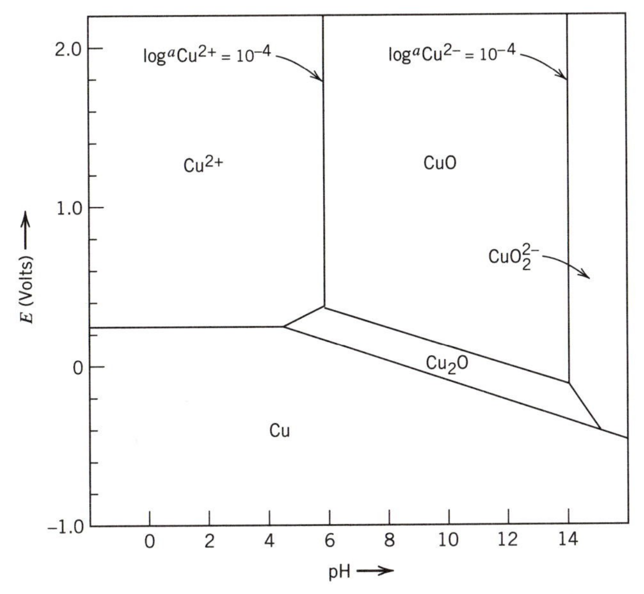

Figure 6.5 Pourbaix diagram for copper in water (selected activity levels for Cu^{2+} and CuO_2^{2-}). (From Ref. 1.)

That means that metallic copper would be corroded under the conditions indicated on the Pourbaix diagram.

6.8 CONCENTRATION CELLS

Suppose a cell is constructed with two identical electrodes, copper, for example. Furthermore, suppose that the copper electrodes are exposed to different copper ion concentrations in the electrolyte. This can be accomplished either through the use of a membrane to separate the two sections of the electrolyte or by use of a device, called a salt bridge, to connect two electrolytes differing in copper ion concentration.

On side I of the cell (Figure 6.6), the concentration of copper ions is Cu^{2+} [I], and the reaction that takes place can be expressed as:

$$Cu \rightarrow Cu^{2+}\ [I] + 2e^- \tag{6.18}$$

On the other side, where the copper ion concentration is Cu^{2+} [II], the reaction is

$$2e^- + Cu^{2+}\ [II] \rightarrow Cu \tag{6.19}$$

The overall reaction is

$$Cu^{2+}\ [II] \rightarrow Cu^{2+}\ [I] \tag{6.20}$$

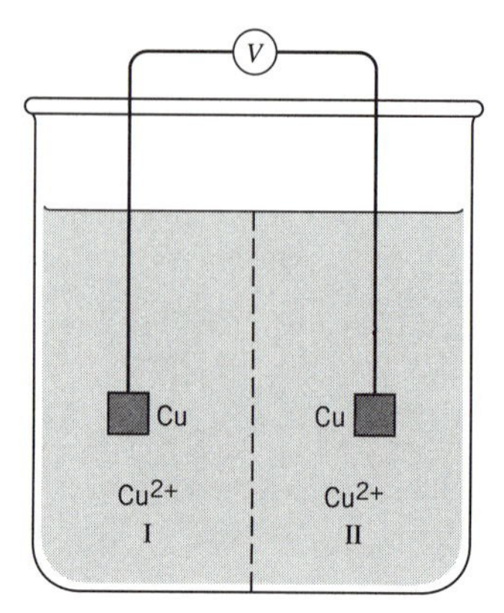

Figure 6.6 Schematic diagram of concentration cell.

Because the standard state for the reaction is the same for both sides (see Eq. 6.10)

$$\Delta G^\circ = 0; \qquad E^\circ = 0$$

Thus the Nernst equation is reduced to

$$E = -\frac{RT}{z\mathcal{F}} \ln \frac{a_{Cu^{2+}[I]}}{a_{Cu^{2+}[II]}} \tag{6.21}$$

From Eq. 6.21, one can see that an electromotive force can be generated between two similar electrodes that are immersed in electrolytes with different ion concentrations. One could think of using such a device as a means for measuring the copper ion concentration. If the activity (related to concentration) of copper ions on one side is known, then a measurement of the voltage would determine the copper ion activity (concentration) on the other side.

One could also think of a cell consisting of a uniform electrolyte and with two dissimilar, copper-based electrodes. This could be accomplished by using pure copper as one electrode and an alloy of copper as the other. The thermodynamic activity of copper in the pure copper electrode is one. In the copper alloy electrode is less than one. The electromotive force of the cell would be a measure of the the difference in thermodynamic activity of the copper in the two electrodes.

In this case Eqs. 6.18–6.20 become:

$$Cu_I \rightarrow Cu^{2+} + 2e^- \tag{6.22}$$

$$Cu^{2+} + 2e^- \rightarrow Cu_{II} \tag{6.23}$$

$$Cu_I \rightarrow Cu_{II} \tag{6.24}$$

$$E = -\frac{RT}{z\mathcal{F}} \ln \frac{a_{Cu_{II}}}{a_{Cu_I}} \tag{6.25}$$

6.9 OXYGEN PRESSURE DETERMINATION

The principle of the concentration cell can also be used to measure differences in thermodynamic activity (pressure) of gases. An example is the cell used to measure oxygen activity or oxygen pressure. The electrolyte for such a cell could be a liquid, but in practice solid electrolytes are used. Remember that the criterion for an electrolyte is that it be an ionic conductor (of the appropriate ion), not an electronic conductor. Ceramics such as solid solutions of zirconia and yttria (ZrO_2 and Y_2O_3) are oxygen ion conductors at high temperatures and do not conduct electrons. A cell can thus be constructed as in Figure 6.7. The voltage of such a cell is:

$$E = E° - \frac{RT}{z\mathcal{F}} \ln \frac{P_{O_2}}{P_{O_2,ref}} \qquad E° = 0$$

$$E = -\frac{RT}{z\mathcal{F}} \ln \frac{P_{O_2}}{P_{O_2,ref}}$$

Thus the measurement of the electrical potential difference between the two electrodes is a measure of the difference in oxygen pressures on the two sides. If one is a reference pressure, such as the atmosphere where the oxygen pressure is 0.21 atm, then the voltage generated by the cell is a measure of the oxygen pressure on the other side. This technique is used to measure oxygen pressure (activity) in the exhaust of internal combustion engines. This measurement is used to control the air–

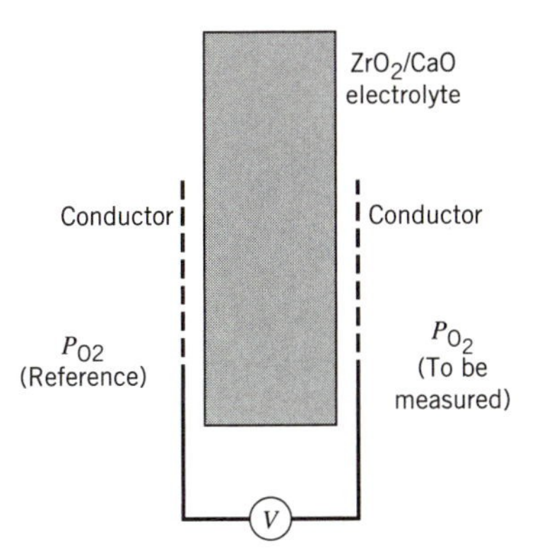

Figure 6.7 Schematic diagram of an oxygen concentration cell.

fuel mixture that goes to the engine, to balance the concentrations in the exhaust of oxidizing and reducing pollutants (carbon monoxide, unburned hydrocarbons, and the oxides of nitrogen) as they enter a catalytic converter. The technique is also used to measure oxygen activity in molten metal baths.

6.10 TEMPERATURE DEPENDENCE OF VOLTAGE

Because the cell voltage is related to the Gibbs free energy change of the reaction being carried out in the cell (Eq. 6.9), the temperature dependence of the cell voltage can be determined by knowing the temperature dependence of the Gibbs free energy function:

$$\Delta G = -Ez\mathcal{F} \qquad (6.26)$$

$$d(\Delta G) = -\Delta S\, dT = -z\mathcal{F}\, dE$$

Thus the temperature coefficient of a cell dE/dT is:

$$\frac{dE}{dT} = \frac{\Delta S}{z\mathcal{F}} \qquad (6.27)$$

$$\frac{dE^\circ}{dT} = \frac{\Delta S^\circ}{z\mathcal{F}} \qquad (6.28)$$

where ΔS is the entropy change for the overall chemical reaction of the cell.

The heat effect or enthalpy change of the electrochemical reaction can be calculated by knowing the cell voltage and the temperature coefficient of cell voltage as follows:

$$\Delta G = \Delta H - T\Delta S$$

$$-Ez\mathcal{F} = \Delta H - Tz\mathcal{F}\frac{dE}{dT}$$

$$\Delta H = -z\mathcal{F}\left[E - T\frac{dE}{dT}\right] \qquad (6.29)$$

$$\Delta H^\circ = -z\mathcal{F}\left[E^\circ - T\frac{dE^\circ}{dT}\right] \qquad (6.30)$$

6.11 ELECTROCHEMICAL POTENTIAL

The concept of electrochemical potential is useful in the analysis of movement (diffusion) of charged atoms or molecules (ions) in an electric field. Consider the movement of n_i moles of an ion from point 1 to point 2 in Figure 6.8. The electric potentials are ϕ_1 and ϕ_2. The chemical potentials of n_i are μ_1 and μ_2, respectively. The charge on the ion is z'.[5]

If n_i moles of an ion i are moved from point 1 to point 2, the work done is:

$$W_{\text{rev}} = z'\mathcal{F}(\phi_2 - \phi_1)n_i + (\mu_2 - \mu_1)n_i \quad \textbf{(6.31)}$$

where μ_i is the chemical potential of i, as defined in Chapter 3, in the absence of an electric field.

Rearranging terms, the work done per mole of i is:

$$\frac{W_{\text{rev}}}{n_i} = (z'\mathcal{F}\phi + \mu)_2 - (z'\mathcal{F}\phi + \mu)_1$$

The term $z'\mathcal{F}\phi + \mu$ is defined as the *electrochemical* potential, and will be noted as $\tilde{\mu}$.

In the *absence* of the electric potentials, the criterion of equilibrium between points 1 and 2, as discussed in Chapter 4, is:

$$W_{\text{rev}} = 0 \qquad \text{hence} \qquad \mu_2 = \mu_1$$

In an electric field, the criterion of equilibrium is:

$$\frac{W_{\text{rev}}}{n_i} = (z'\mathcal{F}\phi + \mu)_2 - (z'\mathcal{F}\phi + \mu)_1 = 0$$

Hence:

$$(z'\mathcal{F}\phi + \mu)_2 = (z'\mathcal{F}\phi + \mu)_1$$

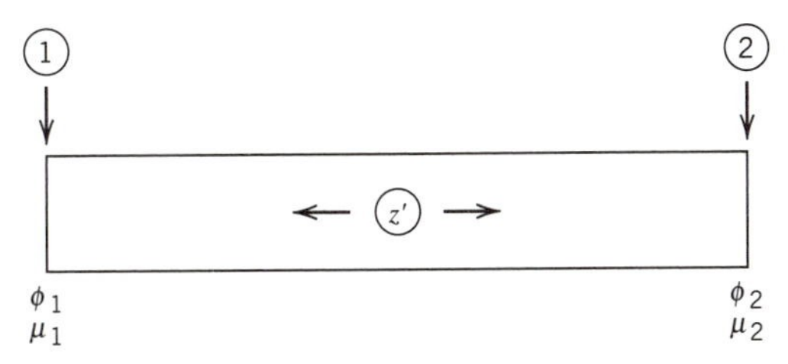

Figure 6.8 Movement of ions in an electric field.

[5]Note that the notation for charge in this section is opposite to the notation used in the rest of this chapter. The charge on the ion, noted as z' (z-prime), is *positive* when the charge is *positive.* In other sections of this chapter the symbol z represents the number of electrons (i.e., negative charges) flowing in an external circuit from point 1 to point 2.

that is,

$$\tilde{\mu}_2 = \tilde{\mu}_1$$

The driving force for movement (diffusion) of species in the *absence* of an electric field is the gradient of the *chemical* potential, grad(μ). In an electric field, the driving force is the gradient of the *electrochemical* potential, grad($\tilde{\mu}$).

REFERENCE

1. Pourbaix, M., *Atlas of Electrochemical Equilibria in Aqueous Solutions,* 2nd ed. National Association of Corrosion Engineers, Houston, TX, 1974.

PROBLEMS

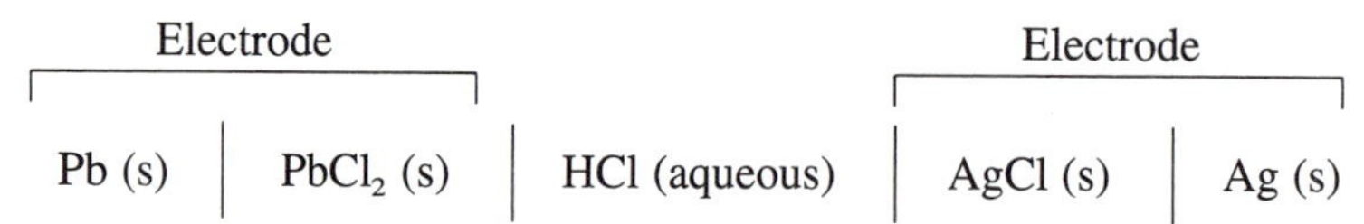

6.1 The emf of the cell represented in the accompanying diagram is 0.490 V at 25°C and the temperature coefficient is -1.84×10^{-4} V/°C. All the components are present as pure solids in contact with an HCl electrolyte.

(a) Write the half-cell reactions.
(b) Write the overall cell reaction.
(c) Calculate the Gibbs free energy change and the entropy change for this reaction.

6.2 The open circuit voltage for an oxygen–hydrogen fuel cell operated with pure gases as reactants and liquid water as the product is 1.229 V at 25°C and 1 atm pressure.

(a) What is the voltage of the cell operated at 20 atm? The reactants and products are all at 20 atm.
(b) What would the voltage be if the product were impure water, (i.e., water containing dissolved salts): higher, lower, or the same?
(c) What would be the voltage of the cell operated at 1 atm with air as the reacting gas instead of pure oxygen?

6.3 A brine (water containing sodium chloride) is to be electrolyzed to hydrogen and oxygen at a temperature of 25°C.

(a) What is the thermodynamic activity of water in the brine at 25°C? (Take pure water as the standard state.)
(b) Will the voltage required to electrolyze the brine be greater or less than 1.229 V if the resulting hydrogen and oxygen are each to be at one atmosphere? How much different?

(c) A small amount of sodium hydroxide is added to change the pH (acidity) of the brine. Will your answer to part b change? If so, in what direction?

NOTE: "Small amount" means enough to change the acidity, but not enough to change the overall salt content of the brine significantly.

DATA

The voltage required to electrolyze pure water is −1.229 V at 25°C.
The vapor pressure of pure water at 25°C is 24 mmHg.
The pressure of water in equilibrium with the brine at 25°C is 23 mmHg.

6.4 An internal combustion engine operated with methanol (CH_3OH) as the fuel has a thermodynamic efficiency of 20%.

(a) How much work is derived from burning one mole of ethanol in the engine?
(b) What is the maximum work that could be derived by reacting the methanol with pure oxygen in a fuel cell at 25°C?
(c) Would the answer in part b be different if air were used as the oxidant? If so, how much?

DATA

For $CH_3OH + \frac{3}{2} O_2 \rightarrow CO_2 + 2H_2O$:

$$\Delta G^\circ = -702.36 \text{ kJ at } 25^\circ\text{C}$$

$$\Delta H^\circ = -726.51 \text{ kJ at } 25^\circ\text{C}$$

6.5 The electromotive force of the following galvanic cell is 0.0324 V at 483°C.

Cd (l) (liquid)	$KCl\text{-}NaCl\text{-}LiCl\text{-}CdCl_2$ (liquid solution)	Cd-Sn (liquid solution) $x_{Cd} = 0.258$

(a) Calculate the relative partial molar free energy for Cd in the liquid Cd-Sn solution.
(b) Calculate the activity of cadmium in the alloy taking pure liquid Cd as the standard state.
(c) Calculate the vapor pressure of Cd over the liquid Cd-Sn solution if the vapor pressure of pure Cd is 9.23 mmHg.

6.6 Two copper samples are immersed in a one molar (1 M) copper sulfate solution. One is stressed to 10,000 psi in tension. The other is unstressed. If the two samples are considered to be electrodes of an electrolytic cell, what electrical potential is developed between the two?

DATA

For copper: Elastic modulus = 16×10^6 psi.
Density is 8.96 g/cm^3.

6.7 **(a)** Write the chemical equation for the dissociation of magnesium chloride into the metal and gaseous chloride.

(b) At 1000 K, what is the pressure of chlorine in equilibrium with liquid metallic magnesium and pure, liquid magnesium chloride?

(c) In an attempt to produce metallic magnesium, pure magnesium chloride is to be treated in an electrochemical cell at 1000 K. What is the minimum voltage required to produce pure liquid magnesium and pure chlorine at one atmosphere pressure?

DATA

The Gibbs free energy of formation of liquid $MgCl_2$ from its elements (liquid Mg and gaseous Cl_2) is given in joules by

$$\Delta G^\circ = -605{,}000 + 125.4T \text{ (K)}$$

6.8 The electrochemical cell represented in the accompanying diagram consists of two electrodes. One is a solid NiO/Ni mixture. The other is a gaseous O_2 electrode. The porous platinum coating acts only as a catalyst.

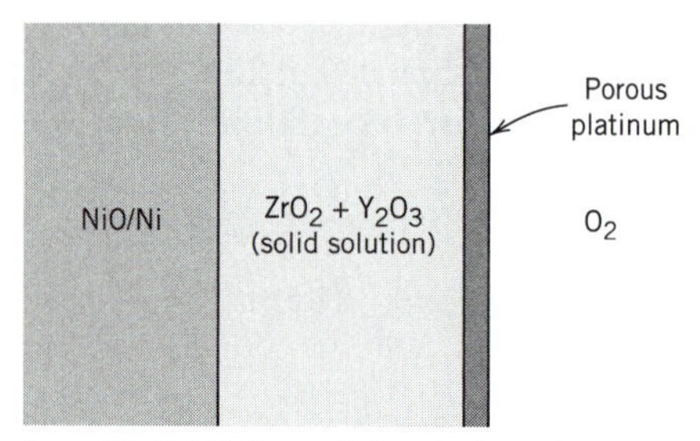

The electrolyte is a ZrO_2/Y_2O_3 solid solution that conducts oxygen ions (O^{2-}) between the oxygen side and the NiO/Ni side.

(a) Calculate the equilibrium oxygen partial pressure at the NiO/Ni electrode ($2Ni + O_2 \rightarrow 2NiO$) at 1000 K.

(b) Identify the cathode and the anode in this electrochemical cell and write the half-cell reactions for the two electrodes. (Specify the direction of the reaction.)

(c) What is the open circuit voltage at 1000 K when the pressure of oxygen on the oxygen electrode side is one atmosphere?

(d) What is the open circuit voltage at 1000 K when the oxygen pressure on the oxygen electrode side is 10^{-10} atm?

DATA

$$\text{For Ni} + \tfrac{1}{2}\,O_2 \rightarrow \text{NiO:}$$

$$\Delta G_f^\circ = -244{,}550 + 98.5\,T$$

where ΔG_f° is in joules and T is in kelvin.

6.9 The electrodes of the zinc–bromine battery shown schematically in the accompanying diagram are pure zinc and liquid bromine. The electrolyte is an aqueous solution of zinc bromide.

(a) Write the half-cell reactions and the overall reaction for the battery.
(b) What is the open circuit voltage of the cell at 298 K when the electrolyte is a one molar solution of zinc bromide?
(c) What is the open circuit voltage at 298 K when the electrolyte concentration has risen to 2 M zinc bromide?
(d) The enthalpy of formation of a one molar solution of zinc bromide from the elements (zinc and bromine) is −397 kJ/mol. What is the temperature coefficent of the cell in volts per kelvin?

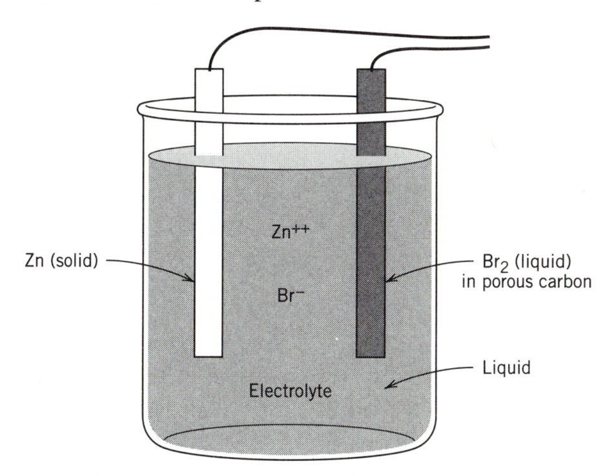

6.10 Lithium-chlorine batteries are being considered as the energy storage system for electric vehicles, even though they have to be operated at elevated temperatures. The reactants in the battery are pure lithium (Li) at one electrode and gaseous chlorine (Cl_2) at the other. Assume that the electrolyte is pure lithium chloride (LiCl). To charge the battery, a voltage is applied to the cell and the lithium chloride is electrolyzed to form metallic lithium and gaseous chlorine (stored in a special way outside the cell). Upon discharge, the metallic lithium and gaseous chlorine react electrochemically to produce lithium chloride.

$$\mathrm{Li(liquid)} + 1/2\ \mathrm{Cl_2\ (gas.)} \underset{\text{charge}}{\overset{\text{discharge}}{\rightleftarrows}} \mathrm{LiCl\ (liquid)}$$

(a) Estimate the open circuit voltage at 1000 K if the electrolyte is pure LiCl.
(b) How many grams of lithium must the battery contain for each kilowatt-hour of energy delivered? (Assume that the energy is delivered at the open circuit voltage.)
(c) If the discharge rate is one kilowatt (at the open circuit voltage), at what rate must heat be added or removed to keep the temperature constant?

(d) If the electrolyte is a solution containing 20% lithium chloride in other salts, and the solution is assumed to be ideal, will the open circuit voltage be different from the one determined in part a?

(e) Will the answer to part c be changed?

DATA

Molecular weights:

Lithium = 6.94 g/mol

Chlorine = 35.45 g/mol

For LiCl $\Delta G_f^\circ = -92{,}000 + 11\,T$ (cal/mol)

6.11 Aluminum oxide (Al_2O_3) is to be electrolyzed to metallic aluminum in an electrochemical cell. The eletrodes of the cell are made of graphite. The products of the reaction are metallic aluminum and carbon dioxide. Metallic aluminum does not react with graphite.

(a) Write the chemical reaction that takes place in the cell.

(b) Calculate the minimum voltage at which the electrolysis may be carried out.

DATA

$$2Al + \tfrac{3}{2}O_2 = Al_2O_3 \qquad \Delta G^\circ = -1{,}676{,}000 + 320T$$

$$C + O_2 = CO_2 \qquad \Delta G^\circ = -394{,}100 - 0.84T$$

Chapter 7

Solutions

The discussion in Chapter 5 on chemical equilibrium dealt with chemical changes among pure components and compounds. Many of the interesting properties of materials and many important chemical reactions take place not among *pure* elements or compounds, but among elements or compounds dissolved in one another as *solutions*. This chapter deals with the thermodynamics of these solutions.

7.1 THERMODYNAMIC ACTIVITY

The concept of fugacity was introduced in Chapter 5 and was defined for gases by Eqs. 5.2 as follows:

$$d\overline{G}_i = RT\, d\,(\ln f_i) \tag{5.2b}$$

The thermodynamic activity (or simply, activity) of a component, i, is defined as

$$a_i \equiv \frac{f_i}{f_i^o} \tag{5.4}$$

where f_i^o is the fugacity of the component i in its standard state. The fugacity of a condensed phase is equal to the fugacity of the vapor phase in equilibrium with it.

If the vapor in equilibrium with the condensed phase is ideal, then the fugacity is equal to the pressure of the vapor.

Although the choice of standard state is arbitrary, in liquid solutions and solid solutions the pure materials at one atmosphere pressure and specified crystal structure are usually taken as the standard state. For example, when dealing with solutions of acetone and water, we choose pure water as the standard state for water and pure acetone as the standard state for acetone. In solid solutions of iron and nickel, the same approach can be used. Pure iron, with specified crystal structure (body-centered cubic or face-centered cubic) can be chosen as the standard state for iron, and pure nickel the standard state for nickel. If that is done, we can write:

$$\overline{G}_i - \underline{G}_{i,\text{pure}} = \overline{G}_i - \underline{G}_i^o = RT \ln a_i \qquad \textbf{(7.1)}$$

An *ideal* solution is defined as one in which the fugacity of component i is equal to the mole fraction of component i multiplied by the fugacity of i in the pure state f_i^p shown in Eq. 7.2a.[1]

$$f_i = x_i f_i^p \qquad \textbf{(7.2a)}$$

If the pure state is taken as the standard state, then:

$$f_i^o = f_i^p$$

$$a_i = \frac{x_i f_i^p}{f_i^o}$$

$$a_i = x_i \qquad \text{(ideal solution)} \qquad \textbf{(7.2b)}$$

This is shown diagrammatically in Figure 7.1. The ideal solution relationship, Eq. 7.2a, is often called Raoult's law. For the component i, the change of Gibbs free energy (per mole of i) upon mixing is:

$$\overline{G}_{i(\text{solution})} - \underline{G}_{(\text{pure})} = RT \ln \frac{f_{i(\text{solution})}}{f_{i(\text{pure})}}$$

If pure material "i" is taken as the standard state, then:

$$\overline{G}_{i(\text{solution})} - \underline{G}_{(\text{pure})} = RT \ln x_i \qquad \text{ideal solution} \qquad \textbf{(7.3)}$$

[1]The ideal solution may also be defined in terms of the enthalpy and entropy of mixing. The enthalpy of mixing of an ideal solution is zero. The entropy of mixing is $-R\sum_i x_i \ln x_i$. These two relationships can be derived from Eq. 7.2b, or vice versa.

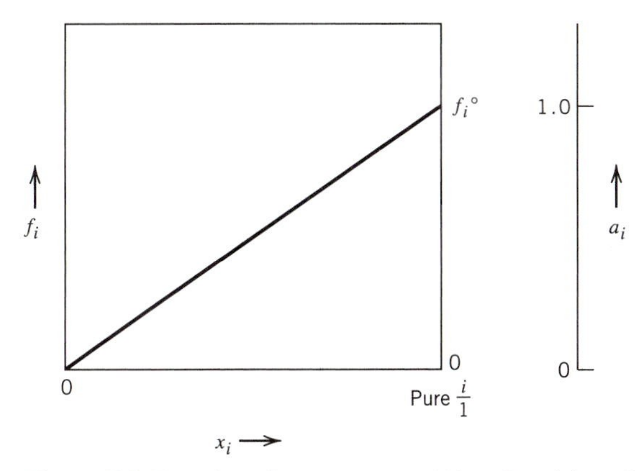

Figure 7.1 Fugacity of component i (f_i) and activity of component i (a_i) as a function of mole fraction (x_i) for an ideal solution.

7.2 PARTIAL MOLAR QUANTITIES

The properties of a material in solution are represented by the partial molar quantities (Section 3.4).

Consider the volume of a solution consisting of materials A and B. The volume of this solution, at constant temperature and pressure, is a function of the amount of A (n_A) and the amount of B (n_B), where n represents the number of moles.

$$V = V\,(n_A,\, n_B)_{T,P}$$

Because volume is a point, or state, function, we may write its total differential dV at constant T and P as

$$dV = \left(\frac{\partial V}{\partial n_A}\right)_{T,P,n_B} dn_A + \left(\frac{\partial V}{\partial n_B}\right)_{T,P,n_A} dn_B \tag{7.4}$$

The quantity $(\partial V/\partial n_A)_{T,P}$ is defined as the partial molar volume of A, and is written as

$$\overline{V}_A = \left(\frac{\partial V}{\partial n_A}\right)_{T,P,n_B} \tag{7.5a}$$

Then:

$$dV = \overline{V}_A\, dn_A + \overline{V}_B\, dn_B \tag{7.5b}$$

The significance of Eq. 7.5a is shown in Figure 7.2. It is the rate of change of the volume of the solution with respect to the moles of B added.[2] The partial molar volume of a component in a solution is the volume change of a very large amount (infinite, in fact) of the solution when one mole of the particular component is added to it, at constant temperature and pressure. To cite a specific example, if one mole of water (18 g, with a volume of 18 cm^3), is added to a very large bath of an ethanol–water solution (70 mol%/30 mol%), the volume of the solution will change by approximately 16 cm^3. The partial molar volume of water in the solution is thus 16 cm^3/mol. Molar volume has been used as an example. The same principle applies to other quantities, such as enthalpy, entropy, and Gibbs free energy.

Equation 7.5b can be integrated along a special path.[3] If a solution consists of n_A moles of A and n_B moles of B, then we can integrate Equation 7.5a along the path in which the ratio of n_A to n_B remains constant. If the composition is constant, then $\overline{V}_A$ and $\overline{V}_B$ are constant, thus:

$$V = \overline{V}_A\, n_A + \overline{V}_B\, n_B \tag{7.6}$$

Dividing Eq. 7.6 by the total number of moles, $n_A + n_B$:

$$\frac{V}{n_A + n_B} = \overline{V}_A \frac{n_A}{n_A + n_B} + \overline{V}_B \frac{n_B}{n_A + n_B}$$

$$\underline{V} = \overline{V}_A\, x_A + \overline{V}_B\, x_B \tag{7.7a}$$

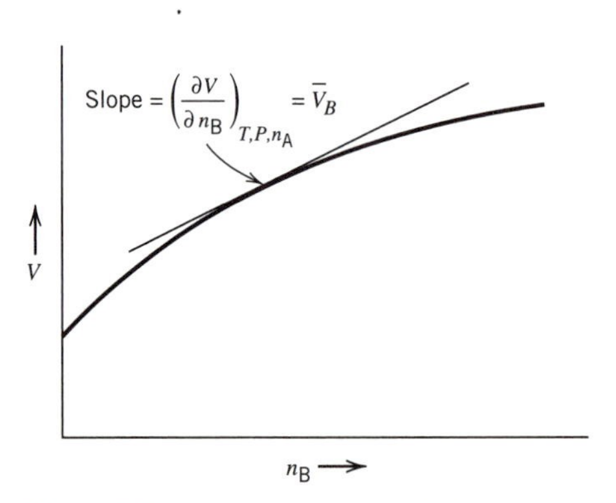

Figure 7.2 Volume of solution A–B as a function of moles of B added to the solution.

---

[2]As a limiting case, consider the partial molar volume of pure B. The *partial molar* volume of pure B is just the molar volume of B. If a small quantity of material B is added to a large mass of B, the volume change of the large mass will be just the molar volume of B multiplied by the number of moles of B added.

[3]Another way of deriving Eq. 7.6 involves the use of Euler's theorem of homogeneous functions (see Section A.17), recognizing that the equation for V is a first-order, homogeneous function of n_A and n_B (but not of T and P).

where x_A and x_B are the mole fractions of A and B in the solution, and $\underline{V}$ its molar volume. The differential form can be derived by dividing eq 7.5b by the total number of moles, n. The result is:

$$d\underline{V} = \overline{V}_A \, dx_A + \overline{V}_B \, dx_B \tag{7.7b}$$

7.3 RELATIVE PARTIAL MOLAR QUANTITIES

Upon mixing A and B to form a solution, the volume changes as follows:

$$\Delta V_{\text{mixing}} = V_M = V_{\text{final}} - V_{\text{initial}}$$

where V_{final} is the volume of the solution and V_{initial} is the volume of the components before mixing. The term V_M is the volume change upon mixing:

$$V_M = n_A \, \overline{V}_A + n_B \, \overline{V}_B - n_A \, \underline{V}_A - n_B \, \underline{V}_B \tag{7.8}$$

$$V_M = n_A \, (\overline{V}_A - \underline{V}_A) - n_B \, (\overline{V}_B - \underline{V}_B)$$

The term $\overline{V}_A - \underline{V}_A$ is called the relative partial molar volume of A, $\overline{V}_A^{\text{rel}}$. We can think of it as the change in volume of pure A as it enters the solution. It is the partial molar volume of A in solution relative to the molar volume of pure A. Thus we may write:

$$V_M = n_A \, \overline{V}_A^{\text{rel}} + n_B \, \overline{V}_B^{\text{rel}} \tag{7.9}$$

For one mole of solution, this becomes

$$\underline{V}_M = x_A \, \overline{V}_A^{\text{rel}} + x_B \, \overline{V}_B^{\text{rel}} \tag{7.10}$$

Equations 7.9 and 7.10 use volume as a representative thermodynamic property. The same forms of the equations apply to Gibbs free energy, entropy, enthalpy, and similar quantities.

The equations for variations of thermodynamic properties with temperature, pressure, and so on (such as $dG = -S \, dT + V \, dP$) apply to solutions as well as to pure components. Thus they apply to the relative partial molar quantities. For example, at constant P ($dP = 0$), and constant composition:

$$d\overline{G}^{\text{rel}} = -\overline{S}^{\text{rel}} \, dT \tag{7.11}$$

If in the initial state $\underline{G}_A = \underline{G}_A^\circ$, then:

$$\overline{G}_A^{\text{rel}} = \overline{G}_A - \underline{G}_A^\circ = RT \ln a_A \tag{7.12}$$

7.4 ENTROPY OF MIXING: IDEAL SOLUTION

For an ideal solution, in which $a_i = x_i$,

$$\overline{G}_i^{\text{rel}} = RT \ln x_i \tag{7.13}$$

The relative partial entropy of component A, from Eqs. 7.11 and 7.13, is

$$\overline{S}_A^{\text{rel}} = -\frac{\partial}{\partial T}(\overline{G}_A^{\text{rel}})_{P,n_B} = \frac{\partial}{-\partial T}(RT \ln x_A)_{P,x_B} \tag{7.14}$$

$$\overline{S}_A^{\text{rel}} = -R \ln x_A \tag{7.15}$$

The molar entropy of mixing for a solution of A + B is

$$\underline{S}_M = x_A \overline{S}_A^{\text{rel}} + x_B \overline{S}_B^{\text{rel}}$$

$$\underline{S}_M = -R [x_A \ln x_A + x_B \ln x_B] \tag{7.16}$$

For ideal solutions with many components, this can be generalized as:

$$\underline{S}_M = -R \sum_i x_i \ln x_i \tag{7.17}$$

The term $\underline{S}_M$ is called the entropy of mixing of an ideal solution. This is consistent with the expression obtained by considering the solution of two ideal gases in Section 3.7, Eq. 3.32.

7.5 ENTHALPY OF MIXING: IDEAL SOLUTION

The enthalpy of mixing of a solution can be obtained by a technique analogous to that demonstrated in Section 7.4 (Eqs. 7.13 and 7.14).

Based on Eq. 5.28:

$$d\left(\frac{\Delta G}{T}\right)_P = \Delta H \, d\left(\frac{1}{T}\right)_P$$

Therefore

$$\frac{d\left(\dfrac{\overline{G}_A^{\text{rel}}}{T}\right)}{d\left(\dfrac{1}{T}\right)} = \overline{H}_A^{\text{rel}} \tag{7.18}$$

Because the relative Gibbs free energy for a component of an ideal solution is:

$$\overline{G}_A^{rel} = RT \ln a_A = RT \ln x_A \qquad \textbf{(7.19)}$$

and we note that the term $\overline{G}_A^{rel}/T$ is not a function of temperature, therefore:

$$\frac{d\,(R \ln x_A)}{dT} = \overline{H}_A^{rel} = 0 = \overline{H}_A - \underline{H}_A \qquad \textbf{(7.20)}$$

We conclude thus that the enthalpy of mixing of an ideal solution is zero.

$$\underline{H}_M = x_A\, \overline{H}_A^{rel} + x_B\, \overline{H}_B^{rel} = 0 \qquad \text{ideal solution} \qquad \textbf{(7.21a)}$$

Similary, the volume change upon mixing, the ΔV of mixing, is also zero for an ideal solution because the term $RT \ln x_A$ is not a function of pressure.

$$\left(\frac{\partial G}{\partial T}\right)_T = V \qquad \text{and} \qquad \left(\frac{\partial \overline{G}_A^{rel}}{\partial P}\right)_T = \overline{V}_A^{rel}$$

$$\frac{d\,(RT \ln x_A)}{dP} = \overline{V}_A^{rel} = 0$$

$$\underline{V}_M = x_A\, \overline{V}_A^{rel} + x_B\, \overline{V}_B^{rel} = 0$$

To summarize, for an ideal solution:

$$\underline{G}_M = RT \sum_i x_i \ln x_i$$

$$\underline{S}_M = -R \sum x_i \ln x_i$$

$$\underline{H}_M = 0$$

$$\underline{V}_M = 0$$

7.6 GRAPHICAL REPRESENTATION

This section demonstrates that if we know a property of a two-component solution as a function of composition, we can determine the partial molar values of that property.

Take, as an example, molar volume. The molar volume of a solution can be determined simply by forming solutions of A and B with different compositons and measuring the resulting specific volume: that is, the volume divided by the total number of moles of the resulting solution. The result would look something like the solid line in Figure 7.3 From Eq. 7.7b, the equation relating the change of specific volume of a solution to changes in composition is:

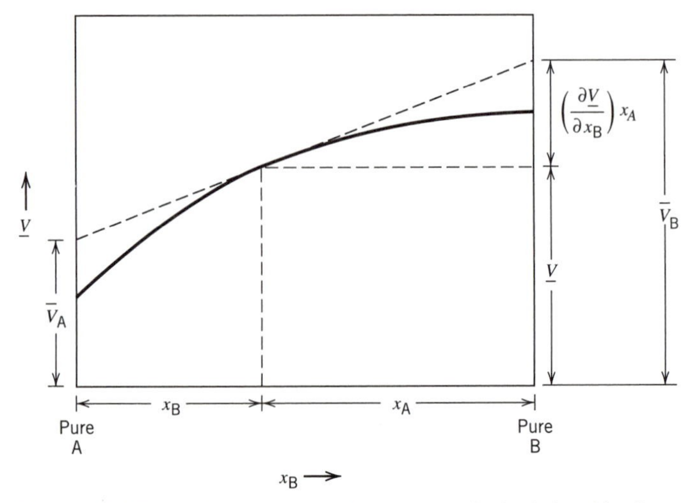

Figure 7.3 Plot of $\underline{V}$ versus x_B to show geometrical relationships in an A–B solution.

$$d\underline{V} = \overline{V}_A\, dx_A + \overline{V}_B\, dx_B \qquad \textbf{(7.7b)}$$

Because $x_A + x_B = 1$, $dx_A + dx_B = 0$, and $dx_A/dx_B = -1$, Eq. 7.7b can also be written as

$$\left(\frac{\partial \underline{V}}{\partial x_B}\right)_{T,P} = \overline{V}_B - \overline{V}_A \qquad \textbf{(7.22)}$$

This gives rise to the following three equations:

$$\text{I. } x_A \left(\frac{\partial \underline{V}}{\partial x_B}\right)_{T,P} = x_A\, \overline{V}_B - x_A\, \overline{V}_A \qquad \textbf{(7.23)}$$

$$\text{II. } x_B \left(\frac{\partial \underline{V}}{\partial x_B}\right)_{T,P} = x_B\, \overline{V}_B - x_B\, \overline{V}_A \qquad \textbf{(7.24)}$$

$$\text{III. } \underline{V} = \overline{V}_A\, x_A + \overline{V}_B\, x_B \qquad \textbf{(7.7)}$$

These three equations can be rearranged as follows:

$$\text{I + III} \qquad \underline{V} + x_A \left(\frac{\partial \underline{V}}{\partial x_B}\right)_{T,P} = \overline{V}_B \qquad \textbf{(7.25)}$$

$$\text{III − II} \qquad \underline{V} - x_B \left(\frac{\partial \underline{V}}{\partial x_B}\right)_{T,P} = \overline{V}_A \qquad \textbf{(7.26)}$$

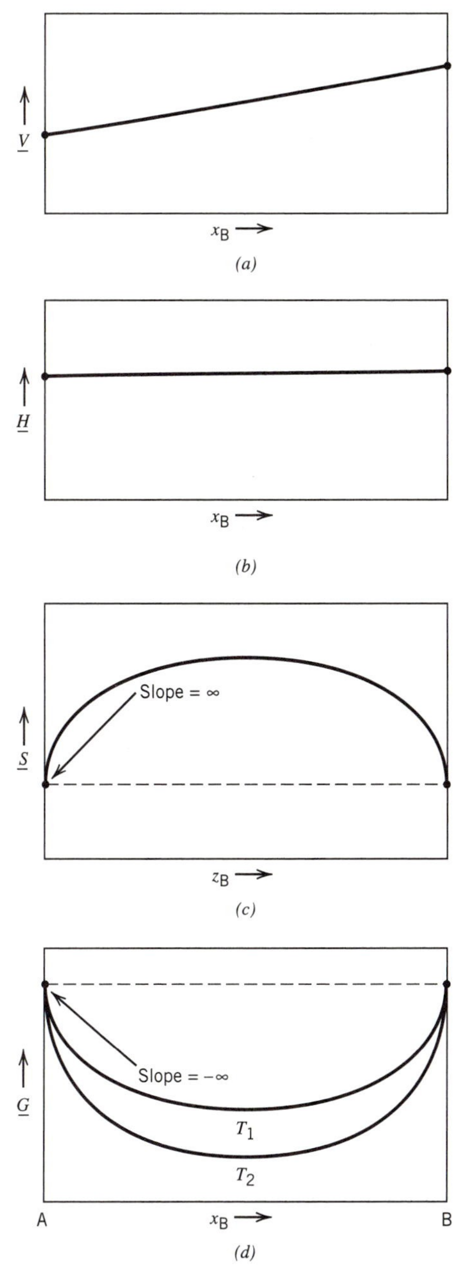

Figure 7.4 Molar properties for an ideal solution A–B: (*a*) volume of mixing, (*b*) enthalpy of mixing, (*c*) entropy of mixing, and (*d*) Gibbs free energy of mixing.

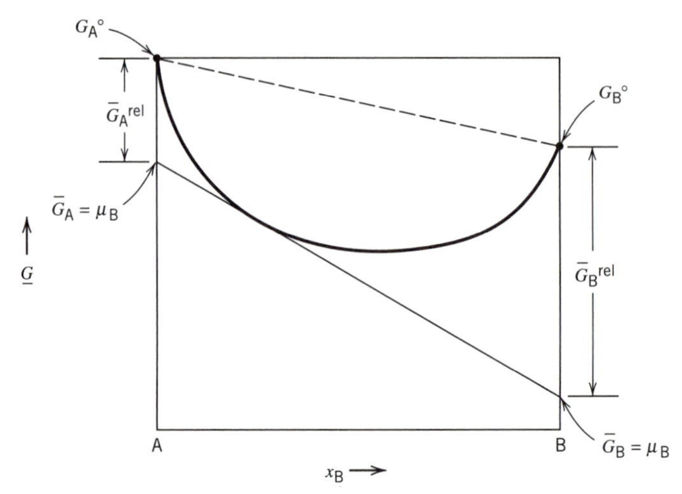

Figure 7.5 Plot of $\underline{G}$ versus x_B to show geometrical relationships between molar Gibbs free energy and mole fraction of B. (Note free choice of level for G_A° and G_B°.)

The relationships in Eqs. 7.25 and 7.26 take on special significance as shown in Figure 7.3. The tangent to the versus x_B curve at composition x_B intersects the A axis at $\bar{V}_A$ and intersects the *B* axis at $\bar{V}_B$.

For *ideal* solutions the graphs of molar volume of mixing, molar enthalpy of mixing, molar entropy of mixing, and molar Gibbs free energy of mixing as a function of composition are shown as Figure 7.4. It can be shown that the slope of specific entropy versus composition curve is infinite at pure B ($x_B = 1$) and at pure A ($x_A = 1$).

The ideal solution was defined by Eq. 7.2. By considering the diagram in Figure 7.4*d*, one can conclude that materials A and B can minimize their Gibbs free energy by forming a solution. The Gibbs free energies of the mixtures of pure A and pure B (not yet in solution) are shown along the dashed line joining the two points on the A and B axes. The Gibbs free energies of the ideal solutions created by A and B are represented by the solid line. At any composition, x_B, the two pure materials can spontaneously form a solution because the change in Gibbs free energy upon so doing is negative.

Figure 7.5 is Figure 7.4*d* redrawn, with the two points representing the specific Gibbs free energies of pure A and B taken at different levels. The values for $\underline{G}_A^\circ$ and $\underline{G}_B^\circ$ can be arbitrarily determined. The conclusion is that $\bar{G}_A^{rel}$ and $\bar{G}_B^{rel}$ are independent of the choice of zero point on the scales for G_A and G_B.

7.7 NONIDEAL SOLUTIONS

In a nonideal solution, the curve of specific Gibbs free energy of the solution as a function of composition may have many forms. Figure 7.6 illustrates one of these. In the region between x_{B1} and x_{B2}, the solution represented by the solid line could minimize its free energy by decomposing into two solutions of composition x_{B1} and

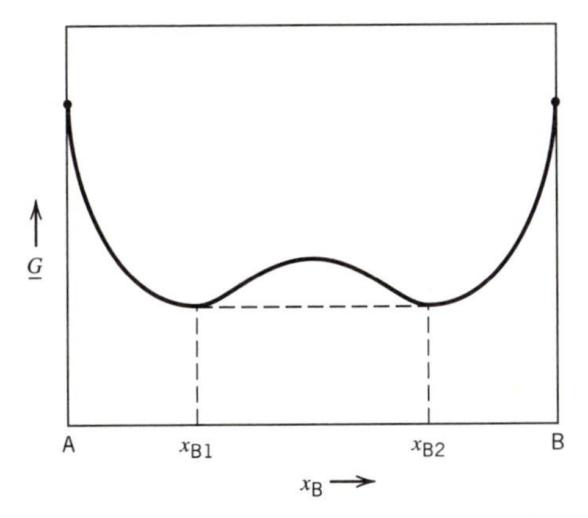

Figure 7.6 Plot of $\underline{G}$ versus x_B for a nonideal solution, showing positive deviations from Raoult's law. (In the region between x_{B1} and x_{B2}, the system can minimize G by forming mixtures of two solutions, x_{B1} and x_{B2}.)

x_{B2}. The Gibbs free energy of the combination (mixture) of the two solutions (x_{B1} and x_{B2}) is shown by the dashed line. A single solution in this region is unstable relative to the mixture of solutions of x_{B1} and x_{B2}. If equilibrium were attained, the single solution would decompose into two solutions because the Gibbs free energy change associated with such a decomposition is negative.

One extreme of such nonideal behavior is the case of two *immiscible* materials; that is, materials A and B form no solutions.[4] In that case the graph of molar Gibbs

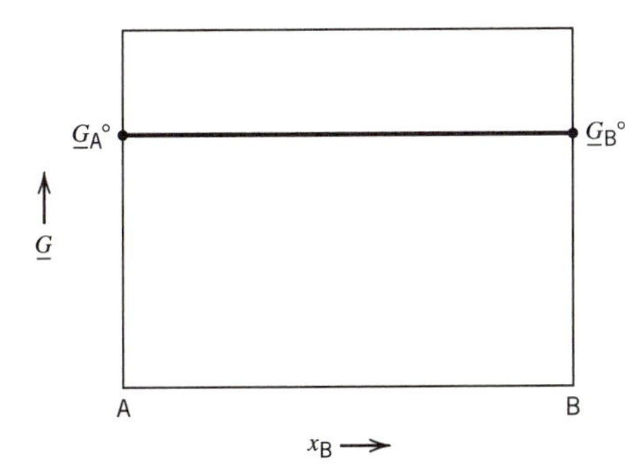

Figure 7.7 Plot of $\underline{G}$ versus x_B for mixtures of A and B, where A and B are immiscible.

[4]Actually, total immiscibility among elements is impossible because the slope of the curve of $\Delta\underline{G}_M$ versus x_B is infinite at the pure material axes. For our purposes we will assume that immiscibility means solubility so low that it is of no practical consequence in calculations or in graphical representations.

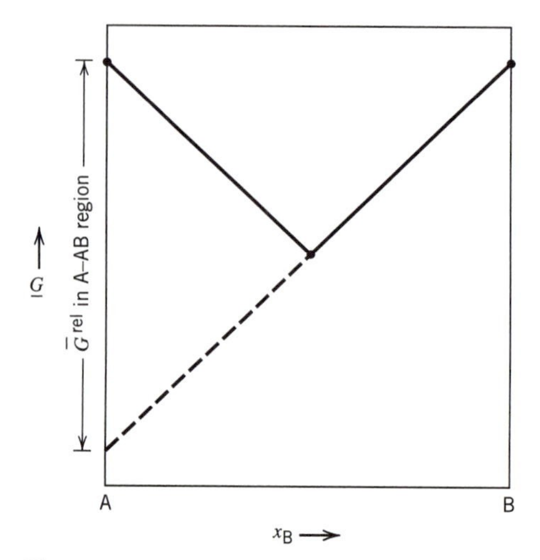

Figure 7.8 Plot of $\underline{G}$ versus x_B in an A–B system when compound AB is formed.

free energy against composition is simply a straight line between the two points on the A and B axes (Figure 7.7).

As another extreme, suppose that the materials A and B form a compound AB, which does not dissolve in either A or B. The Gibbs free energy of combinations of A and B as a function of x_B is shown in Figure 7.8.

If materials A and B form a continuous set of solutions, but are non-ideal, the activity of material B is usually expressed as:

$$a_B = \gamma_B \, x_B \tag{7.27}$$

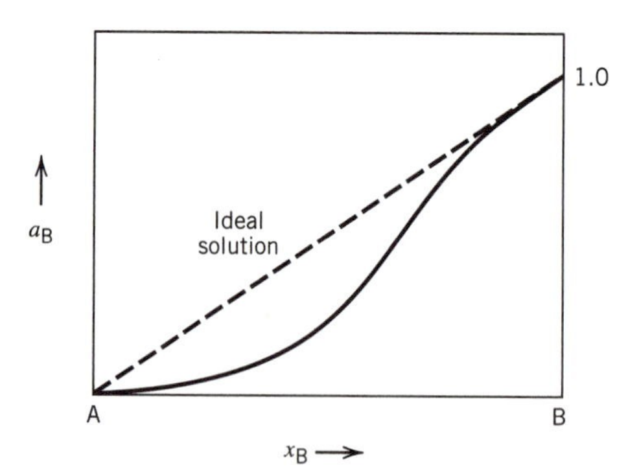

Figure 7.9 Plot of activity versus composition for a nonideal solution.

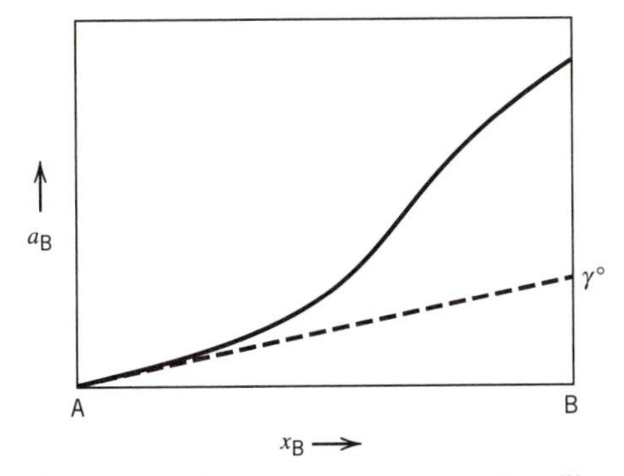

Figure 7.10 Extrapolation from the dilute region in a plot of activity versus composition for a nonideal solution.

Consequently:

$$G_B = RT \ln a_B = RT \ln (\gamma_B \, x_B) \quad \textbf{(7.28)}$$

where γ_B is called the activity coefficient of B in the A–B solution. A diagram of activity of B versus composition could look like the one in Figure 7.9. Note that $\gamma = 1$ for an ideal solution, the one shown in a dashed line in that figure.

In dilute solutions, the activity coefficient of material B, γ_B, is often constant, and is represented by the symbol γ_B°. This is illustrated in Figure 7.10.

7.8 GIBBS–DUHEM RELATION

The molar Gibbs free energy of an A-B solution can be written by analogy to Eq. 7.7:

$$\underline{G} = x_A \, \overline{G}_A + x_B \, \overline{G}_B \quad \textbf{(7.29)}$$

Differentiating Equation 7.29, we obtain:

$$d\underline{G} = \overline{G}_A \, dx_A + \overline{G}_B \, dx_B + x_A \, d\overline{G}_A + x_B \, d\overline{G}_B \quad \textbf{(7.30)}$$

Noting, however, that

$$G = G \, (n_A, \, n_B)_{T,P}$$

$$dG = \overline{G}_A \, dn_A + \overline{G}_B \, dn_B$$

we can write, on a molar basis:

$$d\underline{G} = \overline{G}_A \, dx_A + \overline{G}_B \, dx_B \quad \textbf{(7.31)}$$

Subtracting Eq. 7.31 from Eq. 7.30 yields:

$$x_A \, d\overline{G}_A + x_B \, d\overline{G}_B = 0 \tag{7.32}$$

This equation, known as the Gibbs–Duhem equation, can be put in more useful form as

$$x_A \, d(RT \ln a_A) + x_B \, d\,(RT \ln a_B) = 0 \tag{7.33}$$

or

$$d\,(\ln a_A) = -\frac{x_B}{x_A} \, d\,(\ln a_B) \tag{7.34}$$

This relationship is a particularly important one. It shows that if the activity of one component of a binary solution is known as a function of composition, then the activity of the other can be determined. (See section 7.10.) Some consequences of this follow.

Suppose that in the A–B solution, we know that B behaves ideally. The activity of B is equal to the mole fraction of B if we take pure B as the standard state. Integrating Eq. 7.34 from pure A ($x_B = 0$) to an arbitrary value of $x_B$we have:

$$\int_{-\infty}^{\ln a_A} d\,(\ln a_A) = -\int_{-\infty}^{\ln a_B} \frac{x_B}{x_A} \, d\,(\ln a_B) = -\int_{-\infty}^{\ln x_B} \frac{x_B}{x_A} \, d\,(\ln x_B)$$

$$\ln a_A = -\int_{-\infty}^{x_B} \frac{x_B}{x_A} \frac{dx_B}{x_B} = -\int_{-\infty}^{x_B} \frac{dx_B}{x_A}$$

Switching variables, $dx_A = -\,dx_B$

$$\ln a_A = \int_1^{x_A} \frac{dx_A}{x_A} = \ln x_A$$

$$a_A = x_A$$

We thus come to a conclusion that if material B in a solution of A and B behaves ideally, then A does also. This could have been demonstrated, as well, by dealing with the activity *coefficients* of A and B (from Eq. 7.27) in Eq. 7.34.

$$x_A \, d\,[\ln(\gamma_A \, x_A)] + x_B \, d\,[\ln(\gamma_B \, x_B)] = 0$$

$$x_A \, d\,(\ln \gamma_A) + x_B \, d\,(\ln \gamma_B) + x_A \, d\,(\ln x_A) + x_B \, d\,(\ln x_B) = 0$$

But $x_A \, d\,(\ln x_A) = dx_A$ and $x_B \, d\,(\ln x_B) = x_B$ and

$$dx_A + dx_B = 0$$

Hence:

$$x_A\, d\, (\ln \gamma_A) + x_B\, d \ln (\gamma_B) = 0 \tag{7.35}$$

$$d\, (\ln \gamma_A) = -\frac{x_B}{x_A}\, d\, (\ln \gamma_B) \tag{7.36}$$

Equation 7.36 is another form of the Gibbs–Duhem equation. It can also be used to demonstrate that if component B in a two-component system behaves ideally, then component A does also. If B is ideal; $\gamma_B = 1$, then $d \ln (\gamma_B) = 0$. Based on Eq. 7.36,

$$d\, (\ln \gamma_A) = 0 \text{ or } \gamma_A = \text{constant} \tag{7.37}$$

But at $x_A = 1$, $a_A = 1$ or $\gamma_A = 1$. Hence component A is ideal.

7.9 DILUTE SOLUTIONS AND COLLIGATIVE PROPERTIES

Consider a nonideal solution A–B whose activity–composition diagram is given by Figure 7.10. In solutions of this type there is usually a region in the vicinity of pure A (in the dilute B region) in which the activity of B is a linear function of composition. The activity coefficient of B (γ_B°) is a constant, not necessarily equal to 1. This region is referred to as the Henry's law region for B.

$$\gamma_B = \gamma_B^\circ$$

$$d\, (\ln \gamma_B^\circ) = 0$$

If this is true, the activity coefficient of A in this region must also be constant, based on arguments presented (Eqs. 7.35–7.37). If $d \ln \gamma_B^\circ = 0$, then $d \ln \gamma_A^\circ = 0$. That means that the activity coefficient of A in this region is a constant. But the activity of A equals 1 when the mole fraction of A equals 1 (pure A); hence the activity coefficient of A in this region must be 1.

We thus conclude that if, in a mixture of A and B, material B follows Henry's law, then material A follows Raoult's law: that is, its activity simply equals its mole fraction (Figure 7.11). In most dilute solutions, the activity of *solutes* follows Henry's law up to a few mole percent, which means that the thermodynamic activity of the *solvent* follows Raoult's law. This property of the solvent in dilute solutions is the basis of a series of properties of dilute solutions called *colligative properties.* In a dilute solution:

1. The vapor pressure of the solvent is reduced. If x_S represents the mole fraction of solvent, and x_u the mole fraction of solute, then the vapor pressure of the solvent is:

$$P_S = P_S^\circ\, x_S = P_S^\circ\, (1 - x_u) \tag{7.38}$$

where P_S° is the vapor pressure of the pure solvent.

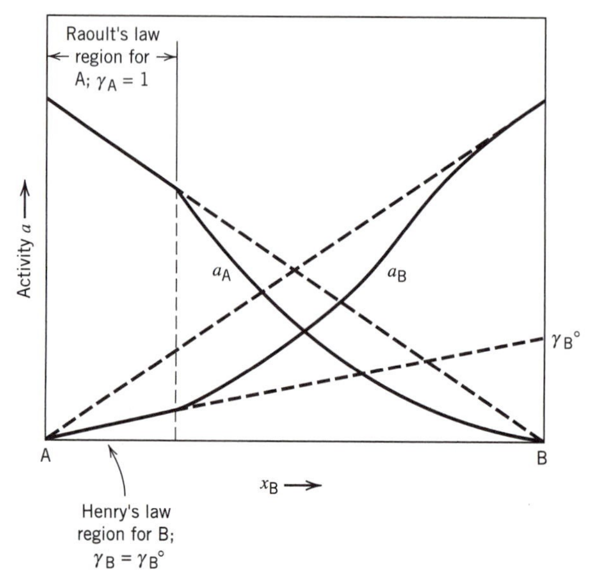

Figure 7.11 Henry's law and Raoult's law regions for the nonideal solution.

2. **2.** The boiling point of the solvent is elevated. If the vapor pressure of the solvent is lowered, one would expect that its boiling point, the temperature at which its pressure is one atmosphere, would be raised. The relationship is

$$\ln x_S = \ln(1 - x_u) = \frac{\Delta H^\circ_{vap}}{R}\left[\frac{1}{T_B} - \frac{1}{T^\circ_B}\right] \tag{7.39}$$

where ΔH°_{vap} is the molar heat of vaporization of the solvent, T_B is the boiling point of the solvent in the solution, and T°_B is the boiling point of the pure solvent.

3. **3.** The freezing point of the solvent is lowered. This will be discussed in Chapter 9 (Section 9.1).
4. **4.** The solvent will display an osmotic pressure (see Section 7.14).

7.10 INTEGRATING THE GIBBS–DUHEM EQUATION

From the Gibbs-Duhem equation (Eq. 7.34), we know that if the activity of B is known throughout an A–B solution, then the activity of A may also be calculated throughout as follows:

$$\int_{-\infty}^{\ln a_A} d(\ln a_A) = \int_{-\infty}^{\ln a_B} -\frac{X_B}{X_A}\, d(\ln a_B)$$

$$\ln a_A = -\int_{-\infty}^{\ln a_B} \frac{x_B}{x_A}\, d(\ln a_B) \tag{7.40}$$

If an algebraic expression for the activity of B (a_B) as a function of mole fraction of B (x_B) is known, then Eq. 7.40 may be integrated using the normal methods of calculus. As an example, let us consider the case of copper–zinc alloys (called brass in the higher copper concentrations). The activity of zinc is easy to measure because zinc is volatile, and it is possible to measure its pressure in equilibrium with zinc-containing liquids or solids. Its vapor pressure over an alloy relative to the vapor pressure over pure zinc is its activity. In the temperature range 1400–1500 K an expression that fits the data for the activity coefficient of zinc is:

$$RT \ln \gamma_{Zn} = -38{,}300 x_{Cu}^2$$

Applying the Gibbs–Duhem equation in the form of Eq. 7.36,

$$d(\ln \gamma_{Cu}) = -\frac{x_{Zn}}{x_{Cu}} d\left(-\frac{38{,}300}{RT} x_{Cu}^2\right)$$

$$d(\ln \gamma_{Cu}) = x_{Zn} \frac{2(38{,}300)}{RT} dx_{Cu}$$

$$dx_{Cu} = -dx_{Zn}$$

$$\int_0^{\ln \gamma_{Cu}} d(\ln \gamma_{Cu}) = -\int_0^{x_{Zn}} \frac{2(38{,}300)}{RT} x_{Zn}\, dx_{Zn}$$

$$\ln \gamma_{Cu} = -\frac{38{,}300}{RT} x_{Zn}^2$$

The foregoing relations illustrate that the activity (or activity coefficient) of one component of a two-component solution can be determined easily if the activity of the other is known in algebraic form. If, however, a_B as a function of x_B is known only empirically, the expression in Eq. 7.40 must be integrated numerically or graphically. The integral on the right-hand side of the equation is the negative of the area under the curve of x_B/x_A (ordinate) versus $\ln a_B$ (abscissa). This integration is made inaccurate by the need to integrate from $a_B = 0$ where $\ln a_B$ is negative infinity.

The alternate form of the Gibbs–Duhem equation (Eq. 7.36) is the basis of a convenient way to avoid the need to integrate from negative infinity (on the abscissa) when integrating the Gibbs–Duhem equation graphically.

$$d(\ln \gamma_A) = \frac{x_B}{x_A} d(\ln \gamma_B) \tag{7.41}$$

$$\int_0^{\ln \gamma_A} d(\ln \gamma_A) = \ln \gamma_A = -\int_{\ln \gamma_B^\circ}^{\ln \gamma_B} \frac{x_B}{x_A} d(\ln \gamma_B)$$

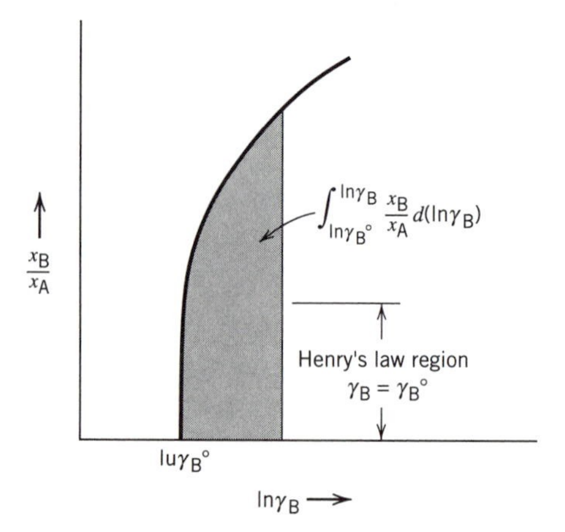

Figure 7.12 Graphical integration of the Gibbs–Duhem equation.

The diagram of x_B/x_A versus the natural logarithm of γ_B is shown in Figure 7.12. The vertical portion of the line represents that region in which the activity coefficient of B is constant in the dilute solution, that is, the Henry's law region.

7.11 REGULAR SOLUTIONS

One example of a nonideal solution whose thermodynamic properties can be described using a simple algebraic function is the regular solution. From Eq. 7.10, the Gibbs free energy of mixing of a solution can be expressed as:

$$\underline{G}_M = x_A\overline{G}_A^{rel} + x_B\overline{G}_B^{rel}$$

$$\underline{G}_M = x_A\, RT \ln (\gamma_A x_A) + x_B\, RT \ln (\gamma_B x_B)$$

$$\underline{G}_M = RT\,(x_A \ln x_A + x_B \ln x_B) + RT\,(x_A \ln \gamma_A + x_B \ln \gamma_B) \qquad \textbf{(7.42)}$$

The first two terms in Eq. 7.42 represent the Gibbs free energy of mixing of an ideal solution. The second two terms, which involve the activity of coefficients of A and B, are referred to as the *excess* Gibbs free energy of mixing—that is, the Gibbs free energy of mixing in excess of the ideal.

$$\underline{G}_M = \underline{G}_M^{ideal} + \underline{G}_M^{xs}$$

where

$$\underline{G}_M^{ideal} = RT\,(x_A \ln x_A + x_B \ln x_B)$$

and

$$\underline{G}_M^{xs} = RT\,(x_A \ln \gamma_A + x_B \ln \gamma_B) \qquad \textbf{(7.43)}$$

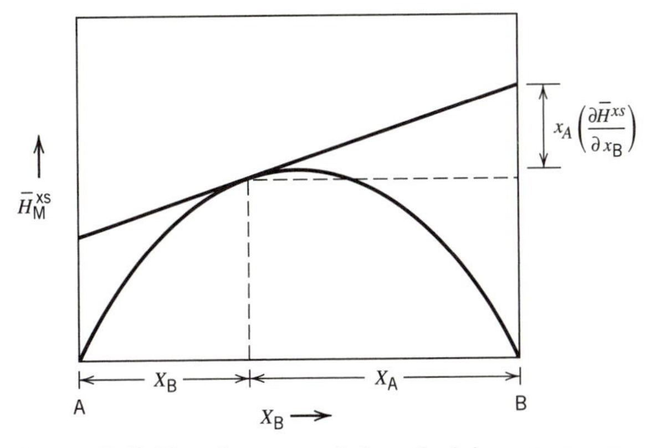

Figure 7.13 Plot of excess enthalpy of mixing versus composition for a nonideal solution.

A regular solution of two components is defined as one in which the activity coefficient of A has the form:

$$\ln \gamma_A = \frac{\omega}{RT} x_B^2 = \frac{\omega}{RT}(1 - x_A)^2 \tag{7.44}$$

In addition, the entropy of mixing in a regular solution is equivalent to the ideal entropy of mixing, that is the excess entropy of mixing is zero. From these two relationships, we can show that

$$\underline{G}^{xs} = \underline{H}^{xs} \qquad \text{regular solution} \tag{7.45}$$

Thus the *excess* enthalpy of mixing (actually the enthalpy of mixing because the enthalpy of mixing of an ideal solution is zero, from Eq. 7.21a) can be derived using Eqs. 7.42 and 7.44

$$\underline{G}_M = RT \ln (x_A \ln x_A + x_B \ln x_B) + \underbrace{\underbrace{x_A\, \omega\, x_B^2 + x_B\, \omega\, x_A^2}_{(x_A + x_B)(\omega\, x_A\, x_B)}}_{\omega\, x_A\, x_B} \tag{7.46}$$

$$\underline{G}_M = RT (x_A \ln x_A + x_B \ln x_B) + \omega\, x_A\, x_B$$

The first term in Eq. 7.46 is just $-T\underline{S}_M$, thus the second term is $\underline{H}_M$.

$$\underline{H}_M = \omega\, x_A\, x_B \tag{7.47}$$

7.12 REGULAR SOLUTIONS: ATOMISTIC INTERPRETATION

In this section, we depart from the methods of classical (macroscopic) thermodynamics and seek an understanding of the algebraic form of the regular solution (Eqs. 7.44 and 7.47) in terms of interactions between the components of the solution, A and B. We will speak of A and B as *atoms,* as in the case of metallic solutions, but they could just as well be *molecules,* as in the case of polymer solutions.

Let us think of the mixture of A and B as a solid solution, with the A and B atoms distributed on the sites of the crystallographic lattice of the solid. The basic assumption of the model will be that the atoms (A and B) are randomly distributed on the lattice sites. This is equivalent to stating that the entropy of mixing of A and B is equal to the entropy of mixing of an ideal solution. (This will be explored further in Chapter 10 on statistical thermodynamics.)

Each atom will be assumed to have an energy interaction only with its nearest neighbors, Z in number (Z is also called the coordination number). The bonding energy of the crystal will be expressed in terms of the energies of the bonds between the atoms, E_{AA} being the bonding energy of two A atoms, for example.

We will mix N_A atoms of A with N_B atoms of B. Before mixing, the number of N_{AA} bonds among the A atoms (in pure A) is $N_AZ/2$. We arrive at this conclusion by counting the number of bonds emanating from each A atom (Z), multiplying by the number of A atoms (N_A), then dividing by 2 because we will have double-counted the bonds. After mixing, the number of AA bonds will be $N_A^2Z/2N_T$, where the total $N_T = N_A + N_B$. To count the number of AA bonds we multiply N_A by two factors: the coordination number Z and the probability that its nearest neighbor is an A atom, N_A/N_T (assuming that A and B atoms are randomly distributed). The same procedure is followed for calculating the number of BB bonds and AB bonds. The resulting energies after mixing are listed in Table 7.1.

The energy of mixing is the final energy less the initial energy.

$$E_M = Z\left[\frac{N_A^2}{2N_T}E_{AA} + \frac{N_B^2}{2N_T}E_{BB} + \frac{N_AN_B}{N_T}E_{AB} - \frac{N_A}{2}E_{AA} - \frac{N_B}{2}E_{BB}\right] \quad \textbf{(7.48)}$$

The mole fraction of A is x_A, which is N_A/N_T. Performing the substitution of mole fractions (x) in Eq. 7.48 yields:

$$E_M = ZN_T\left\{\frac{[x_A^2 - x_A]\,E_{AA} + [x_B^2 - x_B]\,E_{BB}}{2}\right\} + x_A\,x_B\,E_{AB}$$

$$E_M = ZN_T\,x_A\,x_B\,[E_{AB} - \tfrac{1}{2}(E_{AA} + E_{BB})]$$

or

$$E_M = \omega\,x_A\,x_B \quad \textbf{(7.49)}$$

where

$$\omega = ZN_T\,[E_{AB} - \tfrac{1}{2}(E_{AA} + E_{BB})] \tag{7.50}$$

Equation 7.49 for the energy of mixing has the same form as Eq. 7.47 for enthalpy of mixing of ΔH_M. The two are essentially equal for solids because the volume changes influencing the *PV* term in the definition of enthalpy are very small.

From Eq. 7.49 it can be seen that the solution will behave as ideal ($\Delta H_M = 0$) when the A–B binding energy is equal to the average of the A–A and B–B bonding energies. If the A–B binding energy is larger, in absolute value, than the A–A, B–B average, then the enthalpy of mixing will be negative. (Remember that binding energies are negative.)

Let us now express the thermodynamic mixing functions for the solution in terms of the *excess* quantities (xs), the quantities in excess of the ideal mixing functions (in Eq. 7.45):

$$\underline{G}_M^{xs} = \underline{H}_M^{xs} - T\underline{S}_M^{xs}$$

For the regular solution, $\underline{S}_M = 0$, and $\underline{H}_M = \omega\, x_A x_B$. Then:

$$\underline{G}_M^{xs} = \underline{H}_M^{xs} = \omega\, x_A x_B$$

From Figure 7.13

$$\overline{H}_B^{xs} = \underline{H}_M^{xs} + x_A\left(\frac{\partial \underline{H}_M^{xs}}{\partial x_B}\right)_T \tag{7.51}$$

$$\overline{H}_B^{xs} = \omega x_A x_B + x_A\left(\frac{\partial(\omega x_A x_B)}{\partial x_B}\right)_T$$

$$\overline{H}_B^{xs} = \omega x_A x_B + x_A\,\omega\,(x_A - x_B) \tag{7.52}$$

$$\overline{H}_B^{xs} = \omega\, x_A^2$$

Table 7.1 Bond Energies After Mixing Atoms A and B in a Solid Solution

	Number of Bonds	Energy per Bond	Energy
N_{AB}	$N_A \dfrac{N_B}{N_T} Z$	E_{AB}	$\dfrac{N_A N_B}{N_T} Z\, E_{AB}$
N_{AA}	$\frac{1}{2}\dfrac{N_A^2}{N_T}$	E_{AA}	$\frac{1}{2}\dfrac{N_A^2}{N_T} Z\, E_{AA}$
N_{AB}	$\frac{1}{2}\dfrac{N_B^2}{N_T}$	E_{BB}	$\frac{1}{2}\dfrac{N_B^2}{N_T} E_{BB}$

But, for a regular solution, $\overline{S}_B^{xs} = 0$

$$\overline{G}_B^{xs} = \overline{H}_B^{xs}$$

From Eq. 7.43:

$$\overline{G}_B^{xs} = RT \ln \gamma_B$$

Therefore

$$\ln \gamma_B = \frac{\omega}{RT} x_A^2$$

This justifies the algebraic form of the activity coefficient in a regular solution.

7.13 POLYMER SOLUTIONS

In this section we will deal with the thermodynamics of solutions in which the solute is a polymer and the solvent is a material with lower molecular weight. These solutions are different from the ones we have discussed thus far because the molecular weights and the molar volumes of the components are very different. If the molar volumes of the polymer and solvent differ only by a factor of 2 or 3, the relationships we have derived in the preceding sections are still valid. Often, however, the molar volume of the polymers is very many times the molar volume of solvents, and we must, therefore, modify our approach to solution thermodynamics.

To gauge the magnitude of the difference in molar volumes, let us make a sample calculation for the case of linear polymers. A polymer has many repetitive segments (mers). Let n be the number of segments in a polymer chain. Assume that the volume of each segment is equal to the molecular volume of the solvent. The molar volume of the polymer is then n times the molar volume of the solvent. If we designate the polymer with the subscript 2, and the solvent with subscript 1,

$$\underline{V}_2 = n\, \underline{V}_1 \qquad (7.53)$$

For a binary solution consisting of solvent and polymer, the volume fractions of solvent (ϕ_1) and polymer (ϕ_2) are:

$$\phi_1 = \frac{x_1 \underline{V}_1}{x_1 \underline{V}_1 + x_2 \underline{V}_2} = \frac{x_1}{x_1 + n x_2} \qquad (7.54)$$

$$\phi_2 = \frac{x_2 \underline{V}_2}{x_1 \underline{V}_1 + x_2 \underline{V}_2} = \frac{n x_2}{x_1 + n x_2} \qquad (7.55)$$

where x_1 and x_2 are the mole fractions of solvent and polymer, respectively, and n is the number of segments in the polymer.

Take the case of a polymer with 100 segments, $n = 100$. In a solution consisting of 90 mol % solvent ($x_1 = 0.90$), the volume fraction of solvent will be:

$$\phi_1 = \frac{x_1}{x_1 + nx_2} = \frac{0.90}{0.90 + 100(0.10)} = \frac{0.90}{10.9} = 0.083$$

The volume fraction occupied by the polymer, 0.917 (that is, 1.0 − 0.083) is disproportionately larger than its mole fraction 0.10. This leads to the conclusion that Raoult's law for ideal solutions stated in terms of *mole* fraction is not useful in the case of polymers in which the molar volumes of solvent and solute differ so radically. In this case, Raoult's law is stated in terms of *volume* fraction, ϕ_i.

$$a_i = \phi_i \tag{7.56}$$

Based on this definition of ideality, the following expression for the entropy of mixing can be derived based on the Flory–Huggins lattice model (see Refs. 1 and 2; the derivation is covered in Chapter 10):

$$\underline{S}_{\mathrm{M}} = -R\,[x_1 \ln \phi_1 + x_2 \ln \phi_2] \tag{7.57}$$

Note that this expression for the entropy of mixing is not symmetrical about $x = 0.5$ because the volume fractions are not equal to the mole fractions. If the molar volumes of the two components are equal, the expression becomes the same as the one derived for the entropy of mixing (Eq. 7.16).

In an ideal polymeric solution, the enthalpy of mixing is zero and the molar Gibbs free energy of mixing is

$$\underline{G}_{\mathrm{M}} = RT\,[x_1 \ln \phi_1 + x_2 \ln \phi_2] \qquad \text{for ideal polymer solutions} \tag{7.58}$$

To treat nonideality, a model similar to the regular solution is used (Sections 7.11 and 7.12). In this approach, the entropy of mixing is that of an ideal solution. The enthalpy of mixing for a solution with a total of N polymer segments and solvent molecules ($N = N_1 + nN_2$) is:

$$H_{\mathrm{M}} = \tfrac{1}{2}\, Z\, N\, \phi_1\, \phi_2\, \Delta w \tag{7.59}$$

where Δw is an interaction parameter similar to the interaction energy term $[E_{\mathrm{AB}} - \frac{1}{2}(E_{\mathrm{AA}} + E_{\mathrm{BB}})]$ in regular solutions (Eq. 7.50).

The term $\frac{1}{2}Z\,\Delta w$ is set equal to χRT, where χ is a constant. The expression for the enthalpy of mixing, H_{M} is then written as:

$$H_{\mathrm{M}} = \chi RT\, N\, \phi_1\, \phi_2 \tag{7.60}$$

Note that $N\phi_1 = N_1$, hence $H_M = \chi RTN_1\phi_2$. The molar enthalpy of mixing, $\underline{H}_M$ is $H_M/(N_1 + N_2)$. Then:

$$\underline{H}_M = \chi RTx_1\phi_2 \tag{7.61}$$

The molar Gibbs free energy of mixing is then:

$$\underline{G}_M = RT\,[x_1 \ln \phi_1 + x_2 \ln \phi_2 + \chi x_1\phi_2] \tag{7.62}$$

The partial molar Gibbs free energy (chemical potential) of the solvent is:

$$\mu_1 - \mu_1^\circ = RT\left[\ln \phi_1 + \left(1 - \frac{1}{n}\right)\phi_2 + \chi\,\phi_2^2\right] \tag{7.63}$$

and for the polymer it is:

$$\mu_2 - \mu_2^\circ = RT\,[\ln \phi_2 - (n - 1)\,\phi_1 + n\,\chi\,\phi_1^2] \tag{7.64}$$

7.14 OSMOTIC PRESSURE

The conclusion reached in Section 7.9 concerning the ideal behavior of the solvent in dilute solutions allows us to make some interesting observations on the properties of these solutions. One such observation concerns osmotic pressure, the extra pressure required to raise the chemical potential of the solvent in the solution to the value it has when it is pure.

To illustrate, consider an aqueous solution (water containing a solute B), separated from pure water by a semipermeable, flexible membrane. The membrane is permeable to water, but not to the solute. Cellulose acetate, and other polymeric materials can be made into such membranes. In such a system, a pressure difference is observed between the two sides of the membrane. The pressure inside the membrane (with the impure water) is greater than the pressure on the side with pure water. The difference between the two is indicated by the osmotic pressure Π, as illustrated in Figure 7.14.

Let us calculate the osmotic pressure for a dilute solution of B in A. Assuming that the activity of B follows Henry's law, the activity of water (the solvent, x_A) as a function of solute concentration (x_B) is given by:

$$\mu_A - \mu_A^\circ = RT \ln x_A \qquad \text{Raoult's law}$$

$$\mu_A - \mu_A^\circ = RT \ln (1 - x_B) = -\,RT \ln x_B \tag{7.65}$$

because $\ln(1 - x) = -x$ when x is small.

To find the condition under which the pure water and the solution are in equilibrium at a given external pressure and temperature, we calculate the change of chem-

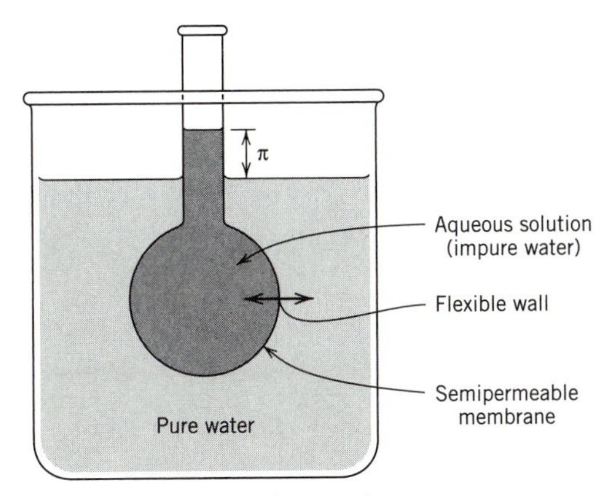

Figure 7.14 Osmotic pressure.

ical potential (molar Gibbs free energy) with pressure for pure water at constant temperature:

$$\mu_A - \mu_A^\circ = \int_{P+\Pi}^{P} \underline{V}_A \, dP = -\underline{V}_A \Pi \qquad \textbf{(7.66)}$$

This is illustrated in Figure 7.15. The osmotic pressure II is:

$$\Pi = RT \frac{x_B}{\underline{V}_A} \qquad \textbf{(7.67)}$$

As an example, let us calculate the osmotic pressure of seawater. Seawater contains approximately 35,000 mg of NaCl per liter. This is equivalent to about 0.6 mol/L. The mole fraction of sodium chloride is thus:

$$x_B = \frac{0.6}{1000/18} = 0.0108$$

Because sodium chloride ionizes completely (into Na^+ and Cl^-), the number of ions (both anions and cations) is double the mole fraction of sodium chloride. Osmotic pressure thus is

$$\Pi = \frac{2\ (0.0108)(8.314)(298)}{18 \times 10^{-6}} = 29.7 \times 10^5 \text{ N/m}^2$$

$$\Pi = 29.7 \text{ atm}$$

Note that the osmotic pressure is that pressure at which the two, pure water at one atmosphere and the aqueous solution at the osmotic pressure, *are in equilibrium.*

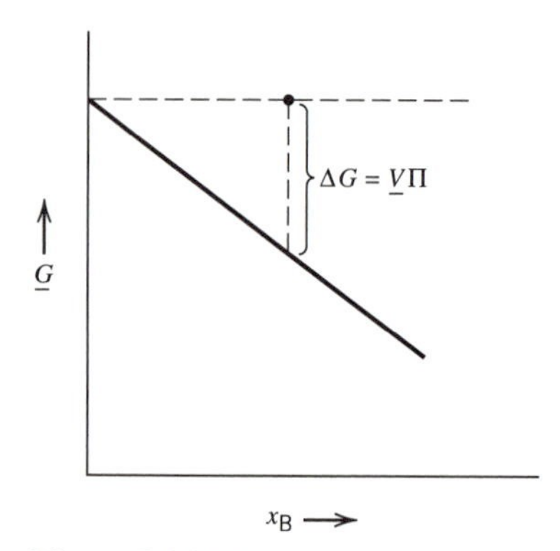

Figure 7.15 Plot of $\underline{G}$ versus x_B for the calculation of osmotic pressure.

If one were to exert on the aqueous solution a pressure greater than the osmotic pressure, water would be forced through the membrane from the seawater side to the pure water side. This process, called reverse osmosis, is used to produce purified water. The pressures in the case of seawater are quite high. The pressure required in the reverse osmosis of brackish water with a lower concentration of impurities is lower.

We have considered water as the solvent. Of course, other liquids can also serve in that capacity. In the case of organic polymer solutions, benzene (C_6H_6) is often used as a solvent. In this case, osmotic pressure determination can be used to find the average molecular weight of the polymer in the solution (number average) if the composition of the solution is known.

For a polymer solution, the chemical potential of the solvent is given by Eq. 7.63. For *dilute* polymer solutions, this expression is reduced to (Ref. 3):

$$\mu_1 - \mu_1^\circ = -RT\frac{\phi_2}{n} + RT\left(\chi - \tfrac{1}{2}\right)\phi_2 \tag{7.68}$$

where the first term represents the chemical potential in the case of ideal solutions, and the second term accounts for nonideality. Let us treat the ideal solution case, where the osmotic pressure is:

$$-\Pi\, V_1 = \mu_1 - \mu_1^\circ = -RT\frac{\phi_2}{n}$$

In dilute solutions, it is reasonable to assume that $\overline{V}_1 = \underline{V}_1$, and that $\phi_2 = nx_2/(x_1 + nx_2) = nx_2 / x_1$. Furthermore, $\underline{V}$ can be taken as $x_1\underline{V}_1$. This results in an expression similar to Eq. 7.67 for osmotic pressure:

$$\Pi = RT\frac{x_2}{\underline{V}}$$

Take as an example a one weight percent solution of a polymer in benzene at 298 K. Let us determine the average molecular weight of the polymer if the osmotic pressure is measured to be 50 mmHg (6660 Pa).

The density of benzene is 0.88 g/cm^3 and its molecular weight is 78 g/mol; hence its specific volume is 88.6×10^{-6} m^3/mol. A 100 g sample of the one weight percent solution of polymer contains one gram of polymer. If its molecular weight is represented by MW, then the number of moles of polymer in the sample is 1/MW. The number of moles of benzene is 99/78. Hence, the mole fraction of polymer is:

$$6660 = \frac{RT}{\underline{V}} x_B$$

Carrying out the calculation in terms of mole fractions:

$$x_B = \frac{1/\mathrm{MW}}{1/\mathrm{MW} + 99/78}$$

but

$$\frac{1}{\mathrm{MW}} \ll \frac{99}{78}$$

$$6660 = \frac{(8.314)\,(298)}{(88.6 \times 10^{-6})} \left[\frac{1}{\mathrm{MW}}\right] \left[\frac{78}{99}\right]$$

$$\mathrm{MW} = 3.3 \times 10^3 \text{ g/mol} \qquad \text{(number average)}$$

Thus osmotic pressure measurements can be used to measure the number-average molecular weight of polymers. In our example, the solution was assumed to be ideal; that is, the term $RT(\chi - \frac{1}{2})\phi_2$ in Eq. 7.68 was ignored. If nonideality is to be taken into account, a series of measurements is made of the solvent with decreasing concentrations of polymer. The ratio of the osmotic pressure to the concentration is then extrapolated to zero concentration, where the assumption of ideality is justified.

REFERENCES

1. Flory, P. J., *Principles of Polymer Chemistry,* Cornell University Press, Ithaca, NY, 1953.
2. Hiemenz, P. C., *Polymer Chemistry,* Dekker, New York, 1984.
3. Young, R. J., and Lovell, P. A., *Introduction to Polymers,* 2nd ed., Chapman & Hall, London, 1991.

PROBLEMS

7.1 The activity coefficient of zinc in liquid brass is given (in joules) by the following equation for temperatures 1000–1500 K:

$$RT \ln \gamma_{Zn} = -38{,}300\, x_{Cu}^2$$

where x_{Cu} is the mole fraction of copper.
Calculate the partial pressure of zinc P_{Zn} over a solution of 60 mol % copper and 40 mol % zinc at 1200 K.
The vapor pressure of pure zinc is 1.17 atm at 1200 K.

7.2 Using the equation given in Problem 7.1 for the activity coefficient of zinc in liquid brass, derive an equation for the activity coefficient of copper using the Gibbs–Duhem equation.

7.3 **(a)** At 900 K, is Fe_3C a stable compound relative to pure Fe and graphite?
(b) At 900 K, what is the thermodynamic activity of carbon in equilibrum with Fe and Fe_3C? Carbon as graphite is taken as the standard state.
(c) In the Fe-C phase diagram, the carbon content of α-iron in equilibrium with Fe_3C is 0.0113 wt %.
What is the solubility of graphite in α-iron at 900 K?

DATA

At 900 K:

$$3Fe + C_{(graphite)} = Fe_3C$$
$$\Delta G^\circ = +\,3463 \text{ J}$$

7.4 From vapor pressure measurements, the following values have been determined for the activity of mercury in liquid mercury–bismuth alloys at 593 K. Calculate the activity of bismuth in a 40 atom % Bi alloy at this temperature.

N_{Hg}	0.949	0.893	0.851	0.753	0.653	0.537	0.437	0.330	0.207	0.063
a_{Hg}	0.961	0.929	0.908	0.840	0.765	0.650	0.542	0.432	0.278	0.092

7.5 For a given binary system at constant T and P, the liquid molar volume of the solution (cm^3/mol) is given by:

$$\underline{V} = 100x_A + 80x_B + 2.5x_Ax_B$$

(a) Compute the partial molar volumes of A and B and plot them, together with the molar volume of the solution, as a function of the composition of the solution.
(b) Compute the volume of mixing as a function of composition.

7.6 For an ideal binary solution of A and B atoms, plot schematically the chemical potential of both species as a function of the composition of the solution. Indicate on the plot the molar Gibbs free energy of pure A and B.

7.7 At 473°C the system Pb–Sn exhibits regular solution behavior, and the activity coefficient of Pb is given by:

$$\log(\gamma_{Pb}) = -0.32(1 - x_{Pb})^2$$

Write the corresponding equation of the variation of γ_{Sn} with composition at 473°C.

7.8 $MgCl_2$ and MgF_2 are two salts that can form solutions. The Gibbs free energy of fusion (J/mol) for both compounds is given by:

For $MgCl_2$:

$$\Delta G = 43{,}905 - 43.644T \qquad \text{Melting point} = 987\ \text{K}$$

For MgF_2:

$$\Delta G = 58{,}702 - 38.217T \qquad \text{Melting point} = 1536\ \text{K}$$

The free energy of mixing (J/mol) for *liquid* mixtures of $MgCl_2$ and MgF_2 is given by:

$$\Delta G_{\text{Mix}} = 2RT\,(x_{MgCl_2} \ln x_{MgCl_2} + x_{MgF_2} \ln x_{MgF_2}) + x_{MgCl_2}\, x_{MgF_2}\,(-2556 + 25\,(x_{MgF_2} - x_{MgCl_2}))$$

Compute the maximum solubility of MgF_2 in liquid $MgCl_2$ at 900°C. $MgCl_2$ does not dissolve in solid MgF_2.

7.9 The thermodynamic properties of Al–Mg solutions at 1000 K are given in the accompanying table.

(a) If one mole of pure liquid aluminum and one mole of pure liquid magnesium, each at 1000 K, are mixed adiabatically, what will be the final temperature of the solution that is formed?

(b) What is the total change in entropy for the process?

DATA

Quantities of Mixing for Liquid Alloys at 1000 *K*

x_{Mg}	ΔG_M (cal/mol)	ΔH_M (cal/mol)	ΔS_M [cal/(mol·K)]	C_P [cal/(mol·K)]
0.1	−800	−300	0.5	7.1
0.2	−1250	−600	0.65	7.18
0.3	−1550	−750	0.8	7.26
0.4	−1700	−850	0.85	7.34
0.5	−1800	−900	0.9	7.42
0.6	−1700	−850	0.85	7.5
0.7	−1550	−750	0.8	7.58
0.8	−1250	−600	0.65	7.66
0.9	−800	−300	0.5	7.74

Chapter 8

Phase Rule

The thermodynamic stability of coexisting elements, compounds, and solutions is an important question for the users and producers of materials. Of course, many useful materials involve thermodynamically unstable structures whose physical stability is maintained by limiting the rate at which the structure approaches equilibrium. Even in those cases, however, it is useful to understand what equilibrium structure would result if equilibrium were to be achieved. In particular, it is important to know how many phases may exist at equilibrium, given a set of specificed physical constraints. The phase rule addresses that question.

8.1 PHASES

A *phase* is a portion of matter that is uniform throughout, not only in chemical composition but also in physical state. Take as an example a mixture of pure ice in pure water. It is common experience that the two may exist in equilibrium, We describe this phenomenon as a two-phase mixture, solid (ice) and liquid (water). We also know that we can have an equilibrium between ice and impure water, such as a salt–water solution. In this case we can also have two phases, ice is one, and the brine (salt–water solution) the other.

If we consider that the ice–brine mixture exists in equilibrium with air (containing water vapor), we would refer to the system as having three phases: ice, brine, and a gas phase containing water vapor and air (oxygen, nitrogen, and other gases in small quantities).

8.2 COMPONENTS

One of the factors that determines the number of phases that may exist at equilibrium is the number of components in a system. The number of components is best defined as the number of chemical species N less the number of independent relationships among them R.

$$C = N - R \tag{8.1}$$

In Chapter 7 we considered the thermodynamic properties of binary solutions, which implies solutions with two components. That is true, according to Eq. 8.1. The solutions are made of two chemical entities, ($N = 2$). In the cases considered, there were no special relations between the two ($R = 0$). Hence, $C = 2$.

Now suppose that the two chemical entities involved, A and B, react to form a compound, AB. In that case there are three chemical entities, A, B, and AB ($N = 3$). But, there is an equilibrium constant for the reaction $A + B = AB$, which relates the thermodynamic activities of the three at a given temperature. If two of the thermodynamic activities are known, the third is determined. That means $R = 1$. The number of components is then two, based on Eq. 8.1. This conclusion will be true even if there are multiple compounds in the A–B system because for each additional compound that increases N, there will be a corresponding equilibrium constant that increases R. Section 8.7 presents several examples of the use of Eq. 8.1 to determine the number of components when chemical reactions are involved.

8.3 SPECIFYING A SYSTEM

To specify a system completely with regard to the phase rule we are about to derive, we must have enough information about that system to define its *intensive* properties, such as the specific volumes of the phases, or the chemical potential μ_i, of the various constituents of the system. For the purposes of the phase rule we do not specify the *amount* of each phase, nor do we take into account surfaces or surface energies.

One way to specify a system is to establish the values of temperature, pressure, and the composition of each phase. Suppose the system of interest has C components. To determine the composition of a phase in the system, we must specify $C - 1$ pieces of information about that phase. For example, in a three-component system, we can specify the composition of phase I by establishing the mole fractions of components 1 and 2 in that phase. The mole fraction of component 3 will then, of course, be determined, because the sum of the mole fractions must be one. Thus the *composition* variables to be specified to have a thermodynamically determined system is the number of phases, P, multiplied by $C - 1$. In addition, two *overall* system variables, such as the temperature and total pressure of the system, must be specified (assuming that no electric or magnetic variables are involved). The total number of variables to be specified (VAR) to define the system is thus:

$$\text{VAR} = P(C - 1) + 2 \tag{8.2}$$

8.4 EQUILIBRIUM CONDITIONS

We know from Eq. 4.6b that for the system to be in equilibrium the chemical potential (partial molar Gibbs free energy) of a component must be the same throughout the system, assuming that there are no significant potential and kinetic energy differences in the system. For component i:

$$\mu_i^1 = \mu_i^2 = \mu_i^3 = \mu_i^4 = \cdots = \mu_i^P \tag{8.3}$$

where μ_i^P is the chemical potential of component 1 in phase P.

If the system contains P phases, Eq. 8.2 yields $P - 1$ *independent* equations $(\mu_i^1 = \mu_i^2, \mu_i^2 = \mu_i^3, \ldots, \mu_i^{P-1} = \mu_i^P)$. If the system contains C components, the total number of relationships, REL, at equilibrium is

$$\text{REL} = C(P - 1) \tag{8.4}$$

8.5 GIBBS PHASE RULE

The degrees of freedom available in the system (F) is the difference between the number of variables required to specify the system (VAR) and the number of relationships required by the equilibrium condition (REL):

$$F = \text{VAR} - \text{REL}$$

$$F = P(C - 1) + 2 - C(P - 1) \tag{8.5}$$

$$P + F = C + 2$$

The value of F gives us the number of system variables that we may freely vary, or arbitrarily fix. The relationship expressed by Eq. 8.5 is called the *Gibbs phase rule.*

8.6 ONE-COMPONENT SYSTEM

Consider a pressure–temperature diagram for a one-component system ($C = 1$), such as pure water (Figure 8.1). In the region containing only one phase, such as point A in the gas region ($P = 1$), the degrees of freedom are:

$$F = C - P + 2 = 2$$

Thus in the gas region, both the temperature and the pressure may be arbitrarily fixed.

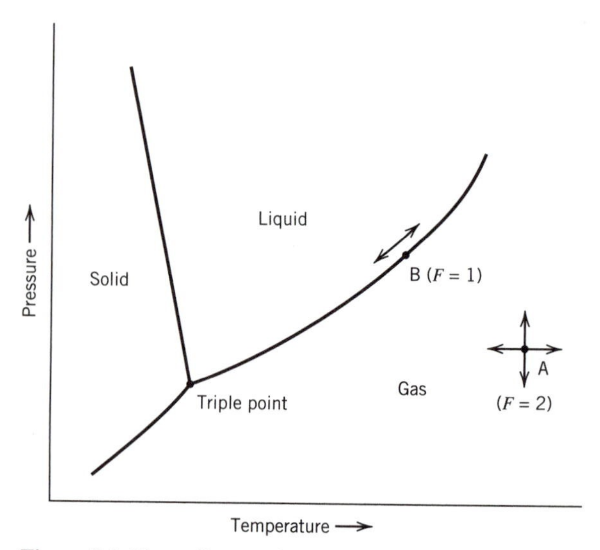

Figure 8.1 Phase diagram for water (one component). In the phase diagram for water, the solid–liquid equilibrium line slopes up and to the left from the triple point because the volume change upon solidification is positive. In most materials, the volume change upon solidification is negative, and the solid–liquid equilibrium line slopes up and to the right.

In two-phase regions, such as Point *B* on the liquid–vapor line, there is only one degree of freedom.

$$F = C - P + 2 = 1 - 2 + 2 = 1$$

On this line, or any other of the two-phase lines, we may arbitrarily fix *either* pressure or temperature. Once that has been done, the other variable is fixed. Thus there is only one degree of freedom.

At a triple point where three phases—solid, liquid, and gas, for example—may exist in equilibrium, there are no degrees of freedom. That is, there is only one combination of temperature and pressure at which all three phases may coexist in a single-component system.

8.7 EXAMPLES

The usefulness of the phase rule in systems with chemical reactions can be demonstrated by running through several examples.

Example 1: Consider the decomposition of calcium carbonate into calcium oxide and carbon dioxide:

$$CaCO_3(s) \rightarrow CaO(s) + CO_2(g)$$

The number of components in this system is determined by counting the chemical species present and subtracting the number of independent relationships available. In this case, there are three chemical species, CaO(s), $CaCO_3$(s), and CO_2(g). There is one relationship among them, the equilibrium constant for the decomposition of calcium carbonate:

$$K_a = \frac{a_{CO_2} a_{CaO}}{a_{CaCO_3}}$$

Thus, the number of "components" $(N - R)$ is two. There are three phases present, calcium carbonate, calcium oxide, and carbon dioxide. Thus the degrees of freedom in this system is limited to one.

$$F = C - P + 2 = 2 - 3 + 2 = 1$$

If the temperature is fixed, then the total pressure of the system (the pressure of carbon dioxide) is also fixed. This is often called the decomposition pressure of calcium carbonate.

Example 2: Suppose that an inert material, such as nitrogen, is added to the system considered in Example 1. The number of chemical species would be increased to four: CaO, $CaCO_3$, CO_2, and N_2. the number of phases is still three: calcium oxide, calcium carbonate, and the gas phase. Thus the degrees of freedom is increased to two. The temperature and the total pressure can each be varied independently. Note that the pressure of carbon dioxide is fixed at a specific temperature, but the *total* pressure of the system can be varied by varying the nitrogen pressure.

Example 3: Consider the equilibrium between solid carbon, carbon monoxide, and carbon dioxide.

$$CO_2(g) + C(s) \rightarrow 2CO(g)$$

There are two phases present, solid carbon and the gas. The number of components present is two because there are three chemical species and one relationship among them (the equilibrium constant for the equation above). Thus the number of degrees of freedom is two. If the temperature and total pressure are fixed, then the system is specified.

Example 4: Ellingham diagrams (Figure 5.7) provide another interesting illustration of the phase rule. If one considers the reaction of the metal, M, with oxygen,

$$M + O_2 \rightarrow MO_2$$

then on the line representing the chemical reaction above in an Ellingham diagram there will be three phases: M, MO_2, and oxygen. The number of components is two, because there are three chemical species and one relationship among them, the equilibrium constant for the reaction. Thus the number of degrees of freedom is one. Specifying the temperature specifies the oxygen pressure, as we have seen.

Example 5: A more complex example involves the decomposition of gaseous ammonia (NH_3):

$$2NH_3 \rightarrow N_2 + 3H_2$$

In this system, there is one phase, the gas. There are two components because there are three chemical species and one relationship among them, the equilibrium constant ($C = N - R = 2$). Thus the number of degrees of freedom is three.

$$F = C - P + 2 = 2 - 1 + 2 = 3$$

With the temperature specified (which removes one degree of freedom), the partial pressures of two of the components in the gas must be specified in order to calculate the third. This can be appreciated by considering the expression for the equilibrium constant:

$$K_a = \frac{a_{N_2} a_{H_2}^3}{a_{NH_3}^2} = \frac{P_{N_2} P_{H_2}^3}{P_{NH_3}^2} \qquad \textbf{(8.6)}$$

Example 6: Suppose in example 5 we had started with pure ammonia. In that case, there is only one component. There are still three chemical species present ($N = 3$), but there are now two relations among them. In addition to the eqiulibrium constant (Eq. 8.6), we also know that the ratio of nitrogen to hydrogen is fixed at 3 because all the nitrogen and hydrogen molecules, in this case, are formed from the decomposition of ammonia. Thus the number of components is one, and the number of degrees of freedom is only two. We can specify the total system by specifying the temperature and the total pressure.

8.8 PHASE RULE FOR CONDENSED SYSTEMS

The thermodynamic properties of condensed systems are relatively insensitive to changes in pressure, as illustrated in Chapter 3. Thus, when dealing with relationships in condensed systems, such as solids, it is common to assume that the pressure of the system is one atmosphere. The pressure variable may be ignored for small pressure variations because the chemical potentials of condensed systems change only slightly with pressure.[1] This reduces by one the number of variables that must be determined to specify a system. The phase rule then becomes:

$$P + F = C + 1 \qquad \text{for condensed systems}$$

---

[1]This is not true, of course, for very large pressure changes considered in some geological studies.

PROBLEMS

8.1 Zinc sulfide (ZnS) is reacted in pure oxygen to form zinc sulfate ($ZnSO_4$).

(a) Write the chemical reaction representing the process.

(b) How many solid phases may exist in equilibrium if pressure and temperature are arbitrarily fixed?

(c) If the temperature is fixed, will the pressure be determined if ZnS and $ZnSO_4$ exist in equilibrium?

8.2 An Fe-Mn solid solution containing 0.001 mole fraction Mn is in equilibrium with an FeO-MnO solid solution and a gaseous atmosphere containing oxygen at 1000 K. How many degrees of freedom does the equilibrium have? What is the composition of the equilibrium oxide solution, and what is the oxygen pressure in the gas phase? Assume that both solid solutions are ideal.

DATA

For Fe:

$$\mathrm{Fe(s)} + \tfrac{1}{2}\,\mathrm{O_2(g)} = \mathrm{FeO(s)} \quad \Delta G^\circ = -259{,}600 + 62.55T$$

For Mn:

$$\mathrm{Mn(s)} + \tfrac{1}{2}\,\mathrm{O_2(g)} = \mathrm{MnO(s)} \quad \Delta G^\circ = -384{,}700 + 72.8T$$

where ΔG° is in joules.

Chapter 9

Phase Diagrams

Chapter 8 contained an example of a one-component phase diagram (Figure 8.1) in which regions of phase stability were plotted for water on a pressure–temperature graph. In a system containing gases, pressure changes are important. In *condensed* systems, however, modest variations in pressure do not appreciably alter phase relationships. If pressure variations are not to be considered, the phase rule is reduced to

$$P + F = C + 1$$

In such condensed phase systems, it is possible to plot phase stability regions for two-component (binary) systems in two dimensions. In two-component systems, $P + F$ is three. At least one phase must be present; hence the maximum number of degrees of freedom that must be accommodated is two. The situation for three-component (ternary) systems is more complex, as demonstrated later in this chapter.

In binary diagrams, composition is generally plotted on the abscissa (horizontal axis) and temperature is plotted on the ordinate (vertical axis). To approach this subject, let us consider the regions of phase stability as a solute (B) is added to a solvent (A).

9.1 FREEZING POINT DEPRESSION

Figure 9.1 represents one line in a phase diagram of a condensed system, the line corresponding to pure A. At the temperature T_m, the two phases, solid and liquid, are in equilibrium.[1] Because we will be interested in the temperature region below T_m, let us take pure solid A as the standard state. For the reaction (or phase transformation) from solid A to liquid A, we can write the Gibbs free energy change as follows:

$$A_s = A_l$$

$$\Delta \underline{G} = RT \ln \frac{f_l}{f_s} = RT \ln \left(\frac{a_{l,pure}}{a_{s,pure}} \right) \tag{9.1}$$

$$\Delta \underline{G} = RT \ln a_{l,pure} \qquad \text{if} \quad a_{s,pure} = 1$$

At the melting temperature T_m, the two phases are in equilibrium; hence the value for the Gibbs free energy change in the reaction is zero. The activity of the liquid is therefore 1, the same as the solid. At temperatures lower than T_m, the value of Gibbs free energy change for the melting of pure A can be written as

$$\Delta \underline{G}_{melting} = \Delta \underline{H}_{melting} - T\, \Delta \underline{S}_{melting} = L - T\, \Delta \underline{S}_{melting} \tag{9.2}$$

The term L (latent heat of fusion) is introduced for the enthalpy of melting to avoid confusion with the notation for mixing. For simplicity, assume that there is no difference in heat capacity between liquid and solid ($\Delta C_P = 0$). In this case, the enthalpy change and entropy change of melting are each independent of temperature. Noting that at the melting temperature, $\Delta G_{melting} = 0$:

$$L = T_m \Delta \underline{S}_{melting} \qquad \text{or} \qquad \Delta \underline{S}_{melting} = \frac{L}{T_m}$$

we can then rewrite Eq. 9.2 as

$$\Delta \underline{G}_{melting} = L \left(1 - \frac{T}{T_m} \right) = \frac{L(T_m - T)}{T_m}$$

$$\Delta \underline{G}_{melting} = RT \ln \frac{a_{l,pure}}{a_{s,pure}} = \frac{L(T_m - T)}{T_m} \tag{9.3a}$$

[1] At the melting temperature T_m, solid A is in equilibrium with liquid A. If both are in equilibrium with A in the vapor at T_m, then the vapor pressures of the liquid and solid are equal. This is because solid and liquid A are in equilibrium with each other. Temperature T_m is the triple-point temperature in Figure 8.1.

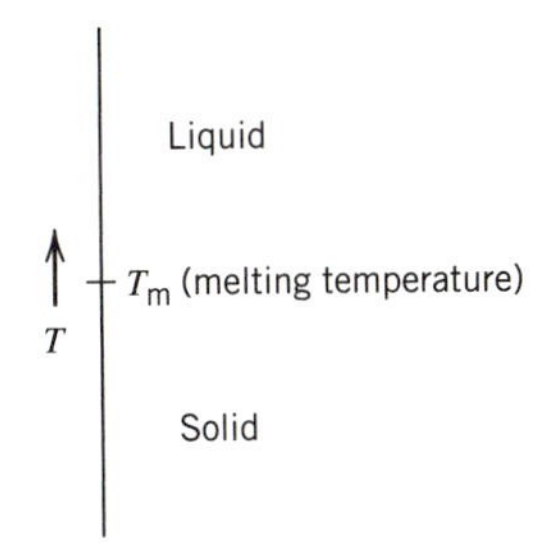

Figure 9.1 Phase diagram of a single-component system.

Based on this equation, it is apparent that the activity of pure liquid A is greater than one at temperatures below T_m, with the solid being considered the standard state. Note that the standard state is defined for each temperature.

Now let us deal with the addition of material B to A. At some temperature T' (below T_m), the activity of A in an ideal A–B solution as a function of composition is shown in Figure 9.2. Consider the case in which A and B are immiscible in the solid state[2], but form ideal solutions in the liquid state. Liquid of composition $x_{A,l}$ (where the activity $a_{A,l} = 1$) is in equilibrium with pure solid A at temperature T'. Consider now the dissolving of pure, liquid A in the liquid solution. For $A_{l,pure} = A_{l,solution}$:

$$\Delta \underline{G} = RT \ln \frac{a_{l,solution}}{a_{l,pure}} \qquad \textbf{(9.3b)}$$

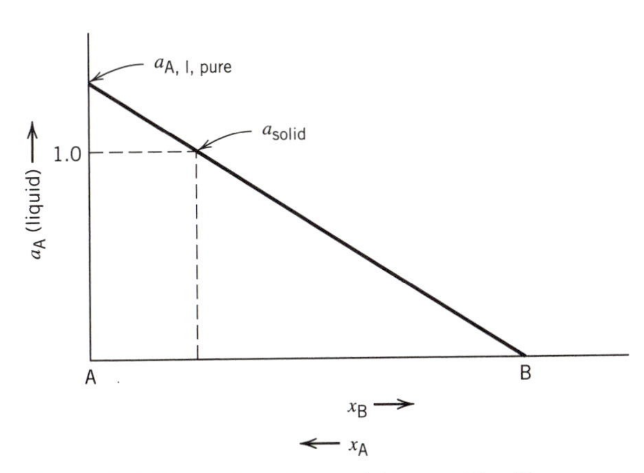

Figure 9.2 Plot of the activity of A_{liquid} at $T < T_{m,A}$ versus composition.

[2]Total immiscibility among elements is impossible in a thermodynamic sense. For our purposes we will assume that immiscibility means solubility so low that it is of no practical consequence in calculations or in graphical representations (see Chapter 7).

The dissolution of pure, solid A in the liquid solution is the sum of the two processes above: the melting of pure A, and the dissolution of pure liquid A in the liquid solution. For:

$$A_{\text{s,pure}} = A_{\text{l,solution}}$$

the Gibbs free energy change is the sum of Eqs. 9.3a and 9.3b. Furthermore, if the liquid solution is in equilibrium with the pure solid, then the $\Delta G = 0$ (Figure 9.2).

$$\Delta \underline{G} = \frac{L(T_m - T)}{T_m} + RT \ln \frac{a_{\text{l,solution}}}{a_{\text{l,pure}}} = 0$$

$$\frac{L(T_m - T)}{T_m} = -RT \ln x_{\text{A,l,solution}} \qquad \text{if the solution is ideal}$$

If T is close to T_m, then:

$$\ln x_{\text{A,l,solution}} = -\frac{L(T_m - T)}{R\,(T_m)^2}$$

Noting that x_A is equal to $1 - x_B$, we have

$$\ln(1 - x_B) = -\frac{L(T_m - T)}{R\,(T_m)^2}$$

where x_B is the mole fraction of B in the solution.

Recall that when a variable z is small, we can write

$$\ln(1 - z) = -z$$

In the region of small x_B, therefore, the relationship between x_B, the composition of the liquid, and the melting point depression, is

$$x_B = \frac{L\,\Delta T}{R(T_m)^2} \tag{9.4}$$

where $\Delta T = T_m - T$, the melting point depression.

This expression can be plotted on a phase diagram in which temperature is the ordinate and composition is the abscissa. A region of such a diagram is shown as Figure 9.3. In the portion of the diagram labeled liquid, the equilibrium phase is a liquid A–B solution. In the two phase region labeled ''L + S'' (liquid plus solid), pure solid A is in equilibrium with a liquid solution. At the temperature T_1, pure solid A is in equilibrium with a liquid solution of composition $x_{B,1}$.

As an example, let us calculate the lowering of the melting point of silver caused

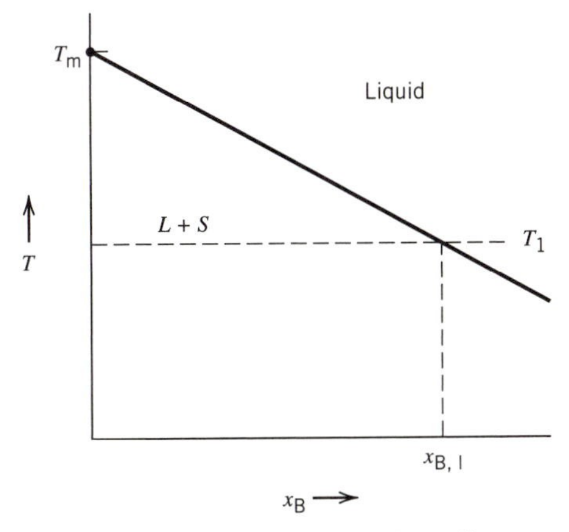

Figure 9.3 Portion of an A–B phase diagram.

by the addition of one mole percent of lead. The conditions assumed in the derivation of Eq. 9.4 are followed by the silver–lead system, although there is a small solubility of lead in solid silver.

For silver: $T_m = 1234$ K and $L = 11{,}300$ J/mol

$$\Delta T = \frac{R(T_m)^2 x_B}{L} = \frac{(8.314)(1234)^2(0.01)}{11{,}300}$$

$$\Delta T = 11.2 \text{ K}$$

Actually, the measured melting point depression is about 10 K for an addition of one mole percent lead. Considering the slight solubility of lead in silver, the agreement between the calculated and measured values is not bad.

Considering the phase rule in the liquid + solid region (condensed phases)

$$F = C - P + 1 = 2 - 2 + 1 = 1 \tag{9.5}$$

Because there are two components, A and B, and two phases, liquid and solid, there is only one degree of freedom. Once the temperature has been specified, the composition of the phases at equilibrium is specified. It is important to note that the relative *amounts* of the phases present (liquid and solid) are *not* determined by the phase rule. Only the *composition* of the phases is determined. We show next that if the overall composition of the A–B combination is given, the quantities of the various phases can be calculated using the lever rule.

9.2 THE LEVER RULE

In the two-phase region illustrated by Figure 9.4, x_B represents the overall composition of a system. At temperature T_1, the phase diagram tells us that the equilibrium

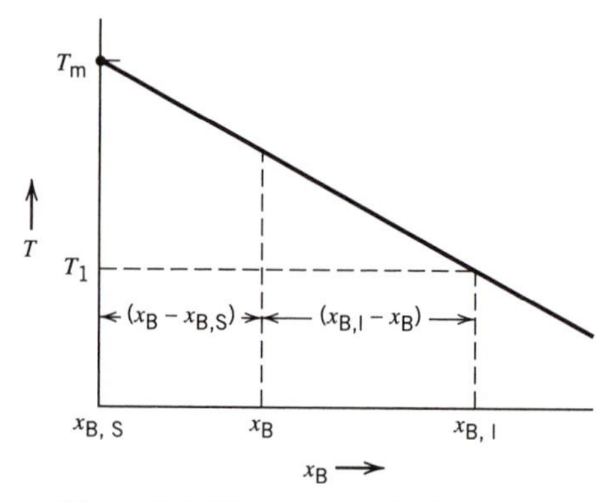

Figure 9.4 Illustration of the lever rule.

liquid composition is $x_{B,l}$. The equilibrium solid composition is $x_{B,s}$. Given the overall composition and the composition of the two phases, we can calculate the relative quantities or fractions of liquid and solid using a mass balance.

If F_l = fraction liquid and F_s = fraction solid, then $F_l + F_s = 1$. Based on a mass balance for B:

$$x_B = F_l\, x_{B,l} + F_s\, x_{B,s}$$

$$(F_l + F_s)x_B = F_l\, x_{B,l} + F_s\, x_{B,s}$$

$$F_l(x_{B,l} - x_B) = F_s(x_B - x_{B,s})$$

The lever rule can be expressed as a ratio of the fraction of liquid to the fraction of solid,

$$\frac{F_l}{F_s} = \frac{x_B - x_{B,s}}{x_{B,l} - x_B} \tag{9.6}$$

or merely as the fraction liquid as follows:

$$F_l = \frac{x_B - x_{B,s}}{x_{B,l} - x_{B,s}} \tag{9.7}$$

As an example, let us calculate the fraction of liquid at 500 K in a lead–tin binary alloy containing 50 mol % tin. At 500 K, the equilibrium composition of the liquid is 59 mol % tin, and the composition of the solid is 24 mol % tin.

$$F_l = \frac{0.50 - 0.24}{0.59 - 0.24} = 0.74$$

At equilibrium, the alloy will be 74% liquid at 500 K, on a *mass* basis. To find *volume* fractions, the molar volumes of the two phases must be known.

9.3 SIMPLE EUTECTIC DIAGRAM

Consider a system in which materials A and B are immiscible in the solid state, but completely miscible in the liquid state. As shown in Section 9.1, the addition of B to A lowers the melting point of A. The reverse is also true. The addition of A to B lowers the melting point of B. This relationship is illustrated in Figure 9.5. Although the linear relationship between composition and temperature derived as Eq. 9.4 may no longer exist, the melting point depressions will continue. When the melting point depression lines intersect, the material will solidify totally into solid A and solid B (Figure 9.5). The temperature at which the two curves intersect, called the *eutectic temperature,* is the lowest temperature at which a liquid solution of A and B may exist at equilibrium with solid A and solid B. The composition at which they intersect is the eutectic composition. At the eutectic point, the phase change may be represented[3] by:

$$\text{liquid} = \text{solid}_A + \text{solid}_B$$

According to the phase rule, there are two degrees of freedom in the liquid region; that is, temperature and composition may be arbitrarily fixed. In the region labeled

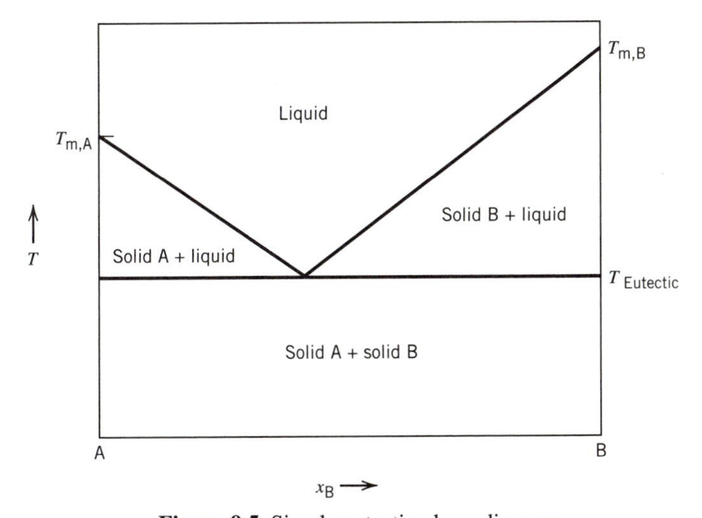

Figure 9.5 Simple eutectic phase diagram.

[3]This type of reaction may also take place entirely in the solid state, in which case it is called a *eutectoid reaction.* An example is shown later, in the reaction at 738°C in the iron–carbon system (Figure 9.17) in which solid γ-iron is converted into α-iron and graphite at equilibrium.

A + liquid, there is only one degree of freedom. Once the temperature has been fixed, the composition of each phase is fixed. The same is true of the region *B + liquid.*

At the eutectic temperature there are no degrees of freedom because there are two components (A and B) and three phases (solid A, solid B, and liquid). Thus $F = 0$. If all three phases (liquid, solid A, and solid B) are present, one must, at equilibrium, be at the eutectic temperature and the liquid will have the eutectic composition.

9.4 COOLING CURVES

If a pure material—pure A, for example—is cooled from a temperature T_m (its melting temperature) to below T_m by removing thermal energy at a constant rate, the temperature of the material as a function of time follows a pattern illustrated in Figure 9.6, assuming that equilibrium is maintained at all times. When material A is above T_m, in the liquid state, the removal of thermal energy lowers its temperature. When the melting point is reached, the removal of thermal energy results in solidification. During solidification, liquid A and solid A are in equilibrium, and the temperature of the system does not change. This condition is called a thermal arrest in the cooling curve. Once all of material A has solidified, the temperature decrease resumes.

The same type of cooling curve, with a thermal arrest at the melting temperature, is observed if one cools a liquid of eutectic composition. As temperature drops, liquid will exist until the eutectic temperature is reached. At that temperature, all of the liquid solidifies into solid A and solid B. When all is solid, the cooling resumes as energy is removed from the system.[4]

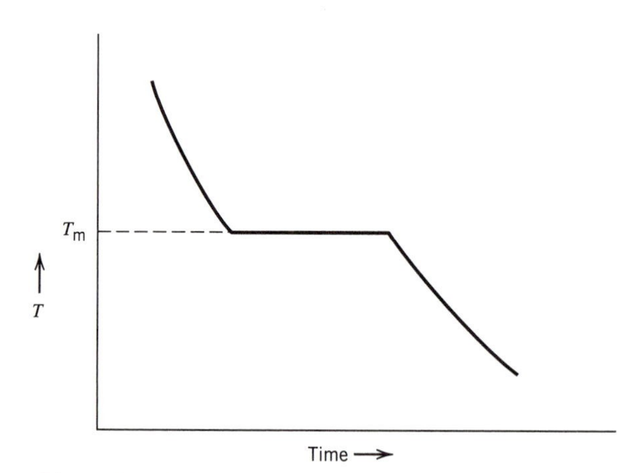

Figure 9.6 Cooling curve for a pure material (or eutectic).

[4]Solidifcation (or melting) at a single temperature is often called congruent solidification (or melting).

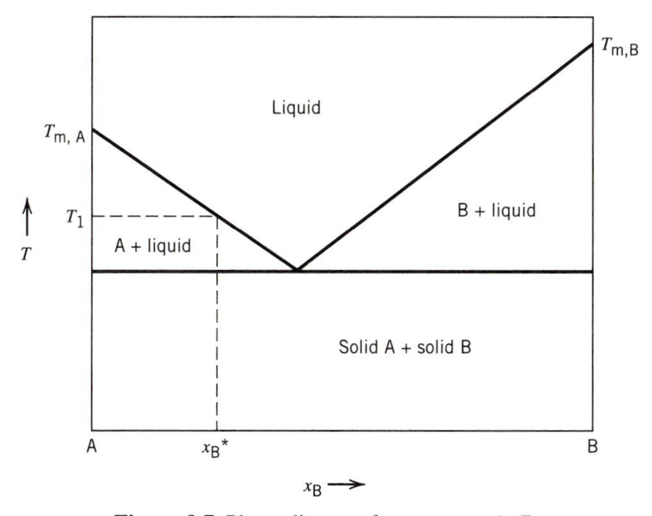

Figure 9.7 Phase diagram for a system A–B.

At compositions other than the eutectic composition, such as composition x_B^* in Figure 9.7, the cooling rate of the material changes when temperature T_1 is reached. At temperatures below T_1 some pure, solid A is formed upon cooling, and the rate of temperature change is diminished (i.e., there is a change in the slope of the cooling curve) because to solidify A, energy must be removed from the system. After the material has reached the eutectic temperature, all the remaining liquid solidifies at a constant temperature, causing a thermal arrest at T_{eu} (Figure 9.8).

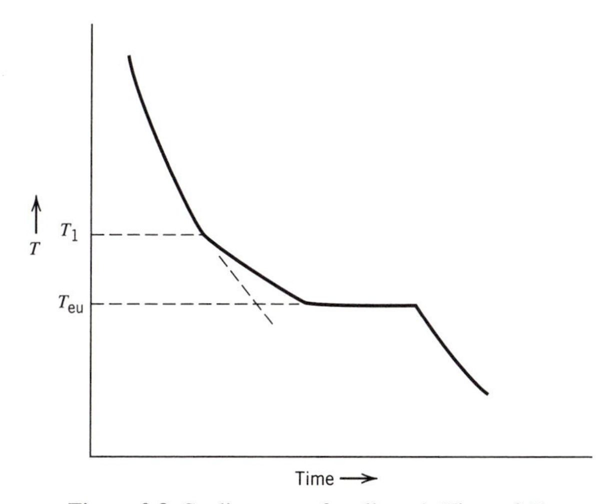

Figure 9.8 Cooling curve for alloy x_B^* (Figure 9.7).

9.5 COMPLETE MISCIBILITY

Consider now a two-component system A–B in which A and B are completely miscible in both solid and liquid states, and form ideal solutions in both. Based on our knowledge of the thermodynamics of ideal solutions (Chapter 7: Sections 7.4 and 7.5) we should be able to predict the equilibrium phase diagram of this system, given information about the melting temperatures and the enthalpies of melting of the two components.

For each pure component, the Gibbs free energy change of melting of each component, the difference in chemical potential between pure liquid and pure solid, is:

$$\Delta \underline{G}_{\text{melting}} = \frac{L(T_m - T)}{T_m} = \mu_l^p - \mu_s^p$$

$$\mu_s^p = \mu_l^p - \frac{L(T_m - T)}{T_m} \tag{9.3a}$$

where T_m = melting temperature
L = enthalpy of melting (latent heat of fusion)
μ_l^p and μ_s^p = chemical potentials of pure liquid and pure solid, respectively

The chemical potential of material A in the liquid solution $\mu_{A,l}$ is

$$\mu_{A,l} = \mu_{A,l}^p + RT \ln a_{A,l} = \mu_{A,l}^p + RT \ln x_{A,l} \qquad \text{(ideal solution)}$$

The molar Gibbs free energy of the liquid solution, is

$$\underline{G}_l = x_A\mu_A + x_B\mu_B = x_A\mu_{A,l}^p + x_B\mu_{B,l}^p + RT(x_{A,l} \ln x_{A,l} + x_{B,l} \ln x_{B,l}) \tag{9.8}$$

The molar Gibbs free energy of the solid solution has a similar form.

If we now elect to choose the pure liquids A and B as the standard states, the terms $\mu_{A,l}^p$ and $\mu_{B,l}^p$ are each zero. Combining the equations above, the molar Gibbs free energies for the liquid and solid solutions, $\underline{G}_l$ and $\underline{G}_s$, are

$$\underline{G}_l = RT(x_{A,l} \ln x_{A,l} + x_{B,l} \ln x_{B,l}) \tag{9.9}$$

$$\underline{G}_s = RT(x_{A,s} \ln x_{A,s} + x_{B,s} \ln x_{B,s}) - x_{A,s}\left[\frac{L}{T_{m,A}}(T_{m,A} - T)\right] - x_{B,s}\left[\frac{L}{T_{m,B}}(T_{m,B} - T)\right] \tag{9.10}$$

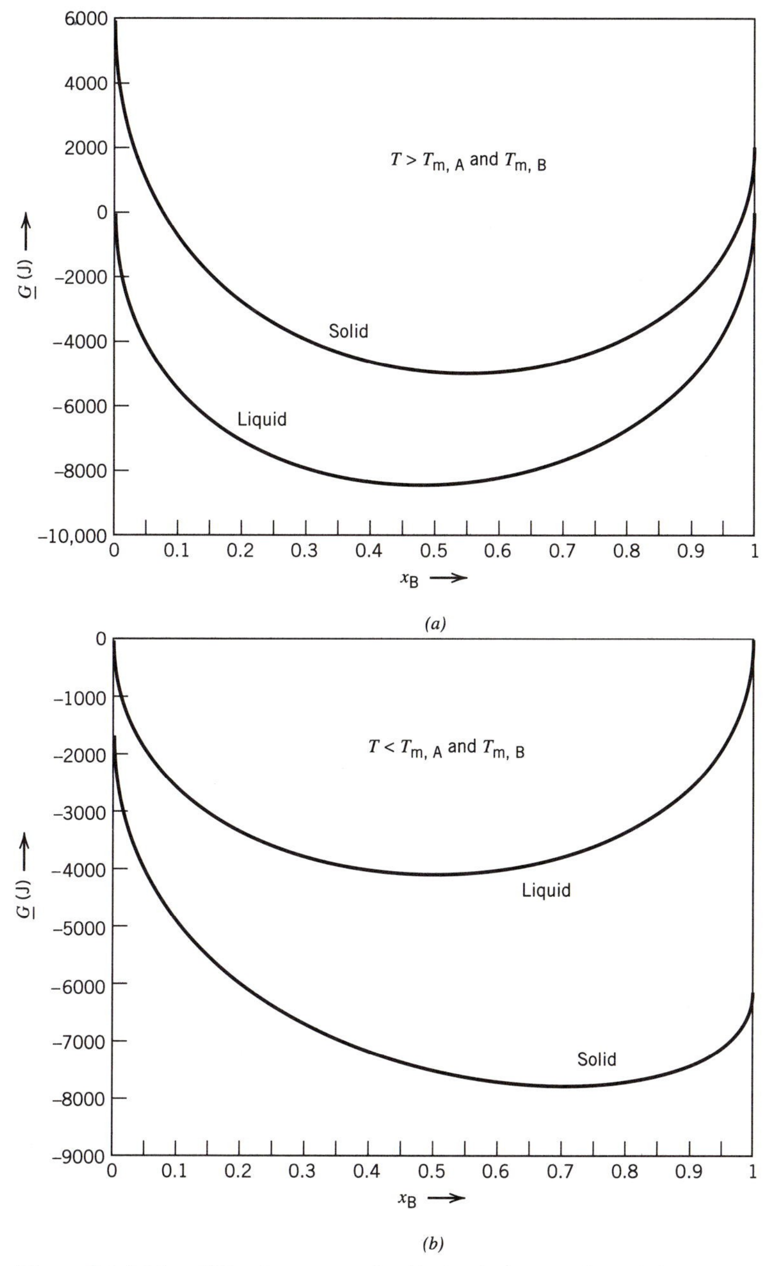

Figure 9.9 Molar Gibbs free energy for ideal solutions (solid and liquid): (*a*) at temperatures above the melting points of A and B, (*b*) at temperatures below the melting points of A and B, (*c*) at the melting point of A, and (*d*) at temperatures between the melting points of A and B.

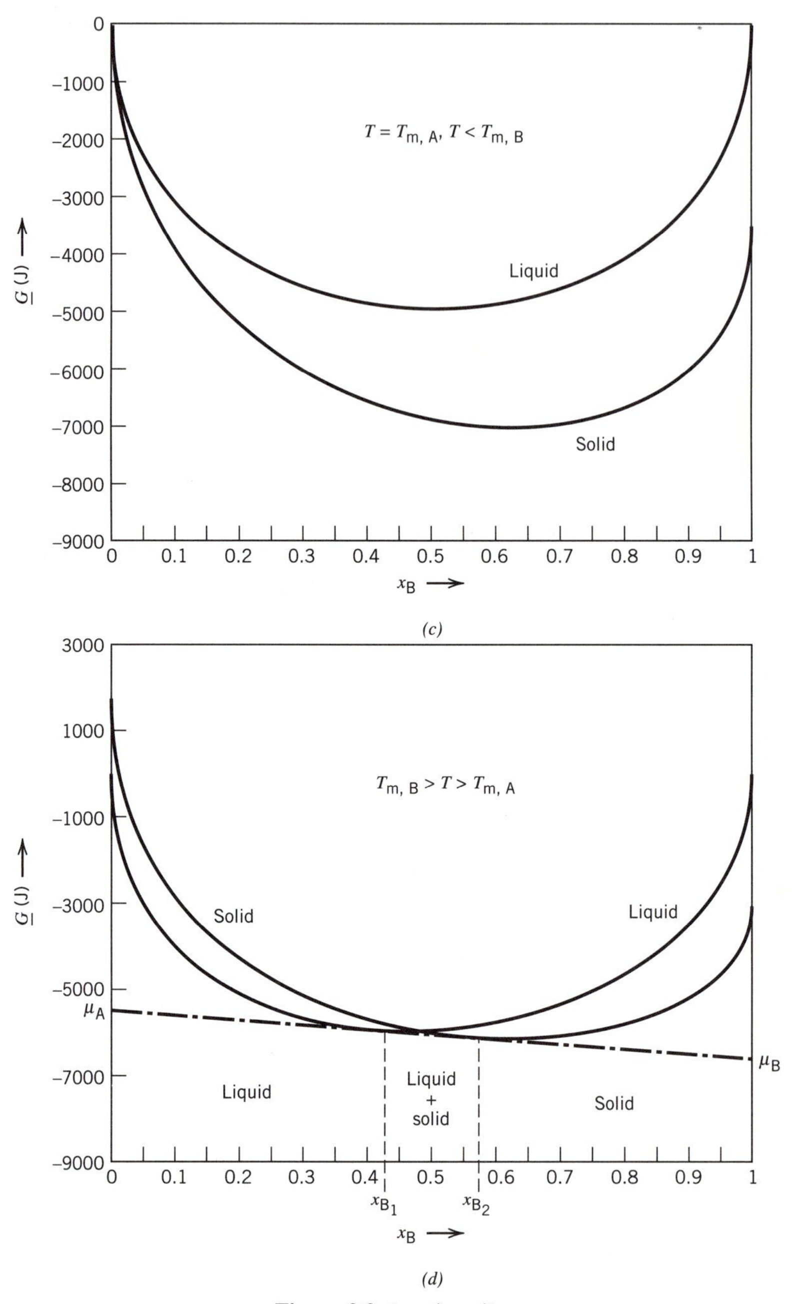

Figure 9.9 (continued)

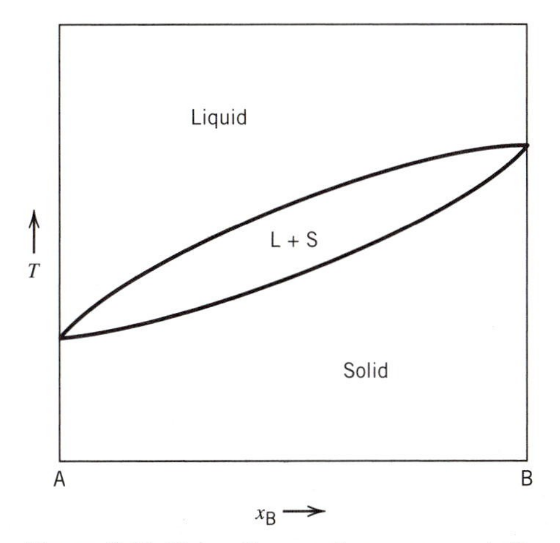

Figure 9.10 Phase diagram for a system A–B: ideal solutions, solid and liquid.

The values of $\underline{G}_l$ and $\underline{G}_s$ are plotted in Figures 9.9*a* through 9.9*d* for various temperatures.[5] The molar Gibbs free energies of the solid and liquid solutions of A and B, at temperatures above the melting points of A and B, are shown in Figure 9.9*a*. The curve for the molar Gibbs free energy of the liquid lies below the curve for the solid across the entire composition range. Hence, liquid exists at equilibrium

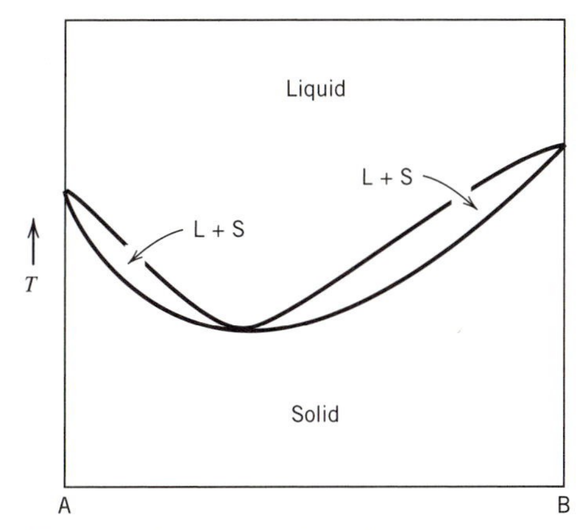

Figure 9.11 Possible phase diagram for a system A–B with complete miscibility of the components in liquid and solid but nonideal behavior.

[5]Values assumed for the melting temperatures of A and B are $T_{m,A}$ = 900 K and $T_{m,B}$ = 1300 K. For both A and B, the value of L/T_m was assumed to be 9 J/K (Trouton's rule).

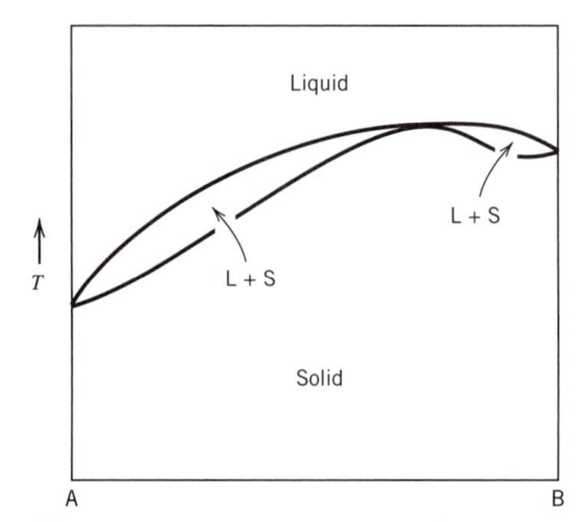

Figure 9.12 Another possible phase diagram for the system A–B described in Figure 9.11.

across the phase diagram, as expected. The reverse is true for temperatures below the melting temperatures of both A and B (Figure 9.9*b*).

At the melting point of A, the curves for liquid and solid intersect at pure A because pure liquid A and pure solid A are in equilibrium at that temperature (Figure 9.9*c*). At temperatures between the melting points of A and B, the Gibbs free energy curves for liquid and solid intersect at a point away from the vertical axes (Figure 9.9*d*).

If, at a given temperature, the molar Gibbs free energy curves for two phases intersect, there is a range of compositions over which the two may exist at equilibrium, $x_{B,1}$ and x_{B2}. The extent of this region can be determined by drawing the common tangent line between the two curves as illustrated in Figure 9.9*d*. The intercepts of the tangent to a Gibbs free energy of mixing curve on the vertical axes $x_A = 0$ and $x_B = 0$ are the chemical potentials of both A and B. The intercepts of the *common* tangent indicate that the chemical potentials of A in both phases are equal; hence the two phases are in equilibrium. The result of a calculation based on this principle is a phase diagram of the type shown in Figure 9.10, with a lenslike stability region for the coexistence of liquid and solid.

If the solid and liquid solutions of A and B are nonideal, then the lenslike liquid–solid phase region can have other forms, such as those in Figures 9.11 and 9.12.

9.6 IMMISCIBILITY

We have discussed examples of complete immiscibility (Section 9.3) and complete miscibility (Section 9.5) in the solid state. There can, of course, be partial miscibility of two components in the solid state. This partial miscibility is usually temperature dependent, with higher miscibility at higher temperatures, unless the individual components undergo a phase change. To examine this phenomenon, let us consider a

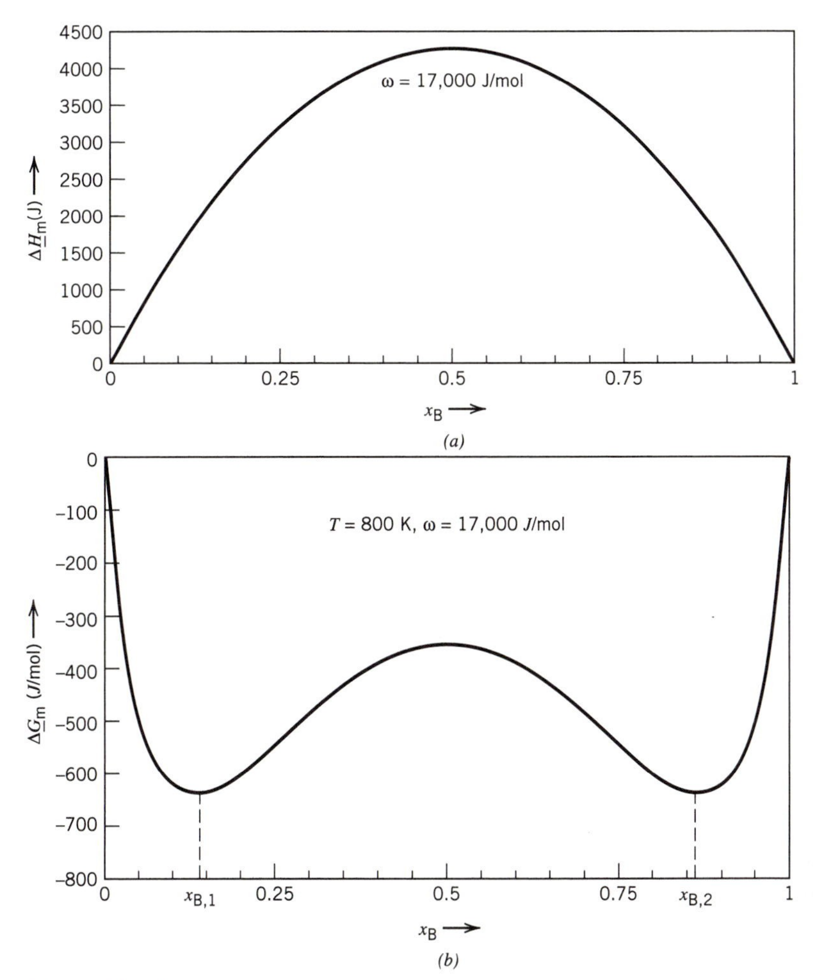

Figure 9.13 Plots of properties of mixing versus composition for regular solutions: (*a*) molar enthalpy and (*b*) molar Gibbs free energy.

regular solid solution of two components. The equation describing the Gibbs free energy of mixing for a regular solution (Eq. 7.46) is:

$$\underline{G}_{\mathrm{M}} = RT(x_{\mathrm{A}} \ln x_{\mathrm{A}} + x_{\mathrm{B}} \ln x_{\mathrm{B}}) + \omega x_{\mathrm{A}} x_{\mathrm{B}}$$

The first term in the equation, $RT(x_{\mathrm{A}} \ln x_{\mathrm{A}} + x_{\mathrm{B}} \ln x_{\mathrm{B}})$, is the one that applies to the Gibbs free energy of mixing for an ideal solution. The other term $\omega x_{\mathrm{A}} x_{\mathrm{B}}$, represents the nonideality of the mixture. If the ω term is positive, then heat of mixing is positive (Figure 9.13*a*), and the molar Gibbs free energy of the solutions of A and

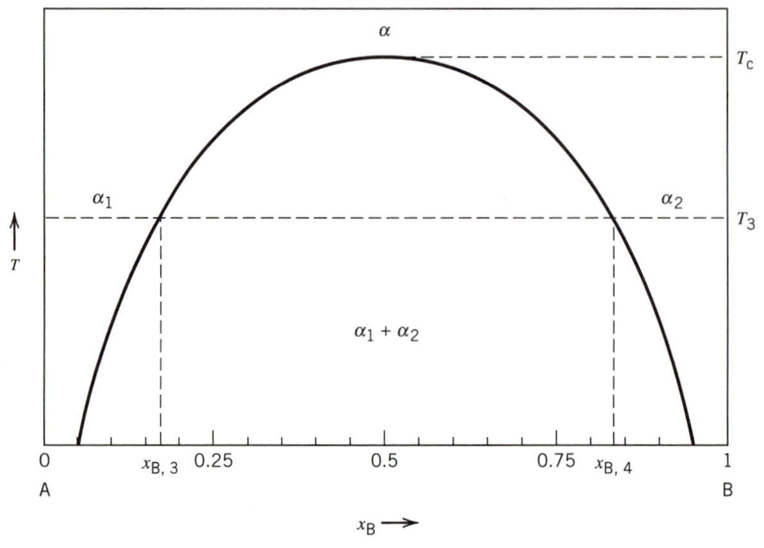

Figure 9.14 The miscibility gap in solid A–B solutions.

B can have minima as illustrated in Figure 9.13*b*. Mixtures of A and B, with the overall composition falling between x_{B1} and x_{B2} can minimize their molar Gibbs free energies by forming two solutions of composition x_{B1} and x_{B2} (i.e., becoming partially immiscible). The locations of x_{B1} and x_{B2} are determined using the common tangent method described in the preceding section. In the case of regular solutions, whose Gibbs free energy of mixing is symmetrical around the midpoint composition, we can locate the minima in the Gibbs free energy function by finding the compositions at which the partial derivative with respect to x_B is equal to zero (recall $x_A = 1 - x_B$):

$$\left(\frac{\partial G_M}{\partial x_B}\right)_T = RT \ln\left(\frac{x_B}{1 - x_B}\right) + \omega(1 - 2x_B) = 0 \qquad \textbf{(9.11)}$$

The solution of this equation, when plotted against temperature, yields a line (Figure 9.14). In this phase diagram, the single-phase region above T_c indicates that the two materials, A and B, are completely miscible above that temperature. At temperatures below T_c, the solution separates into two separate phases, α_1 and α_2. At points under the miscibility gap, the phase compositions vary with temperature. For example, at T_3 the phase compositions $x_{B,3}$ and $x_{B,4}$ are in equilibrium.

9.7 SPINODAL POINTS

Another feature of phase behavior is related to the points of inflection in the curves of Gibbs free energy of mixing with respect to composition: that is, when their second derivative is equal to zero (Eq. 9.12). These inflection points, called spinodal

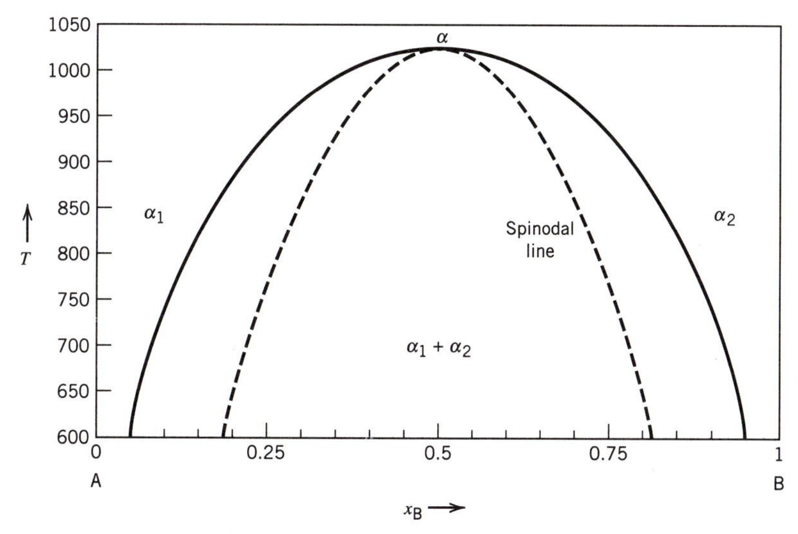

Figure 9.15 The miscibility gap, showing spinodal line.

points, have a special significance in the study of phase transformations. The locus of spinodal points can be indicated as a dashed line in phase diagrams as in Figure 9.15.

$$\left(\frac{\partial^2 \underline{G_M}}{\partial x_B^2}\right) = RT\left(\frac{1}{x_A} + \frac{1}{x_B}\right) - 2\omega \qquad \text{or} \qquad x_A x_B = \frac{RT}{2\omega} \tag{9.12}$$

To appreciate the importance of the spinodal curve consider the region to the right of the spinodal point in Figure 9.16, where the Gibbs free energy of mixing curve is concave downward. In this region the solution may begin the process of

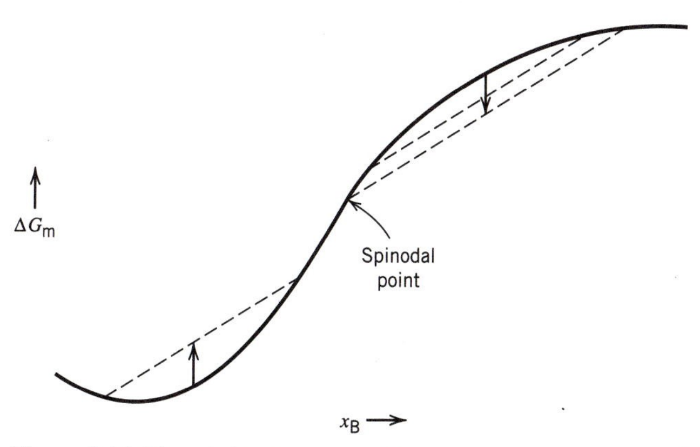

Figure 9.16 The relation between Gibbs free energy of mixing and composition on either side of spinodal point.

decomposition into the equilibrium phases by incremental changes in composition without increasing the total Gibbs free energy of the system. A different situation exists in the region to the left of the spinodal point (inflection point). Here, as the material separates into two phases, the Gibbs free energy of the system must increase before it can finally decrease (Figure 9.16). This difference in path for the Gibbs free energy during decomposition results in a difference in phase transformation behavior. To the left of the spinodal point the transformation is discontinuous. To the right it is not.

The top of the miscibility gap (critical mixing) is the point at which, in Eq. 9.12, $x_A = x_B$, and $T_c = \omega/2R$.

9.8 PERITECTIC PHASE DIAGRAMS

Section 9.3 discussed a simple eutectic diagram (Figure 9.5). Other types of phase behavior are also observed in two-component systems. One of these involves the peritectic transformation in which a liquid phase and a solid phase can combine to form an entirely new solid phase.

$$\text{liquid} + \text{solid}_\alpha = \text{solid}_\beta$$

This is illustrated in Figure 9.17, the carbon–iron system. In this system, when a liquid solution with less than 0.51 wt % carbon is cooled, it first separates into a

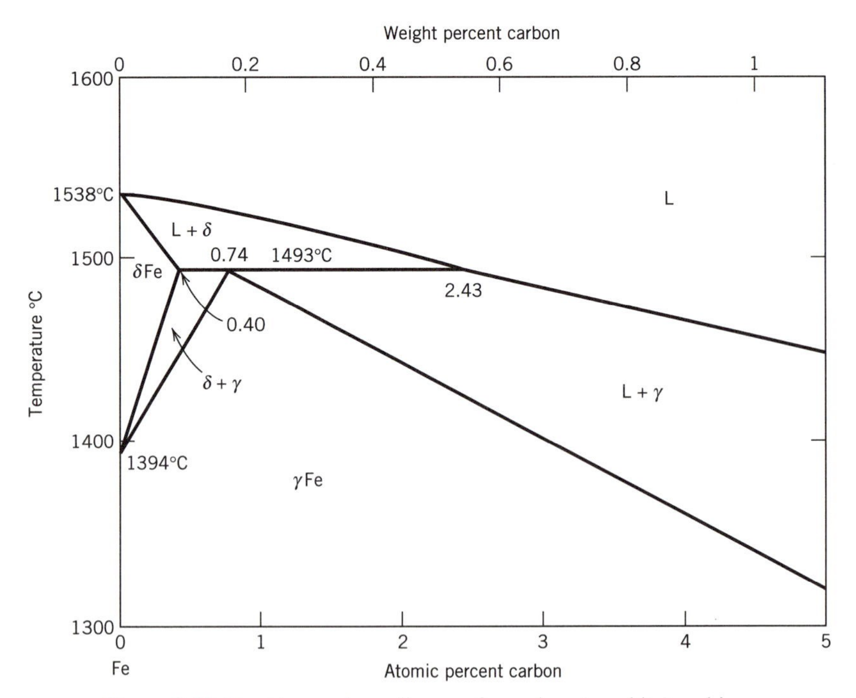

Figure 9.17 The binary phase diagram for carbon (graphite) and iron.

liquid and a solid, labeled δ. When the peritectic temperature is reached, the liquid and the δ phase combine to form a new solid phase indicated by the γ phase region. The lever rule can be used to determine the relative amounts of the two phases present for a given temperature and overall composition.

Peritectic reactions can occur completely in the solid state, in which two solid phases form a third ($\alpha + \beta = \gamma$). These are called peritectoid reactions.

9.9. COMPOUNDS

Two materials, A and B, may in addition to forming liquid and solid solutions, also form compounds (AB, A_2B, AB_2, etc.). Phase diagrams in which compounds form have the general appearance of the system shown schematically in Figure 9.18. Actual phase diagrams of material combinations can show behavior represented by combinations of eutectics, peritectics, and compound formation. It is beyond the scope of this text to deal with these phase diagrams. Good treatments of the subject are given in Refs. 1–5.

9.10 TERNARY DIAGRAMS

For condensed systems, the Gibbs phase rule is given by

$$P + F = C + 1$$

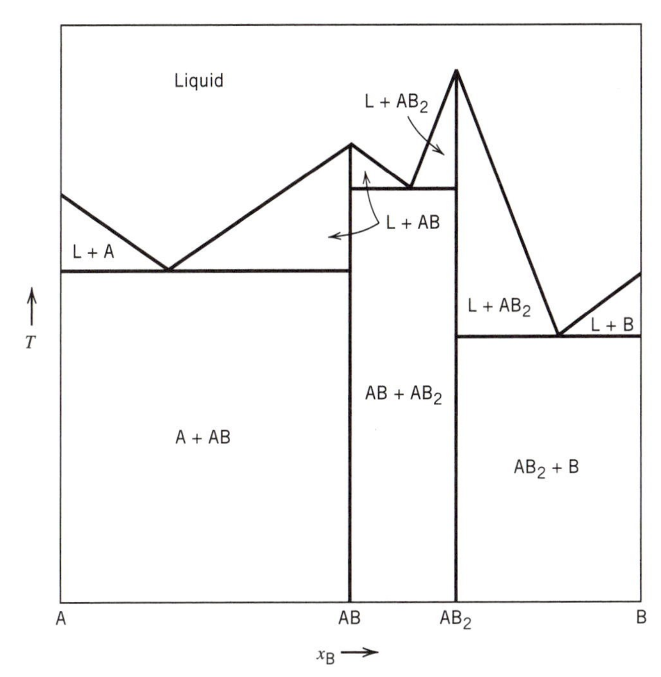

Figure 9.18 Schematic phase diagram for a system A–B containing compounds AB and AB_2.

In a binary system ($C = 2$), a phase diagram describing phase relationships can be displayed on a plane. With a minimum of one phase present ($P = 1$), there are two degrees of freedom. Thus only two dimensions were required to display the system: temperature and composition.

Many useful material systems are composed of more than two components. When three components are present, they are called *ternary systems.* When there are four components, we use the term *quaternary systems.* The display of phase relationships in quaternary systems and higher is beyond the scope of this text, but see Refs. 1 and 2. In this section we examine the diagrams used to display phase relationships in ternary systems.

Because there are three components in a ternary system ($C = 3$), and the minimum number of phases is one ($P = 1$), we need at least three dimensions to describe our system ($F = 3$). Composition is described in two dimensions and temperature on the third. Only two dimensions are needed to describe composition in a three-component system, because specifying the fraction of two components determines the third. The sum of the three compositions must be one.

Because diagrams representing three-dimensional figures are difficult to draw in two dimensions, phase relationships in ternary systems are generally presented as a series of isothermal sections through the three-dimensional structure required to describe them. We can think of these as a series of diagrams printed on sheets of paper or plastic, stacked up vertically. Each sheet of paper contains the phase relationship at a particular temperature.

The isothermal sections show the phases present at various compositions. Because there are three components, only two of them are independent. The concentrations (or mole fractions) of two of the components are plotted on two axes. The two axes may be orthogonal, as shown in Figure 9.19. This type of graph is usually used to describe dilute solutions where the pure solvent (material A) is in the lower left hand corner. Various concentrations of solutes B and C are described on the two axes.

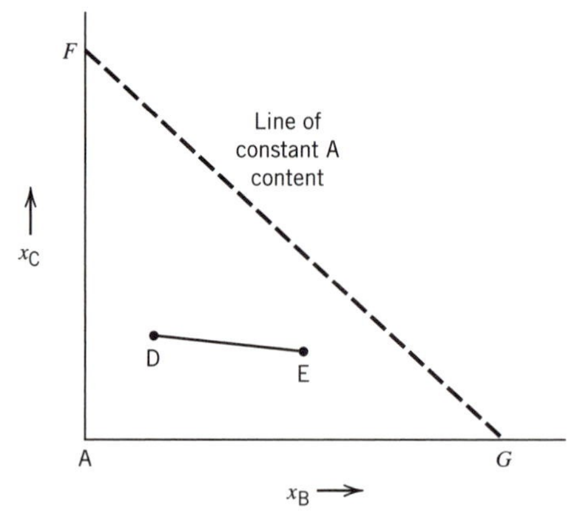

Figure 9.19 Ternary diagram axes: dilute solutions.

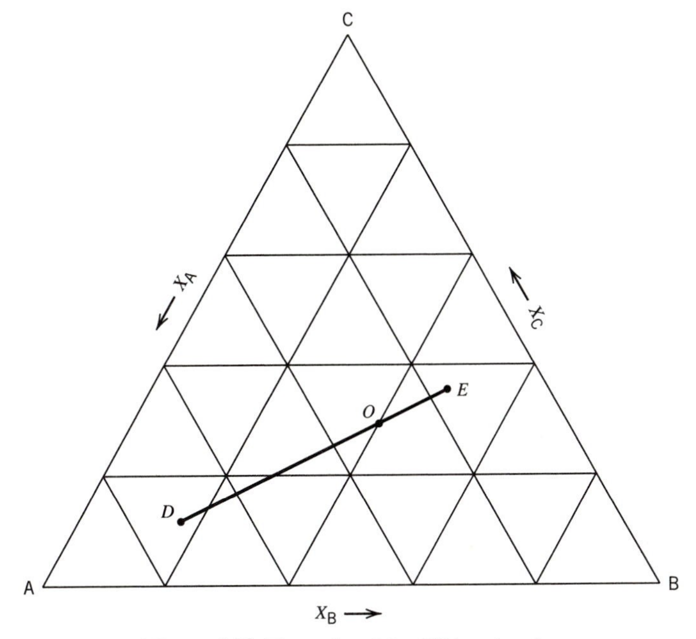

Figure 9.20 Example of the Gibbs triangle.

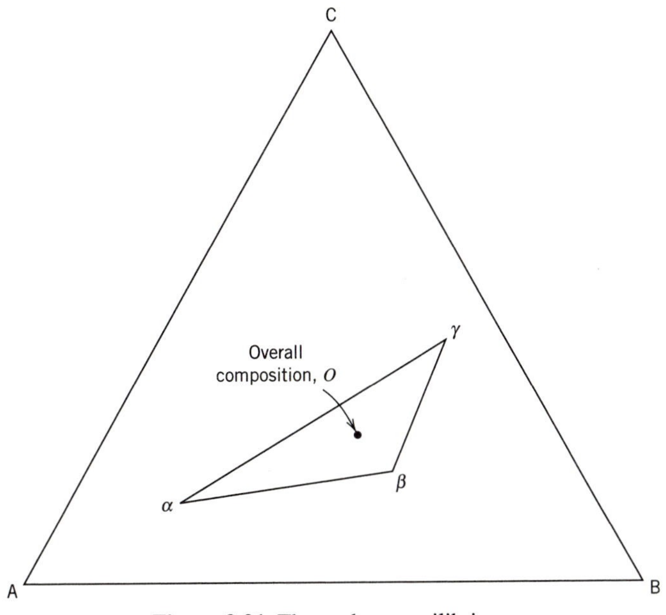

Figure 9.21 Three-phase equilibrium.

The line joining a point on the x_B axis (point G) and a point on the x_C axis (point F) is a line of constant solute concentration if $x_B = x_c$. The composition of combinations of materials D and E (Figure 9.19) lies on the line joining the two points.

Most ternary systems are better described using the triangular format called a Gibbs triangle, shown in Figure 9.20. In this representation, the components A, B, and C are shown at the three corners of the triangle. As in the preceding case, only two independent variables are required to define a position on the plane. One can think of this triangular system as a special case of the system shown in Figure 9.19 with the A–C axis simply tilted over to 60 degrees, instead of in the orthogonal position. The composition of materials represented in the triangle is read by referring to the three axes, labeled x_A, x_B, and x_C. Lines of constant A content are parallel to the line connecting the points representing pure B and pure C. The point O in Figure 9.20 represents a composition of approximately 30% A, 30% C, and 40% B.

As in the case of orthogonal axes, the compositions of materials produced by mixtures of two other materials fall on a straight line between the two. In Figure 9.20 all the combinations (alloys) of D and E fall on the line between the two. Furthermore, the lever rule derived in Section 9.2 applies to these systems. If points E and D represent two phases, and point O represents the overall composition of the mixture of E and D, the fraction of E, F_E, in the mixture is the length of the OD line divided by the length of the DE line:

$$F_E = \frac{\overline{OD}}{\overline{DE}}$$

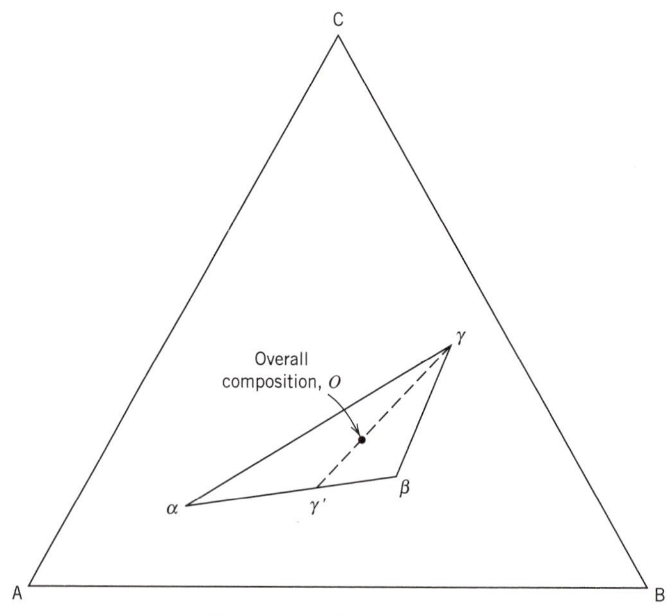

Figure 9.22 The lever rule.

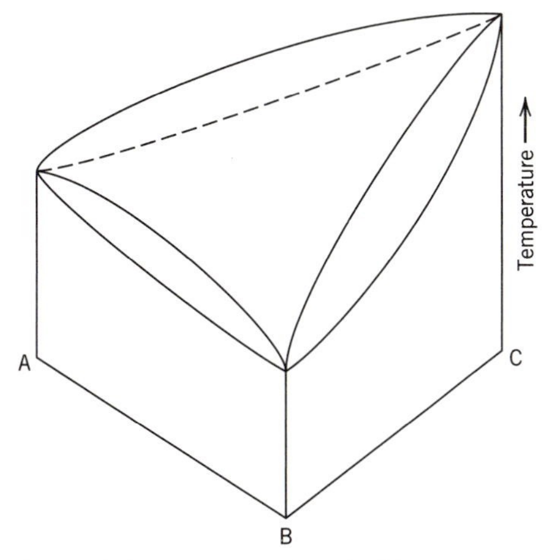

Figure 9.23 A completely miscible system.

In an isothermal section, a three-phase field may exist. The sum of P and F is four ($P + F = C + 1$). With one degree of freedom removed by the isothermal condition, we may have three phases present. In this case there will be no degrees of freedom, and the composition of those phases will be fixed. Figure 9.21 represents such a situation. In this three-phase field, we can solve for the fraction of each phase (F_α, F_β, and F_γ) present at equilibrium if we know the composition of the phases ($x_{A,\alpha}$, $x_{B,\alpha}$, $x_{C,\alpha}$, $x_{A,\beta}$, etc.) and the overall composition of the mixture (x_A, x_B, and x_C):

$$x_A = F_\alpha x_{A,\alpha} + F_\beta x_{A,\beta} + F_\gamma x_{A,\gamma}$$

$$x_B = F_\alpha x_{B,\alpha} + F_\beta x_{B,\beta} + F_\gamma x_{B,\gamma}$$

$$x_C = F_\alpha x_{C,\alpha} + F_\beta x_{C,\beta} + F_\gamma x_{C,\gamma}$$

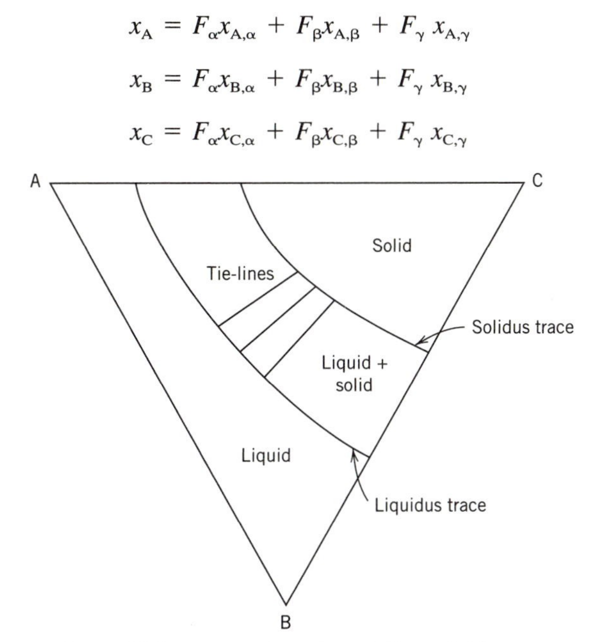

Figure 9.24 Isothermal section of Figure 9.23.

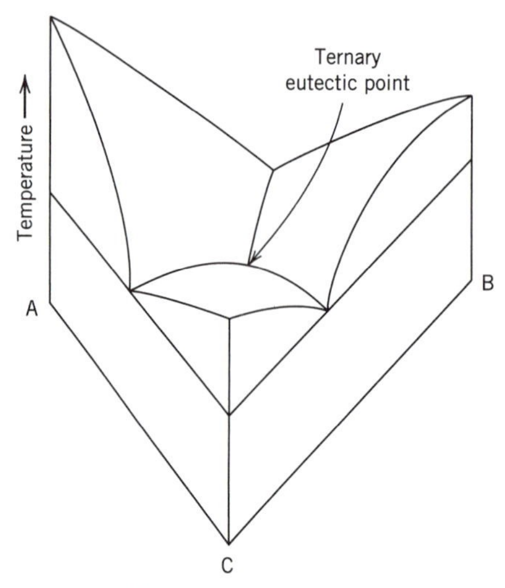

Figure 9.25 Ternary diagram for a eutectic system with no solid solubility.

The three equations may be solved for the three unknown fractions, F_α, F_β, and F_γ.

The lever rule also applies between two points. In Figure 9.22, the fraction of γ phase is the length of the O–γ' line divided by the length of the γ–γ' line.

$$F_\gamma = \frac{O - \gamma'}{\gamma' - \gamma}$$

The simplest ternary diagram is one representing an ideal solution in which each of the three components is completely miscible in the others in both solid and liquid states (Figure 9.23). Isothermal sections of this system at temperatures above the melting temperature of the highest melting component, C, show simply all liquid. Below the melting point of the lowest melting component, B, the section shows all solid.

The temperature at which solidification begins upon cooling for any composition is called the *liquidus* temperature. The temperature at which a solid begins to melt upon heating is the *solidus* temperature. The liquidus and solidus points form surfaces in the three-dimensional diagrams.

At a temperature below the melting point of C, but above the melting points of A or B, a section of the type in Figure 9.24 is observed. The line between the regions labeled ''Liquid'' and ''Liquid + Solid'' is the trace of the liquidus surface. The line between the regions labeled, ''Solid'' and ''Liquid + Solid'' is the trace of the solidus. In the region labeled ''Liquid + Solid,'' the compositions of the liquid and solid in equilibrium are joined by tie-lines, which join the compositions of the solid and liquid that are in equilibrium. These lines need not necessarily extrapolate through the corners of the Gibbs triangle.

In another type of ternary diagram, one involving eutectics (Figure 9.25), the three materials are assumed to be insoluble in the solid state, but completely soluble

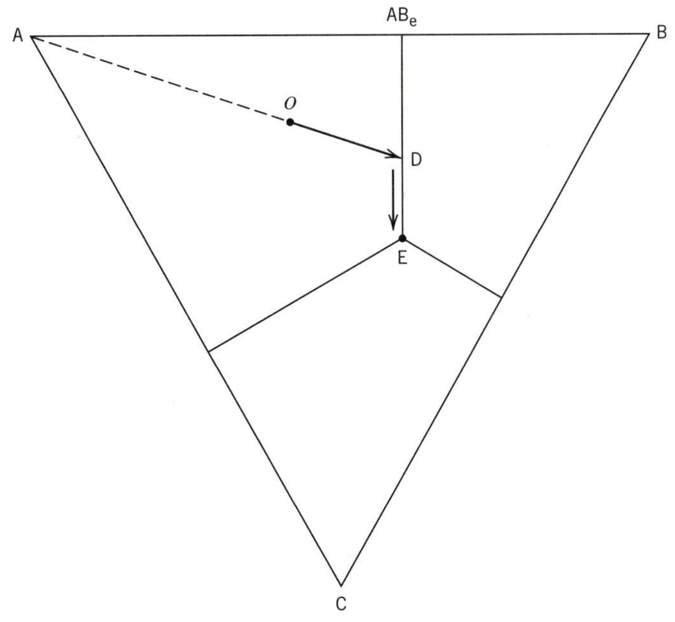

Figure 9.26 Solidification path of material from Figure 9.25.

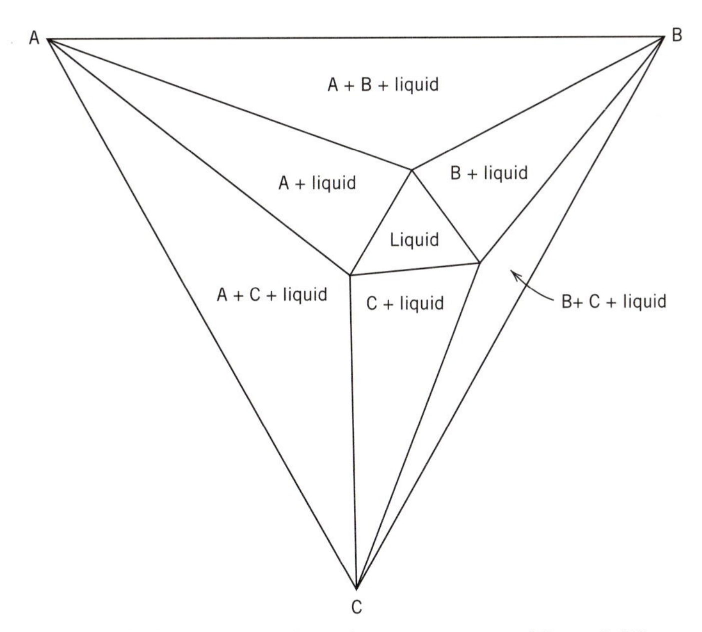

Figure 9.27 Isothermal section of a ternary system (Figure 9.25) at a temperature above the ternary eutectic temperature, but below binary eutectic temperatures.

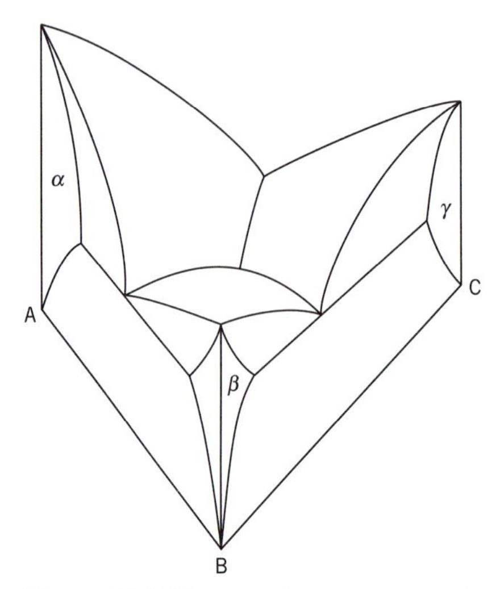

Figure 9.28 Diagram of a ternary eutectic system with solid solubility.

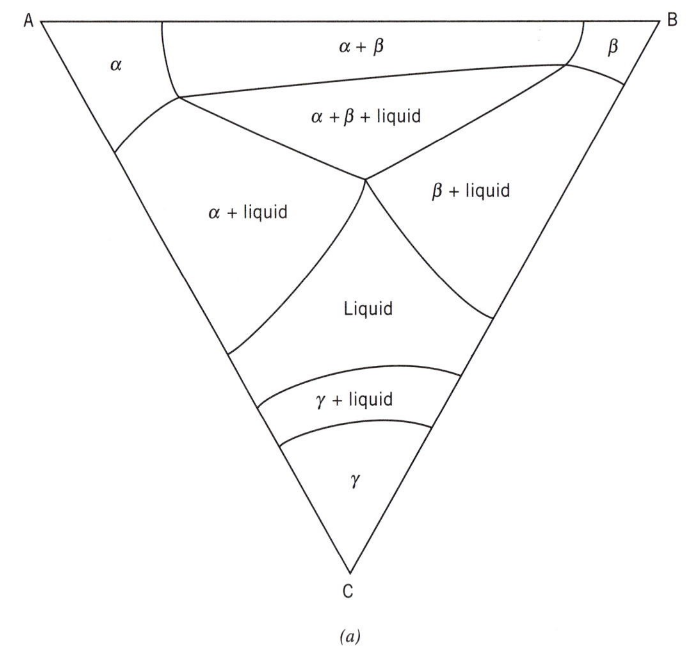

(a)

Figure 9.29 Isothermal sections of the ternary eutectic of Figure 9.28: (*a*) at a temperature above the ternary eutectic and below one binary eutectic temperature (A–B), (*b*) at a temperature above the ternary eutectic temperature but below all binary eutectic temperatures, and (*c*) at a temperature below the ternary eutectic temperature.

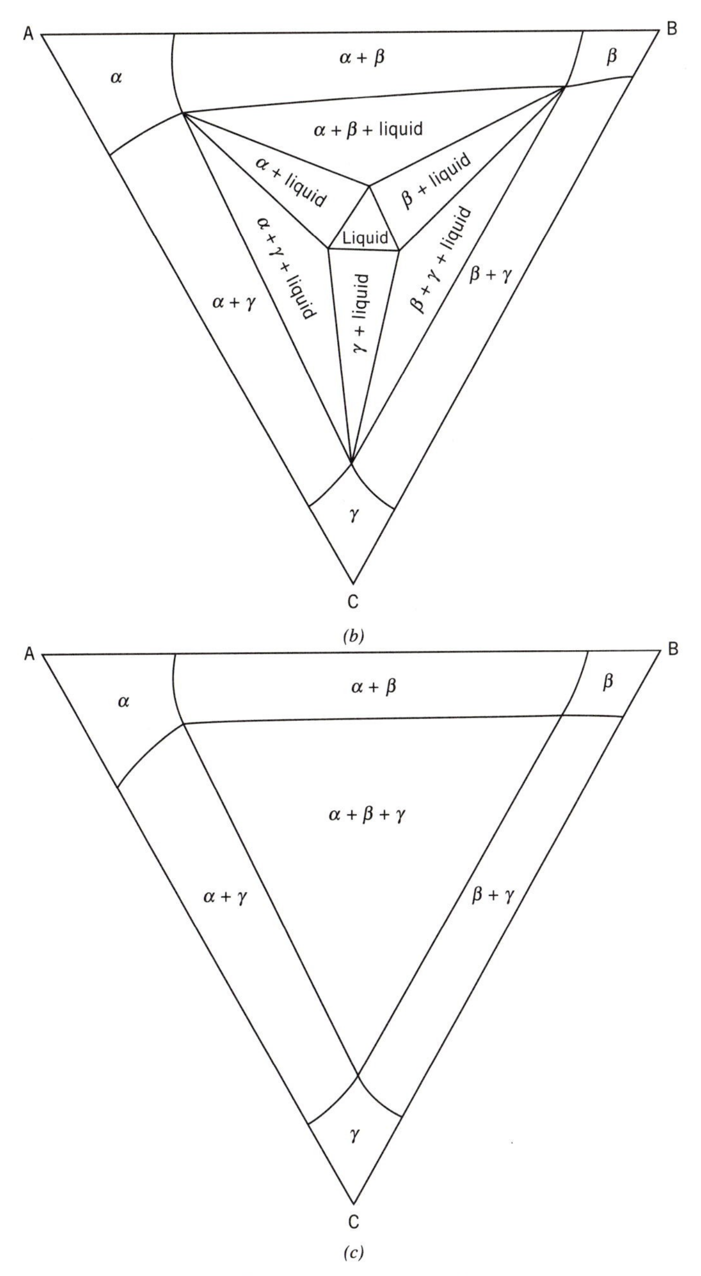

Figure 9.29 (continued)

in the liquid state. Figure 9.26 is a view looking down on the diagram. (Note that this is not an isothermal section, but a projection of the liquidus surface.) In this diagram, a material with the overall composition O will begin to solidify when the temperature reaches the point at which a vertical line at *O* intersects the liquidus surface. The material that will solidify is pure A. Upon further cooling, more A solidifies, and the liquid composition proceeds along line *OD* until the liquid composition intersects the line $(AB)_e$–E, the line representing the saturation in both A and B[6]. At this point both solid A and solid B are formed as the binary A–B eutectic. The liquid now follows the AB_e–E line to the ternary eutectic point where the remaining liquid solidifies as the ternary eutectic.

Figure 9.27 is an isothermal section of the system shown in Figure 9.25 at a temperature below the melting temperatures of the pure components (A, B, and C), and the binary eutectic temperatures, but above the ternary eutectic temperature.

Figure 9.28 represents a ternary three-eutectic system, but one in which there is some solid solubility among the three components. Figures 9.29*a* through 9.29*c* are some isothermal sections of the system.

The ternary systems illustrated are some of the least complicated. Systems involving compounds and peritectics are more complex. Good discussions of these systems are to be found in Refs. 1, 4, and 5.

REFERENCES

1. West, D. R. F., *Ternary Equilibrium Diagrams,* 2nd ed., Chapman & Hall, London, 1982.
2. Lupis, C. H. P., *Chemical Thermodynamics of Materials,* North Holland, Amsterdam, 1983.
3. Marsh, J. S., *Principles of Phase Diagrams,* McGraw-Hill, New York, 1935.
4. Rhines, F. N., *Phase Diagrams in Metallurgy,* McGraw-Hill, New York, 1956.
5. Prince, A., *Alloy Phase Equilibria,* Elsevier, New York, 1966.

PROBLEMS

9.1. (a) If an alloy of 50 atom % copper and 50 atom % silver is brought to equilibrium at 600°C at one atmosphere pressure, what phase or phases in the accompanying Ag-Cu phase diagram are present?

(b) Apply the phase rule to the situation in part a. How many degrees of freedom does the system have?

(c) Assume that the system described in part a is brought to a new equilibrium at 700°C. Describe the physical changes you expect to occur in the system. (You need not limit yourself to the phase rule to discuss this.)

---

[6]The extrapolation of the line *OD* passes through the A corner of the Gibbs triangle in this case because pure, solid A is being formed. The overall composition point *O*, must lie on the line between the points representing the composition of the two phases present.

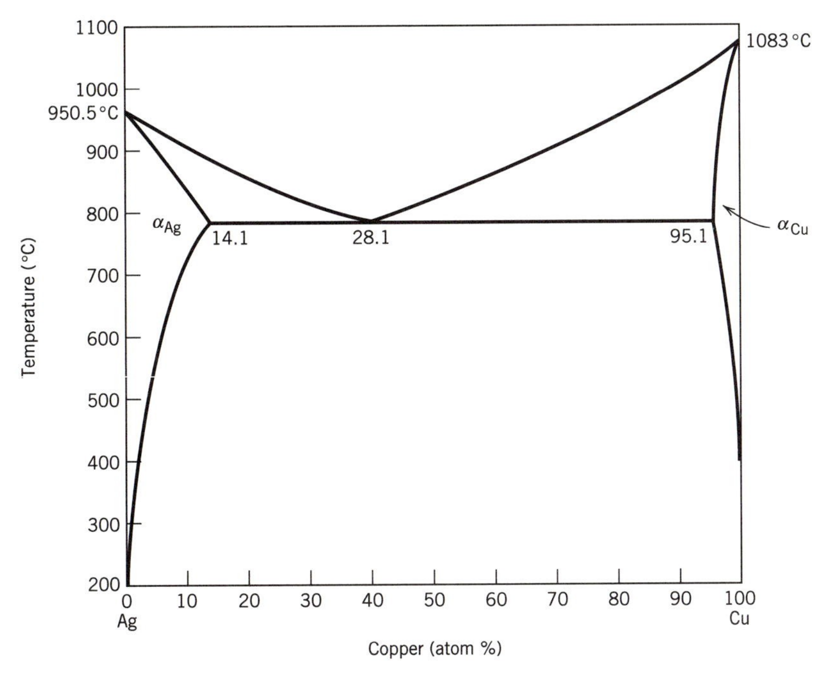

9.2 Use the Ag-Cu phase diagram from Problem 9.1.

(a) Estimate the activity of silver and of copper in a 50-50 atom % alloy at 500, 600, and 700°C.

(b) Does this system obey Raoult's law? Henry's law? Is it a regular solution?

9.3. Using the activities of Ag-Cu as estimated at 600°C in Problem 9.3, determine whether either silver or copper will oxidize in air in this 50-50 alloy at 600°C under the following conditions.

DATA

For silver:

$$4Ag + O_2 \rightarrow 2Ag_2O$$

$$\Delta G^\circ = -58{,}600 + 122.2T$$

For copper:

$$4Cu + O_2 \rightarrow 2Cu_2O$$

$$\Delta G^\circ = -334{,}700 + 144.8T$$

where standard Gibbs free energy (ΔG°) is in joules

9.4 One mole of solid Cr at 1600°C is added to a large quantity of Fe-Cr liquid solution (in which $x_{Fe} = 0.8$), which is also at 1600°C. If Fe and Cr form Raoultian solutions, calculate the enthalpy and entropy changes in the solution resulting from the addition. Assume that the heat capacity difference between solid and liquid Cr is negligible. The enthalpy of fusion of Cr is 21,000 J/mol at its melting temperature, 1900°C.

9.5 The partial pressure of oxygen in equilibrium with pure liquid lead and pure liquid PbO at 1200 K is 3.83×10^{-9} atm. If SiO_2 is added to the liquid PbO such that equilibrium P_{O_2} for the pure Pb–liquid PbO–SiO_2 solution couple is decreased to 9.58×10^{-10} atm, calculate the activity of PbO in the lead silicate melt.

9.6 In the accompanying eutectic equilibrium phase diagram of temperature versus mole fraction of B for the A–B system shown, note that the pressure for the diagram is constant at 1 atm. Consider an alloy containing 40 wt % of B.

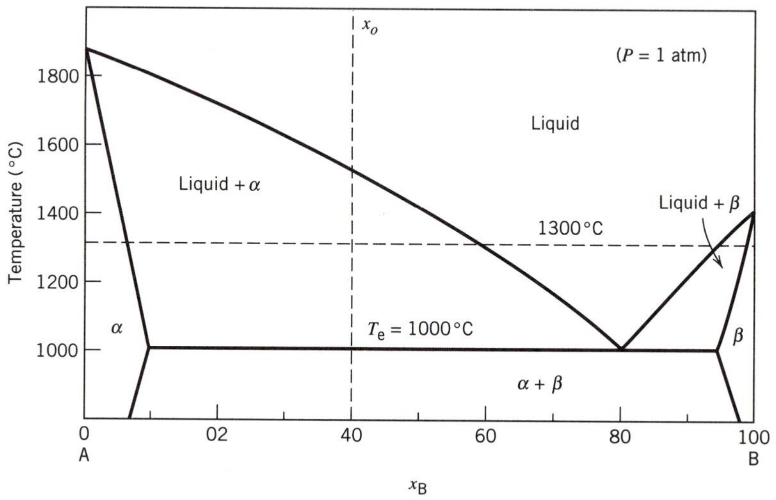

In the table, indicate which of the phases are present in the 40% alloy and give the composition of each (x_B) and the fraction present of each for the temperatures shown.

Temperature (°C)	Phase	Composition x_B (mol %)	Fraction Present, f
1300	Liquid		
	α		
	β		
1000 + (Just above T_e)	Liquid		
	α		
	β		
1000 − (Just below T_e)	Liquid		
	α		
	β		

9.7 How much is the melting point lowered as a result of the solution of 0.160 g of oxygen in 100 g of silver? Neglect the solubility of oxygen in solid silver.

Oxygen dissolves in monatomic form.

DATA

Pure silver melts at 860.8 K.
The heat of fusion of silver is 2690 cal/mol (mole units).

Molecular weights:
Silver = 108 g/mol
Oxygen = 16 g/mol

9.8 The phase behavior of materials A and B can be described using the accompanying phase diagram. Assume that A and B form ideal solutions in the liquid state and in the solid state.

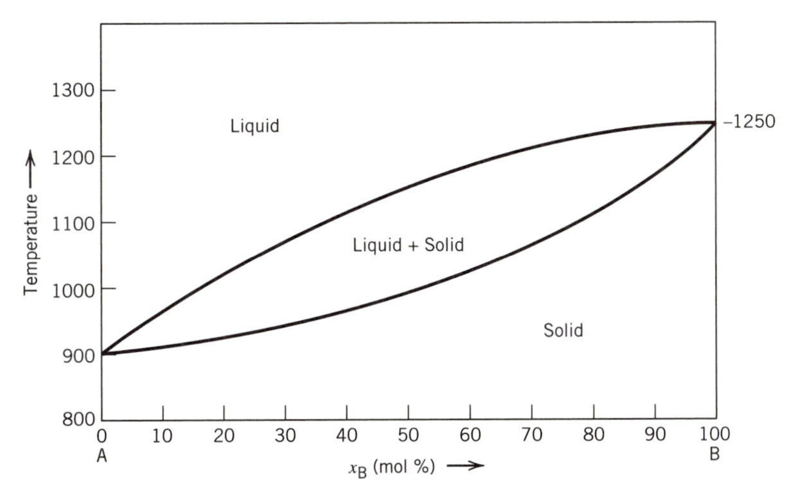

(a) If a solution containing 50 mol % B is cooled from 1300 K, what is the composition of the first solid to form? What is the composition of the last liquid to solidify?

(b) For this 50 mol % B solution, estimate the fraction solid and liquid in equilibrium at 1000 K.

(c) Sketch the activity–composition relationship for material B in the solutions at 1300 K, 1100 K, and 800 K. In each case give the standard state you have chosen.

9.9 An alloy composed of 80 atom % rhodium and 20 atom % rhenium is being slowly cooled from 3500°C during processing. Equilibrium is maintained at each temperture. Use the accompanying Rh-Re phase diagram to answer parts a–c.

(a) At what temperature does the first solid form and what is the composition of that solid?

(b) At what temperature does the last liquid solidify, and what is the composition of the last liquid?

(c) Which phases exist at 2000°C and what is their composition? Give the fraction of each phase. How may degrees of freedom are there in this equilibrium? Explain your answer.

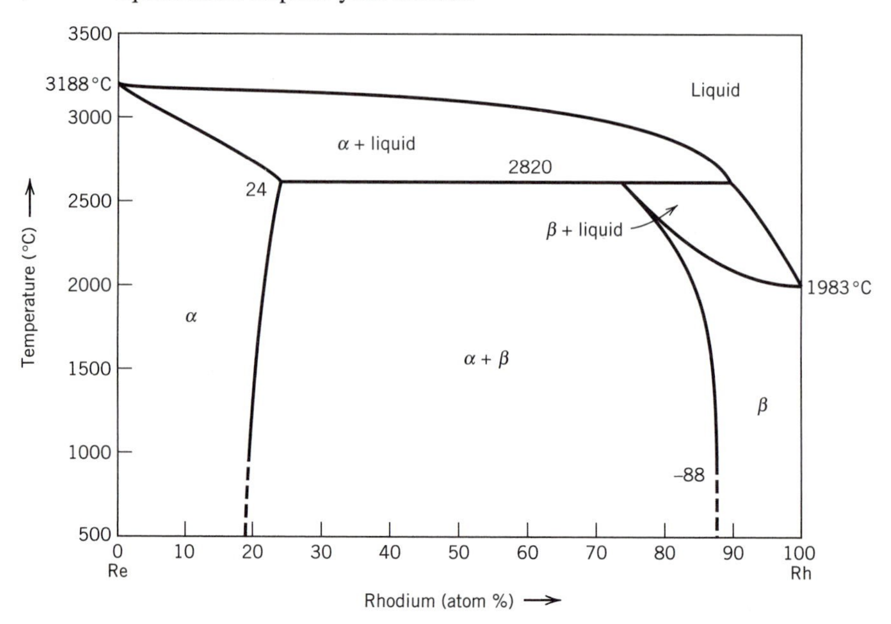

9.10 The accompanying diagram represents the liquidus surface in the ternary phase diagram for $BaCl_2$-NaCl-KCl·$CaCl_2$. Assume for the purposes of this problem that there is no solid solubility of the compounds in one another.

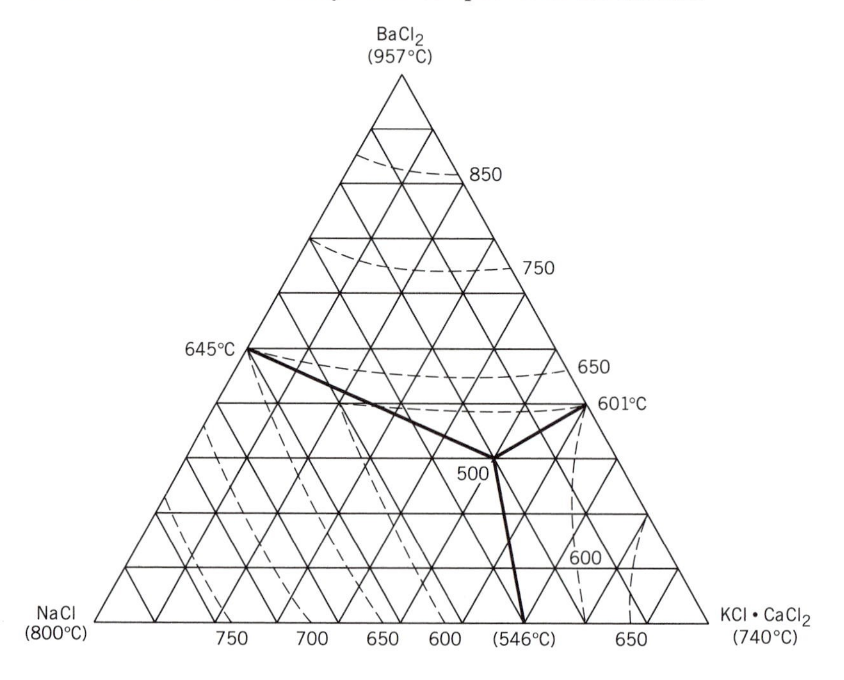

(a) On the diagram trace the path of the liquid composition when a material consisting of 60% $BaCl_2$, 20% NaCl, and 20% $KCl{\cdot}CaCl_2$ is cooled from 900°C. Assume that equilibrium is maintained at all temperatures.

(b) Sketch two blank ternary diagrams and draw in the isothermal sections at 750 and 600°C.

(c) What will be the fraction liquid when the liquid reaches the ternary eutectic temperature (500°C), but none has solidified as a ternary eutectic? That is, what will be the fraction of ternary eutectic in the material when it is all solidified?

9.11 The enthalpy of mixing (J/mol) of Au-Pt solid solutions can be approximately modeled with the following relation:

$$\Delta \underline{H}_{M} = 25{,}000 x_{Au}\, x_{Pt}$$

Since the enthalpy of mixing is positive, solid solutions of Au and Pt will exhibit phase separation at low temperature. Compute the temperature above which single-phase solid solutions of Au and Pt can exist for any composition.

9.12 (a) Five hundred grams (500 g) of an alloy consisting of 45 atom % platinum and 55 atom % silver is kept at 800°C until equilibrium is reached. Which phase(s) is (are) present? What is its (their) composition? How much is present of each phase?

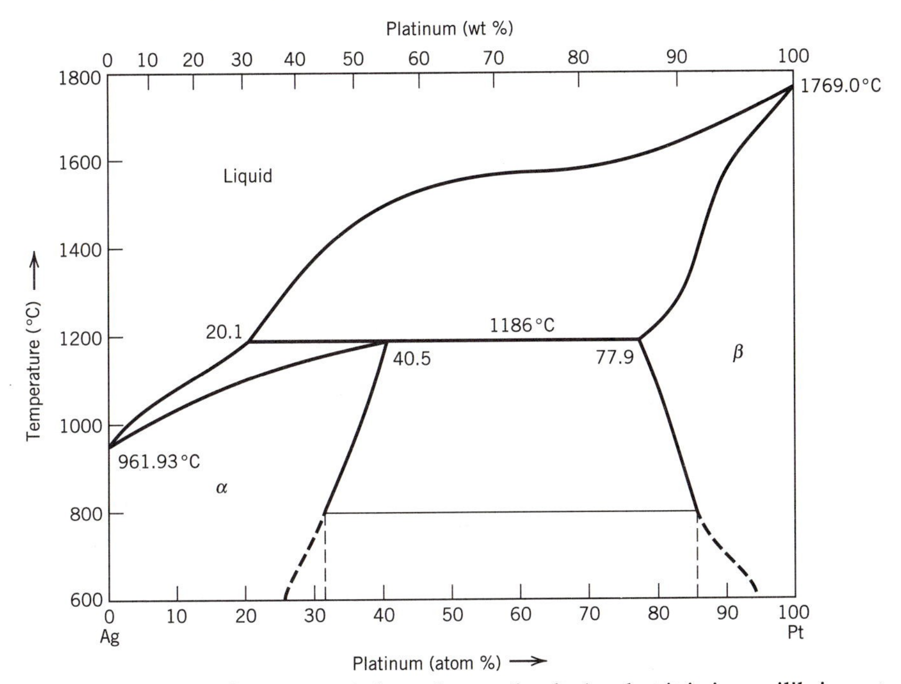

(b) The alloy from part a is heated very slowly (so that it is in equilibrium at all times).
At what temperature does liquid first appear in the alloy?

What is the composition of that liquid?
Give the reaction that takes place at 1186°C upon heating and describe what happens in the material.

(c) I build an electrochemical cell with Ag-Pt alloys, using some fused salt as electrolyte. The cell operates at 600°C. You can assume that Ag behaves in a Raoultian manner in the Ag-rich solid solution (α phase). What cell voltage would I get between an alloy with $x_{Pt} = 0.5$ and an alloy with $x_{Pt} = 0.3$?

(d) Work part c again but reverse the proportions of the alloy (i.e., pure Ag, $x_{Pt} = 0.5$).

9.13 Use the accompanying cadmium–zinc phase diagram for this problem involving an alloy of 80 atom % Zn and 20 atom % Cd, which is cooled slowly.

(a) At what temperature will the alloy start to solidify?
(b) What is the composition of the solid at 300°C?
(c) What is the composition of the last liquid to solidify?
(d) How much eutectic structure will there be?
(e) Draw a picture of the microstructure of the completely solidified material.

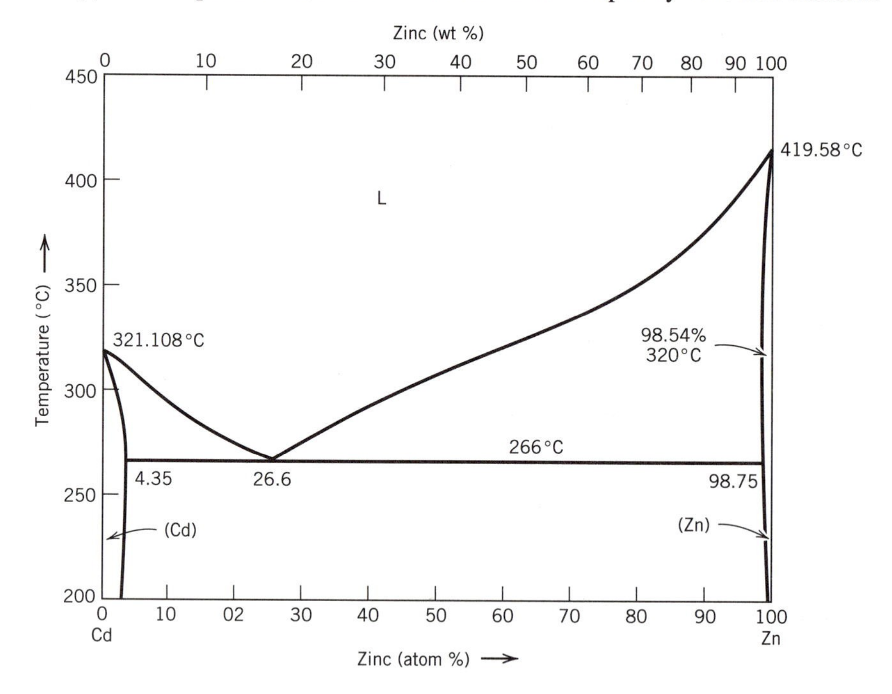

9.14 What happens to the Gibbs free energy of an alloy as it solidifies during a eutectic transformation: Does it

increase? _______ remain constant? _______ decrease? _______

9.15 In a system A–B–C, a ternary alloy of composition 30 wt % B and 30 wt % C consists at a particular temperature of three phases of equilibrium compositions as follows:

Liquid phase: 50% A, 40% B, 10% C
α solid solution: 85% A, 10% B, 5% C
β solid solution: 10% A, 20% B, 70% C

(a) Calculate the proportions by weight of liquid, α, and β present in this alloy.

(b) For the same temperature, deduce the composition of the alloy consisting of equal proportions of α and β phases of the compositions stated above, but with no liquid phase.

9.16 Use the accompanying phase diagrams for the NaCl-NaF-NaI system to work this problem.

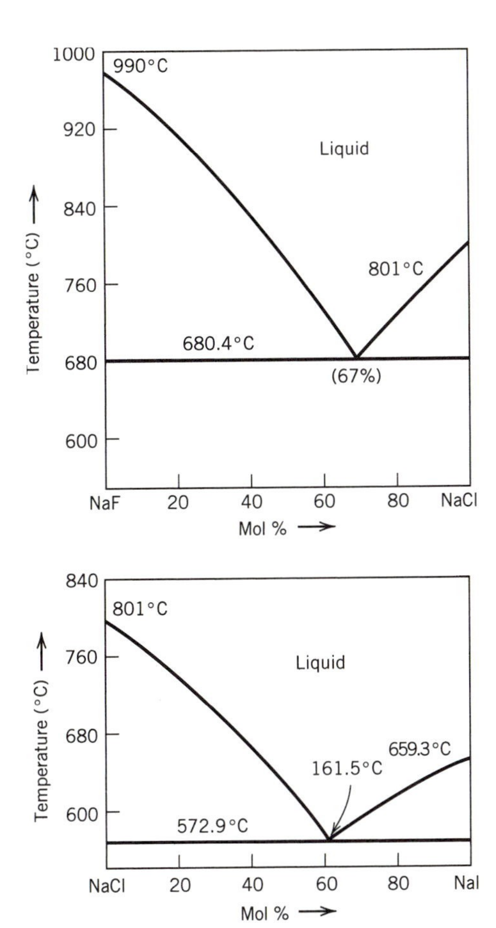

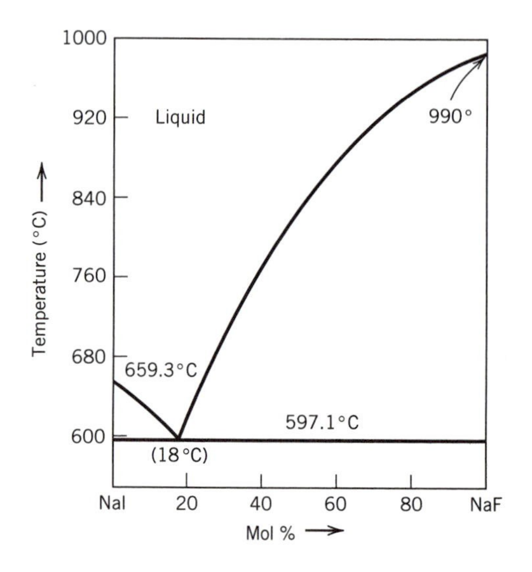

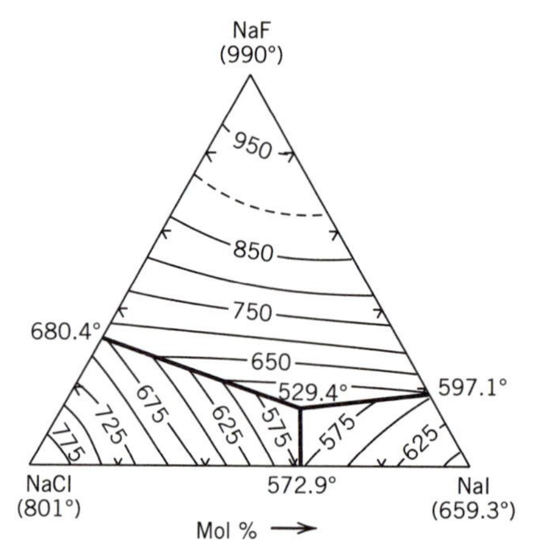

(a) Draw an isothermal section for the NaCl-NaF-NaI system at 850°C. Label all phase fields.

(b) Draw an isothermal section for the NaCl-NaF-NaI system at 650°C. Label all phase fields.

(c) List the solid phases that form, and their order of appearance, for equilibrium solidification of a melt that contains 20 mol % NaF, 18 mol % NaI, and 62 mol % NaCl.

9.17 In the accompanying liquidus projection for an A–B–C system solid solubility is *not* negligible, but the limit of solubility of each of the pure components is about 5 mol %.

(a) Sketch and label an isothermal section for 750, 555, and 300°C.

(b) List the phases that appear, in their order of appearance, for the solidification of a melt containing 50 mol % C and 30 mol % B.

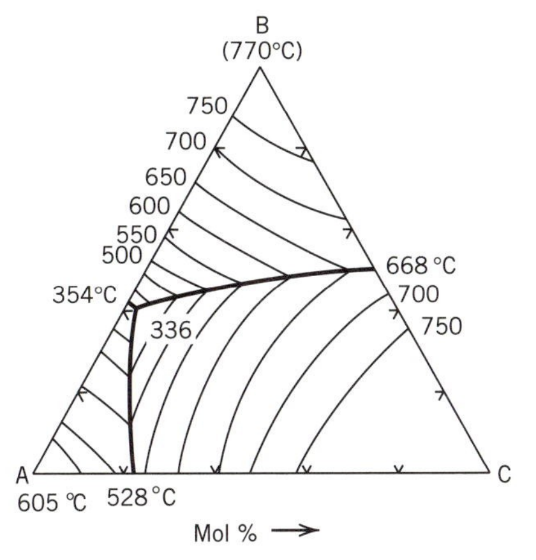

9.18 Part of the Cu-Ag-P system is represented by the accompanying liquidus projection, taking the compound Cu_3P as one of the components. Each of the binary systems contains a eutectic, and the ternary system also contains a eutectic.

Solid solubility in the system is negligible at room temperature. For the purposes of this problem, assume solid solubility to be zero.

Consider an alloy containing 60 wt % Cu and 4 wt % Cu_3P.

(a) List the phases that form, in their order of appearance, for equilibrium solidification of this alloy.

(b) Sketch the isothermal section at 800°C.

(c) Estimate the weight fraction of liquid present for the alloy at 750°C.

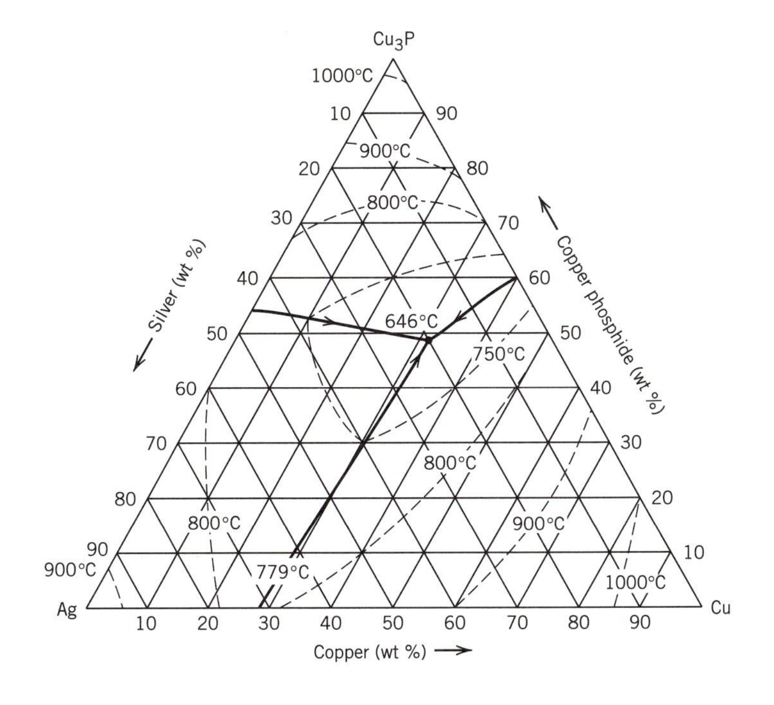

Chapter 10

Statistical Thermodynamics

There are two approaches to the study of thermodynamics, macroscopic and microscopic. Macroscopic thermodynamics, exemplified by the approach used in Chapters 1 through 9, is concerned with the relative changes among the macroscopic properties of matter, such as heats of transformation, pressure, temperature, heat capacity, density, and vapor pressure. Except in the study of chemical reactions, where it is recognized that elements combine in simple proportions to form compounds, macroscopic thermodynamics does not require any knowledge of the atomistic nature of matter. Microscopic thermodynamics, on the other hand, attempts to compute absolute values of thermodynamic quantities based on a *statistical* averaging of such properties of individual atoms and molecules as the mass and volume of atoms, molecular bond strengths, and vibration frequencies. Because of this statistical averaging, microscopic thermodynamics is usually called *statistical thermodynamics.* Macroscopic thermodynamics can ignore the existence of molecules, molecular complexity, and quantum mechanics. Statistical thermodynamics relies heavily on quantum mechanics, and a knowledge of molecular motion and structure.[1]

---

[1]The treatment of statistical thermodynamics used in this chapter relies on the ideas of quantum mechanics, such as quantized energy levels. But, the idea of quantization is not essential to statistical thermodynamics. The ideas that form the basis of classical (non–quantum mechanical) statistical thermodynamics predate the advent of quantum mechanics.

The statistical approach to thermodynamics adds another dimension to the study of thermodynamics. Temperature, in macroscopic thermodynamics, is at first considered to be a quantity inferred from the measurement of some physical property, such as the expansion or contraction of a column of mercury in a thermometer. In the treatment of the second law of thermodynamics, temperature is defined as a conjugate to the entropy function. Statistical thermodynamics will show us another way of looking at temperature. It will also provide a basis for understanding the third law of thermodynamics, which asserts that the entropy of pure, stoichiometric, perfectly ordered crystals is zero at a temperature of absolute zero.

There are many satisfying philosophical insights to be gained by a thorough study of statistical thermodynamics, as recorded in several excellent full-length texts (Refs. 1–4). Rather than aiming for a full-fledged understanding of the philosophical underpinnings of the subject, this chapter introduces enough of the basic ideas of statistical thermodynamics to enable us to calculate some macroscopic properties of materials from the behavior of the individual constituents of the system (atoms and molecules), and to give us a view of what is happening at the microscopic level.

A detailed picture of phenomena at the microscopic level should yield additional information, beyond the calculation of the equilibrium, time-independent quantities we have discussed so far in our study of thermodynamics. It should form the basis for the study of rates of movement: mass diffusion, thermal conductivity, electrical conductivity, chemical reaction rates, and other kinetic phenomena.

A basic idea embodied in statistical thermodynamics is that even when a material is in equilibrium on a macroscopic scale, it is dynamic on a microscopic scale. For example, the macroscopic properties of a gas at equilibrium under constant pressure and temperature do not change with time. The macroscopic thermodynamic state variables (U, V, etc.) are fixed. On the microscopic scale, however, the atoms or molecules are in motion, and their configuration changes constantly. One objective of this chapter is to describe the distribution of velocities in gases. We need not establish such a detailed statistical picture of molecular motion, however, to gain some interesting information about the movement of molecules in ideal gases. As an example, let us calculate an average rate of motion (the root-mean-square velocity) of the atoms of a monatomic ideal gas, such as helium, when the gas is at macroscopic equilibrium at pressure P and temperature T.

10.1 AVERAGE VELOCITY OF GAS MOLECULES

Consider a monatomic, ideal gas at equilibrium, with its pressure and temperature fixed. From a classical point of view, it is static. Looking at the situation from a microscopic point of view, however, the pressure that the gas exerts on its container is generated by the collision of the molecules of the gas with the walls of the container. The impulse on the walls is generated by the change in momentum of the gas atoms when they strike the wall and bounce back from it.

The gas atoms, of course, move in three dimensions. But, for the purpose of the calculation, it is sufficient to consider just the motion of atoms in the x direction. Take the root-mean-square velocity in the x direction to be $\overline{v}_x$. In Figure 10.1, a

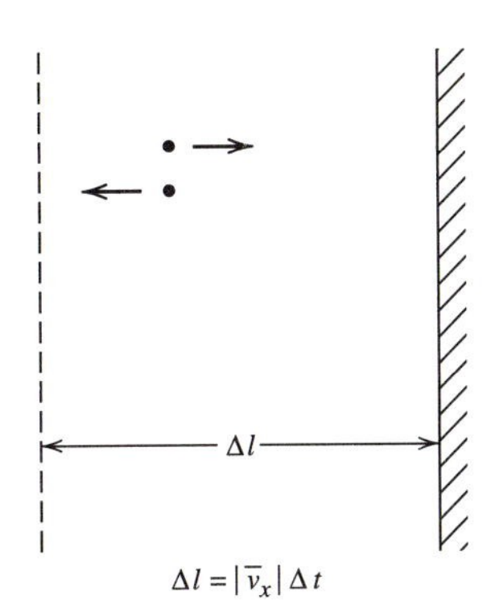

Figure 10.1 Schematic definition of collision volume.

volume is defined by a representative cross-sectional area A and a length Δl, which is the product of the root-mean-square velocity $\bar{v}_x$ and an interval of time Δt. This volume is $A\bar{v}_x\,\Delta t$. Only half the total atoms in the volume will strike the wall in the time period Δt, because half of them will be traveling away from the wall. The number of atoms per unit volume is n_T/V. The root-mean-square momentum of an atom moving toward the wall in Figure 10.1 is $m\bar{v}_x$; after it has struck the wall and rebounded elastically, its momentum will be $-m\bar{v}_x$. The change in momentum is $2m\bar{v}_x$ because the direction of motion is reversed.

The force required to change the momentum of an atom is exerted by the wall the atom strikes. From the principles of mechanics, the change of momentum of a body is equal to the impulse of the force acting on the body. The impulse of a force is the integral of the force over the time interval the force acts. In our case the impulse is $F\,\Delta t$, where F is the average force acting during the time period Δt in response to all the atoms that strike it:

$$F\,\Delta t = A\bar{v}_x\,\Delta t\,\frac{n_T}{V}\,(2m\bar{v}_x)\left(\frac{1}{2}\right) \tag{10.1}$$

where $A\bar{v}_x\,\Delta t(n_T/V)$ is the number of particles in the specified volume, $2\,m\bar{v}_x$ is the change in momentum per particle, and the "$\frac{1}{2}$" term arises because only half the atoms in the volume (Figure 10.1) are moving toward the wall.

The pressure on the wall is defined as the force applied on it divided by the area, and is equal to

$$P = \frac{F}{A} = m\bar{v}_x^2\,\frac{n_T}{V} \tag{10.2}$$

The total number of atoms n_T, is the number of moles n multiplied by Avogadro's number, N_A.

$$n_T = n\,N_A \qquad PV = mn\,N_A\,\overline{v}_x^2 \tag{10.3}$$

From the ideal gas law, $PV = nRT$.
Hence,

$$mn\,N_A\,\overline{v}_x^2 = nRT \qquad m\,\overline{v}_x^2 = \frac{R}{N_A}T = kT \tag{10.4}$$

where k is Boltzmann's constant.

If we now consider velocity in all directions, x, y, and z, we can write

$$\overline{v}^2 = \overline{v}_x^2 + \overline{v}_y^2 + \overline{v}_z^2 \tag{10.5}$$

Assuming that the motion of the atoms is independent of direction—that is, that the probability of motion in each direction is equal—we have

$$\overline{v}_x^2 = \overline{v}_y^2 = \overline{v}_z^2 \tag{10.6}$$

Combining Eqs. 10.5 and 10.6, we have

$$\overline{v}^2 = 3\,\overline{v}_x^2$$

Thus we have an equation relating the root-mean-square velocity of the atoms to the temperature.

$$\tfrac{1}{3}\,m\overline{v}^2 = kT \tag{10.7}$$

Note that the average energy, $\langle E \rangle$, of the molecule is $\frac{1}{2}\,m\overline{v}^2$. Thus

$$\langle E \rangle = \tfrac{1}{2}\,m\overline{v}^2 = \tfrac{3}{2}\,kT \tag{10.8}$$

The specific internal energy per mole is Avogadro's number multiplied by the average energy per particle:

$$\underline{U} = N_A\,\langle E \rangle \qquad \underline{U} = N_A\,\tfrac{3}{2}\,kT = \tfrac{3}{2}\,RT \tag{10.9}$$

Combining with the definition of the heat capacity at constant volume, we can show that $C_V = \frac{3}{2}R$.

$$C_V = \left(\frac{\partial \underline{U}}{\partial T}\right)_V = \frac{3}{2}R \tag{10.10}$$

This rough model, assuming average velocities for molecules, thus yields a value for the constant volume heat capacity (C_V) of a monatomic ideal gas. The constant pressure heat capacity for this gas is just $C_P = C_V + R = \frac{5}{2} R$.

To gain some idea of the rate of motion of molecules in air, let us calculate the root mean square of velocity for oxygen at 300 K. The value of $\overline{v}^2$ from Eq. 10.7 is:

$$\overline{v}^2 = \frac{3kT}{m}$$

$$\overline{v}^2 = \frac{3(1.38 \times 10^{-23})(300)}{\left(\frac{32}{6.022 \times 10^{23}}\right)(10^{-3})} = 23.4 \times 10^4 \text{ m}^2/\text{s}^2 \qquad \textbf{(10.11)}$$

$$\overline{v} = 483 \text{ m/s}$$

The root mean square of velocity is approximately 483 meters per second, or in English units, 1100 miles per hour. We have demonstrated, thus, that although the gas is at macroscopic equilibrium, its molecules are in very rapid motion.

10.2 MACROSTATES AND MICROSTATES

Equation 10.11 yields an estimate of an *average* rate of motion of atoms in a gas (root-mean-square velocity), but it does not tell us about the distribution of atomic velocities. This distribution and its consequences are derived using the methods of statistical thermodynamics.

To establish the basis for the study of statistical thermodynamics, one based on probabilities, we will distinguish between a macrostate and a microstate. A *macrostate* of a system is what, in macroscopic thermodynamics, is called a "state," and which is characterized by a few state variables, such as temperature, volume, and internal energy. A *microstate* of a system characterizes the state of all of the particles in the system. For example, a gas in an *isolated system* at *equilibrium at a given volume* is in one *macrostate*. To specify a *microstate* of this gas requires that we specify the position and velocity of all of the molecules in the system within the limits imposed by the uncertainty principle.

A system in one macrostate passes, with time, through many microstates. In fact it passes very rapidly through many microstates because atoms and molecules move and change direction rapidly. An atom in a solid, for example, vibrates at a frequency on the order of 10^{13} times per second. Gas molecules, as we have seen, have velocities of several hundreds of meters per second. The picture of a material as a rapidly changing system leads us to the realization that when we observe a property of a system, we are really seeing the average of this property in all of the microstates that the material passed through during the observation time. An important premise in statistical thermodynamics is that a system in a given macrostate can exist in every microstate consistent with the constraints of the macrostate. The *ergodic hypothesis* of microscopic thermodynamics states that *the time average of the properties of a system is equivalent to the instantaneous average over the ensemble of the microstates available to the system.*

The pressure in a gas results from the force the molecules exert on the container wall when they strike and bounce back from it. In the previous section, we calculated the *average* pressure assuming that all of the atoms were traveling at the root-mean-square velocity. If we were able to measure the *instantaneous* gas pressure in very short time intervals, we would observe a wildly oscillating pressure. What, in macroscopic thermodynamics, is called "pressure" is the average of this instantaneous pressure over time (Fig. 10.2).

To compute the macroscopic average of a property we need to know:

1. the property of each microstate,
2. which microstates the system can be in, and
3. the probability that the system will be in a given microstate.

To illustrate the difference between macrostates and microstates, consider the following examples. The first involves the throwing of two standard, six-sided dice. Each side of each die is labeled with a number, 1 through 6. The macrostate is the sum of the numbers on the two die faces. A microstate is a specific way of achieving that macrostate. For example, there are six ways (1-6, 2-5, 3-4, 4-3, 5-2, and 6-1) of throwing a seven. The macrostate is "seven." There are six microstates in that macrostate. There is only one microstate in the "12" macrostate (6-6).

As a second illustration, consider a system consisting of four distinguishable spheres, labeled A, B, C, and D. These spheres will be distributed between two containers, I and II. The macrostate is defined by the *number* of spheres in each of the two boxes. A microstate is a specific way of achieving the macrostate. For example, one macrostate is three spheres in box I and one sphere in box II. That macrostate can be achieved in four different ways. The sphere in box II can be either sphere A or B or C or D. In each case, the remaining spheres would be in box I.

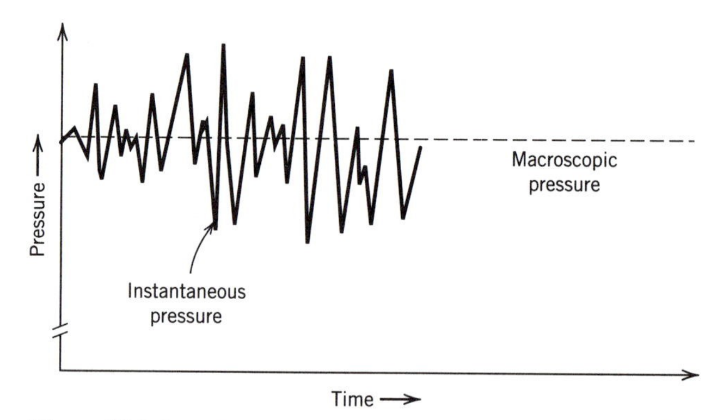

Figure 10.2 Instantaneous gas pressure as a function of time. (The magnitude of the variation of instantaneous gas pressure is exaggerated for emphasis.)

Table 10.1 Distributions of Four Spheres (A, B, C, and D) in Two Boxes I and II

			$\left[\frac{CD}{AD}\right]$		
			$\left[\frac{BD}{AC}\right]$		
		$\left[\frac{BCD}{A}\right]$	$\left[\frac{BC}{AD}\right]$	$\left[\frac{A}{BCD}\right]$	
		$\left[\frac{ACD}{B}\right]$	$\left[\frac{AD}{BC}\right]$	$\left[\frac{B}{ACD}\right]$	
		$\left[\frac{ABD}{C}\right]$	$\left[\frac{AC}{BD}\right]$	$\left[\frac{C}{ABD}\right]$	
	$\left[\frac{ABCD}{O}\right]$	$\left[\frac{ABC}{D}\right]$	$\left[\frac{AB}{CD}\right]$	$\left[\frac{D}{ABC}\right]$	$\left[\frac{0}{ABCD}\right]$
Box I	4	3	2	1	0
Box II	0	1	2	3	3
Ω	1	4	6	4	1

Table 10.1 lists the ways in which these spheres could be distributed in boxes I and II. The quantity Ω represents the number of different ways that the macroconfiguration can be achieved, and is sometimes called the thermodynamic probability.[2]

For the situation just described, the expression for Ω in the case of two boxes is:

$$\Omega = \frac{N!}{N_I!N_{II}!} \tag{10.12}$$

where

N_I = the number of spheres in box I

N_{II} = the number of spheres in box II, and

$N = N_I + N_{II}$.

[2]This is inconsistent with the usual definition of probability, in which the sum of the probabilities equals one. The two can be reconciled by dividing the microstates per macrostate by the total number of possible microstates.

In general, for i boxes, the term Ω is:

$$\Omega = \frac{N!}{\Pi_i N_i!} \tag{10.13}$$

where $\Pi_i N_i!$ is the product of all of the $N_i!$ terms.

To reiterate, stating the total number of particles in each box yields what is called a *macrostate*. Each way of realizing a given macro distribution is called a *microstate*. Each macrostate may be realized by a number of microstates.

10.3 ISOLATED SYSTEMS AND THE BOLTZMANN HYPOTHESIS

Consider an isolated system with N particles in a volume, V, with a fixed energy, E. Each microstate of the system must have the same energy, E. According to the premise stated in Section 10.2, all microstates are equally probable. That is, over a long period of time, the system spends an equal amount of time in each microstate. The average properties of the system are determined by the time average of the microstates.

Let us now consider the collection of possible microstates to be an ensemble. The accepted term in statistical thermodynamics for this collection of microstates with constant energy, volume, and number of particles is a *microcanonical ensemble.*

The Boltzmann hypothesis states that the entropy of a system is linearly related to the logarithm of Ω.

$$S = k \ln \Omega \tag{10.14}$$

The general form of the Boltzmann equation can be rationalized by considering two independent systems and comparing the total entropy of the two with the total probability function for the two. The entropy of the two taken together, S, is simply the *sum* of the two entropies, $S_1 + S_2$, because entropy is an extensive property in the macroscopic sense. The situation for probability is different. For each microstate in the first, all the microstates in the second must be counted. Thus total number of microstates in the two taken together is the *product* of the number in each, $\Omega = \Omega_1 \times \Omega_2$. Boltzmann hypothesized that entropy is proportional to a function of the thermodynamic probability (macrostates per microstate). Because of the additive nature of entropy and the multiplicative nature of thermodynamic probability, the function that relates the two had to be logarithmic. That is a rationalization of the Boltzmann equation (Eq. 10.14). The Boltzmann hypothesis is best viewed, however, simply as a brilliant insight that has stood the test of many experimental verifications.

To explore one manifestation of this hypothesis, let us reconsider the entropy of mixing of two components, derived in Chapter 3. This time let us consider the entropy of mixing of an ideal *solid* solution consisting of atoms labeled 1 and 2. In an ideal solution, the positioning of atoms on a lattice is random.[3]

For one mole of atoms (the sum of N_1 and N_2 is Avogadro's number of atoms, N_A) the function Ω is

$$\Omega = \frac{N_A!}{N_1!N_2!} \tag{10.15}$$

One way of thinking about this is to consider that there are N_A ways of introducing the atoms onto the lattice. The first atom can be chosen from among the N_A atoms in N_A different ways. The second may be chosen in $N_A - 1$ ways, the third in $N_A - 2$ ways, and so forth. The total number of different ways of putting all of the N_A atoms on the lattice is

$$N_A(N_A - 1)(N_A - 2) \cdots (3)(2)(1) = N_A!$$

Not all these different configurations are distinguishable. To account for this, we must divide by the number of different ways that the N_1 atoms can be distributed on their sites. Using an argument similar to the one above, rearranging the atoms labeled 1 on the lattice can be done in $N_1!$ ways. A similar relationship applies for N_2. The entropy of this configuration, S_M, the entropy of mixing N_1 and N_2,[4] is:

$$\underline{S}_M = k \ln \Omega = k \ln \frac{N_A!}{N_1!\, N_2!} \tag{10.16}$$

$$\underline{S}_M = k[\ln N_A! - \ln N_1! - \ln N_2!]$$

To evaluate $\ln X!$ we make use of the Stirling approximation (Section A.17, on factorials, in the volume appendix):

$$\underline{S}_M = k[N_A \ln N_A - N_A - N_1 \ln N_1 + N_1 - N_2 \ln N_2 + N_2]$$

$$\underline{S}_M = k[(N_1 + N_2)\ln N_A - N_1 \ln N_1 - N_2 \ln N_2]$$

$$\underline{S}_M = -k\left[N_1 \ln\left(\frac{N_1}{N_A}\right) + N_2 \ln\left(\frac{N_2}{N_A}\right)\right]$$

---

[3]In an ideal solution there is no enthalpy or energy of mixing. Hence, all the configurations are at the same energy level, and all are equally probable.

[4]The term S_M is also called the configurational entropy of the system because it is related to the uncertainty in the placement of the particles in the system relative to one another. The thermal entropy of the system (Section 2.9) is related to the uncertainty in the position of the particles due to thermal motion. The total entropy of the system is the sum of the two.

$$\underline{S}_M = -kN_A\left[\frac{N_1}{N_A}\ln\left(\frac{N_1}{N_A}\right) + \frac{N_2}{N_A}\ln\left(\frac{N_2}{N_A}\right)\right]$$
$$\underline{S}_M = -R[x_1\ln x_1 + x_2\ln x_2]$$

where x_i is the mole fraction of i and R is the gas constant.

To generalize, the entropy of mixing for one mole of an ideal solid solution of many (i) components, starting with Eq. 10.13 is:

$$\underline{S}_M = -R\sum_i x_i \ln x_i \qquad \textbf{(10.17)}$$

The Boltzmann hypothesis thus leads to the same equation we derived using the principles of classical thermodynamics (Section 7.4, Eq. 7.17).

10.4 THE THIRD LAW

It is interesting to contrast the entropy of mixing of a random (ideal) solid solution with the "mixing" of the materials 1 and 2 in a perfect compound, that is, one with no imperfections in the placement of atoms. The arrangement of the atoms in such a compound is shown schematically in Figure 10.3. The entropy of mixing for this compound is zero because the location of atoms on the lattice points is fixed.[5] Once the atom in the upper left-hand corner has been selected (in this schematic representation of a crystalline compound), it follows that the next one to it on the right and also the one below it must be atom 2, and so forth. This restriction results in a value of Ω equal to 1 because the second atom selected (to the right of the first) is not drawn from the total population N_A; it must be drawn from N_2. The next one to the right is drawn from a pool of $(N_1 - 1)$ atoms. The next is drawn from a pool of $(N_2 - 1)$, the next from $(N_1 - 2)$, and so forth. The numerator of the probability function Ω is thus $N_1!N_2!$. The denominator, which accounts for the indistinguish-

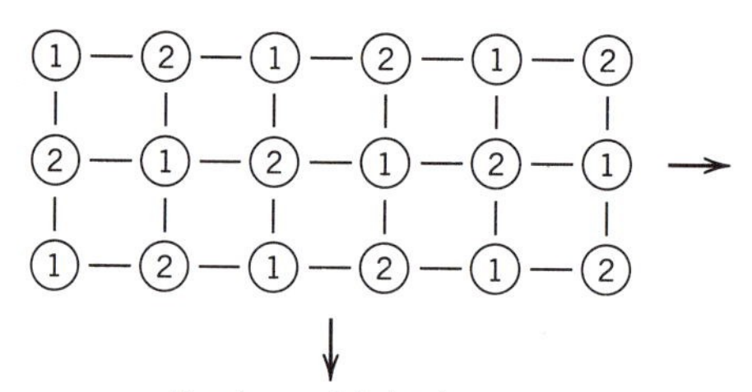

Figure 10.3 Arrangement of atoms in a perfect compound.

[5]This ignores the negligible randomness in the selection of the first atom, the one to occupy the upper left-hand corner.

ability of the "1" atoms on the "1" sites and the "2" atoms on the "2" sites, is also $N_1!\,N_2!$, as in Eq. 10.15. Hence the Ω function is one, and the entropy of mixing, $k \ln \Omega$, is zero.

This relationship gives some insight to the basis of the third law of thermodynamics. If two components (atoms) at absolute zero of temperature are combined to form a *perfectly ordered* crystal of the compound AB, the entropy change is zero. If, however, there is some randomness in the placement of the atoms (or molecules) in the crystal at absolute zero, the entropy change will not be zero. As an upper limit, the entropy at absolute zero of temperature will be the entropy of mixing of an ideal solution (Section 10.3).

10.5 SYSTEMS AT CONSTANT TEMPERATURE

In Section 10.2 we discussed the distribution of microstates and the properties of an *isolated* system, that is, one at *constant energy.* Consider now a closed system, with a fixed number of particles N and in a fixed volume V, immersed in a large constant temperature bath in which heat flow to or from the system is possible. The temperature of the system will be held constant, but the energy may fluctuate around some average value, just as the pressure of a system fluctuates around its average value (Figure 10.2). We can now think of an ensemble of systems, each at a temperature T but whose energy is allowed to vary. Such an arrangement is called a *canonical ensemble.*

An expression for the entropy of this canonical ensemble is given by

$$S = -k \sum_i P_i \ln P_i \tag{10.18}$$

where P_i is the probability that the system will be in microstate i, and the sum is the sum over all the microstates.

The logic behind this statement for entropy, sometimes called the Gibbs formulation, is discussed in many texts on statistical thermodynamics (see, e.g., Refs. 1–4). The Gibbs formulation can be shown to be identical to the Boltzmann formulation (Eq. 10.14). In a microcanonical ensemble (isolated system), each microstate is equally probable. Hence:

$$P = \frac{1}{\Omega}$$

$$S = -k \sum_{i=1}^{i=\Omega} \frac{1}{\Omega} \ln \left(\frac{1}{\Omega}\right)$$

$$S = -k\Omega \left[\frac{1}{\Omega} \ln \left(\frac{1}{\Omega}\right)\right]$$

$$S = k \ln \Omega$$

Either of the two statements, Gibbs or Boltzmann as appropriate, can be used to calculate equilibrium distributions and conditions by adopting the following principle:

> *The state of knowledge we must assume is the one that maximizes* S *relative to the information given.*

With systems at constant energy (microcanonical ensembles), this means that the macrostate with the most microstates is the most probable distribution and is the one that will be observed at equilibrium. The veracity of this statement is not particularly apparent from the example given in Section 10.2, with only four spheres distributed into two boxes. In that case the most probable distribution, two spheres in each box, existed in only six of the 16 possible cases. But this ''most probable'' distribution becomes much more probable than any other distribution as the number of spheres, or particles, increases. The probability of observing an appreciably different distribution is inversely proportional to the square root of n, where n is the total number of particles in the system in Ref. 4. Typically we will be dealing with systems containing on the order of 10^{23} particles. Thus, the probability of observing a distribution different from the one predicted by the principle stated above is minuscule.

10.6 INFORMATION AND ENTROPY

To illustrate the use of the principle just stated for the calculation of equilibrium distributions and conditions in a very simple example, consider the situation in Figure 10.4. One particle is to be located in one of the m boxes. In this case, all of the boxes are at the same energy level. What is the probability that the particle will be found in any one box? Intuitively, the answer is 1 divided by m. Any formal analysis that we apply to the situation should yield the same answer.

The principle stated above requires that we maximize the function S relative to the constraint $\Sigma_i P_i = 1$, where P_i is the probability of finding the particle in the ith box. The common and very useful method for maximizing functions relative to constraints is the Lagrange method of undetermined multipliers described in Section A.19. In this case the Lagrange method is applied as follows:

$$\frac{S}{k} = -\sum_i P_i \ln P_i$$

To maximize S/k (or minimize $-S/k$), we write:

$$\delta\left(-\frac{S}{k}\right) = 0 = \delta\left(\sum_i P_i \ln P_i\right)$$

$$\sum_i (P_i\, \delta \ln P_i + \ln P_i\, \delta P_i) = 0$$

$$\sum_i \delta P_i + \sum_i \ln P_i\, \delta P_i = 0$$

□ □ □ □ □ → m

Figure 10.4 Case of m possible locations for a particle.

Because the sum of the probabilities is one (see below):

$$\sum_i \delta P_i = 0$$
$$\sum_i \ln P_i \, \delta P_i = 0 \tag{10.19}$$

The constraint:

$$\sum_i P_i = 1$$

means that

$$\sum_i \delta P_i = 0 \tag{10.20}$$

Using a Lagrange multiplier of 1 for Eq. 10.19 and λ for Eq. 10.20, we have

$$\sum_i (\ln P_i + \lambda) \, \delta P_i = 0 \tag{10.21}$$

Recognize that the values of δP_i can be varied arbitrarily, subject to only one constraint: their sum must be zero. Thus their coefficient $(\ln P_i + \lambda)$ must be zero.

$$\ln P_i + \lambda = 0$$
$$\ln P_i = -\lambda \tag{10.22}$$
$$P_i = e^{-\lambda}$$

From Eqs. 10.20 and 10.22:

$$\sum_{i=1}^{i=m} e^{-\lambda} = 1; \qquad me^{-\lambda} = 1; \qquad e^{-\lambda} = \frac{1}{m}$$
$$P_i = \frac{1}{m} \tag{10.23}$$

The probability, as we expected intuitively, is $1/m$. The entropy of the situation as we described it is

$$S = -k \sum_{i=1}^{i=m} P_i \ln P_i = -k \sum_{i=1}^{i=m} \frac{1}{m} \ln \left(\frac{1}{m}\right) \qquad \textbf{(10.24)}$$
$$S = k \ln m$$

Thus, there is an entropy associated with the uncertainty, or randomness, in the location of the particle, the greater the uncertainty (m), the greater the entropy.

This aspect of entropy provides a useful quantitative measure of information and can be used to define the information content of a message. Suppose the entropy of a situation before the receipt of a message is S_b. After the receipt of a message, the entropy is reduced to S_a. The information contained in the message is the reduction in entropy, $S_a - S_b$. As an example, consider a situation in which a letter is located in any one of eight postal boxes. In this case $m = 8$. The entropy of the situation, a measure of the randomness *before* the message was received (S_b), is:

$$S_b = k \ln 8$$

Suppose a message is received indicating that the letter is in an even-numbered box. At this point, m is no longer 8; it is reduced to 4, because the message can be only in box 2, 4, 6, or 8. The entropy (S_a) in that situation is

$$S_a = k \ln 4$$

The reduction of entropy by the message is considered a measure of the information delivered by that message. In particular, the information in the message (I) is the reduction in entropy, the negative of ΔS:

$$\Delta S = S_a - S_b = k \ln 4 - k \ln 8 = -k \ln 2$$
$$I = -\Delta S = k \ln 2$$

10.7 BOLTZMANN DISTRIBUTION

Consider a situation similar to the one in Section 10.6, except that the boxes contain N particles, at different energy levels. Let us assume also that there is *no limit on the number of particles that may exist at any energy level.*[6]

In the situation shown in Figure 10.5, the task is to find the distribution of N particles among the m energy levels that will maximize the entropy S, subject to the

---

[6]This is different from the analysis required when the Pauli exclusion principle applies (i.e., where the number of particles, typically electrons, in each state is limited). These cases are considered later in Section 10.13.

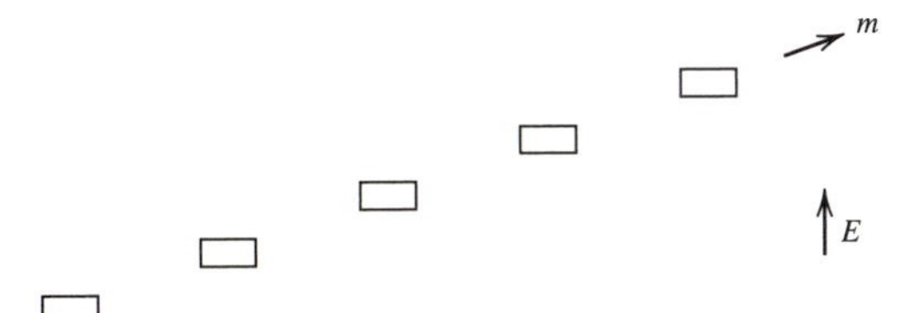

Figure 10.5 The m locations at different energy levels.

constraints that the energy $E = \Sigma_i N_i E_i$ and $N = \Sigma_i N_i$. The probability P_i, which is equal to the number of particles in state i divided by the total number of particles (N_i/N), is the same as the probability that any specific particle will be found in the state i.

$$\text{Maximize: } S = -k \sum_i P_i \ln P_i \tag{10.25}$$

$$\text{Constraint: } \langle E \rangle = \sum_i P_i E_i \tag{10.26}$$

$$\text{Constraint: } 1 = \sum_i P_i \tag{10.27}$$

where $\langle E \rangle$ is the average energy or expected energy of a particle.

Using the method of the Lagrange undetermined multipliers, we have the following tabulation:

	Multiplier	
$\sum_i \ln P_i \, \delta P_i = 0$	1	**(10.28)**
$\sum_i E_i \, \delta P_i = 0$	β	**(10.29)**
$\sum_i \delta P_i = 0$	$-\ln \alpha$	**(10.30)**

The result is

$$\sum_i (\ln P_i - \ln \alpha + \beta E_i)\delta P_i = 0$$

Therefore

$$\ln P_i - \ln \alpha + \beta E_i = 0 \tag{10.31}$$

$$P_i = \alpha e^{-\beta E_i}$$

Noting that

$$\sum_i P_i = 1$$

$$\alpha \sum_i e^{-\beta E_i} = 1 \qquad \textbf{(10.32)}$$

$$\alpha = \sum_i e^{-\beta E_i}$$

we can write

$$P_i = \frac{e^{-\beta E_i}}{\sum_i e^{-\beta E_i}}$$

$$P_i = \frac{e^{-\beta E_i}}{Z} \qquad \textbf{(10.33)}$$

The term $\Sigma_i e^{-\beta E_i}$ is called the *partition function.* In most texts on the subject, it is represented by the letter Z. It is the sum of the $e^{-\beta E_i}$ terms over all the states.[7] The entropy, in terms of the partition function and the average energy, can be given as

$$S = -k \sum_i P_i \ln P_i = -k \sum P_i(-\ln Z - \beta E_i)$$

$$S = k \sum_i P_i \ln Z + k \sum_i \beta P_i E_i$$

Noting Eqs. 10.26 and 10.27, and recognizing that β and ln Z do not depend on the index i, we can write

$$S = k \ln Z + k\beta \langle E \rangle \qquad \text{or} \qquad S = k \ln Z + k\beta U \qquad \textbf{(10.34)}$$

At this point, if we compare this statement of the entropy with the term derived from classical thermodynamics, we can arrive at the conclusion that the term β is related to temperature as follows:

$$\beta = \frac{1}{kT} \qquad \textbf{(10.35)}$$

The derivation of this relationship is given as the chapter Appendix.

---

[7]The letter Z is derived from the German word for sum of states, Zustandsumm.

The result is that the probability P_i that a particular particle will be found in the ith energy level is given by

$$P_i = \frac{\exp(-E_i/kT)}{Z} \qquad \textbf{(10.36)}$$

where $Z = \Sigma_i \exp(-E_i/kT)$.

To check Eq. 10.36 against a relationship derived using classical thermodynamics, consider the situation in an isothermal column of gas of height h. The potential energy of an atom at the *top* of the column relative to the *bottom* is the difference in potential energy, mgh (Figure 10.6).

Consider two spaces of equal volume in the column, one at the top and one at the bottom. The probability of finding any atom in the top (P_1) relative to the probability of finding it in the bottom (P_0) is

$$P_1 = \frac{\exp(-E_1/kT)}{Z} \quad \text{and} \quad P_0 = \exp\left(\frac{-E_0}{kT}\right) \qquad \textbf{(10.37)}$$

$$\frac{P_1}{P_0} = \exp\left(\frac{-E_1 - E_0}{kT}\right) = \exp\left(-\frac{mgh}{kT}\right)$$

Because this ratio of probability (Eq. 10.37) applies to all the atoms, the number of atoms in the volume at the top N_1 divided by the number of atoms in the volume at the bottom N_0 is

$$\frac{N_1}{N_0} = \exp\left(-\frac{mgh}{kT}\right)$$

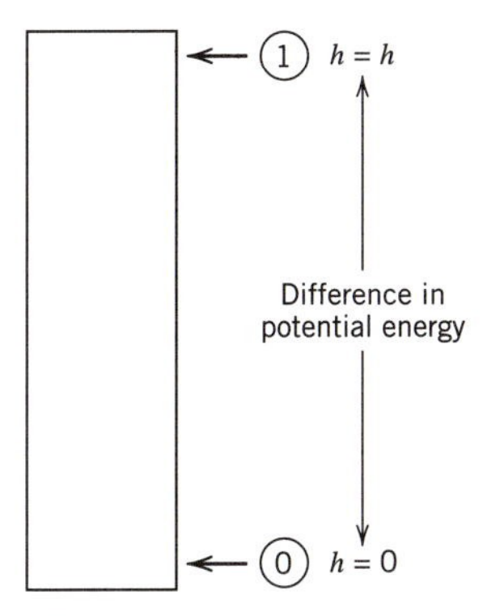

Figure 10.6 Isothermal column.

If the gas is ideal, the pressures in the top and bottom will be proportional to the number of atoms in equivalent volumes at constant temperature.

$$\frac{\text{pressure}_1}{\text{pressure}_0} = \exp\left(-\frac{mgh}{kT}\right) \tag{10.38a}$$

where m is the mass of an atom. If M is the mass of a mole, the equation is written as:

$$\text{pressure}_1 = \text{pressure}_0 \times \exp\left(-\frac{Mgh}{RT}\right) \tag{10.38b}$$

This is exactly the barometric equation that was derived from classical thermodynamic principles (Section 4.2).

10.8 DEGENERACY

At any energy level E_i there may be a number of states g_i each having the same energy, but possessing other distinguishing characteristics. This factor, g_i, is called the degeneracy or the statistical weight of the energy level. We must be careful in our statements to be clear on whether we are asking about the probability that a particle will be at a specified energy *level,* or whether it will be in a specific *state* within that level. The probability that it will be in a specific *state* is:

$$P_i = \frac{\exp(-E_i/kT)}{Z}$$

where

$$Z = \sum_i \exp\left(-\frac{E_i}{kT}\right) \qquad \text{summed over all the } \textit{states}$$

The probability that a particle will be in a specified energy *level,* is

$$P_j = \frac{g_j \exp(-E_j/kT)}{Z} \tag{10.39}$$

where

$$Z = \sum_j g_j \exp\left(-\frac{E_j}{kT}\right) \qquad \text{summed over all the levels} \tag{10.40}$$

Equations 10.39 and 10.40 may be used interchangeably. One can sum over each possible state, considering all the substates within one energy level to be independent, in which case Eq. 10.39 applies. Alternatively, if one accounts for the number of substates in each energy level, Eq. 10.40 applies, and we sum over energy levels. The usefulness of the latter form will be observed in Section 10.12, when we derive the Maxwell–Boltzmann distribution of energy levels or velocities in an ideal gas, or when, in metal physics, we consider the question of electronic energy levels in a conduction band.

10.9 PARTITION FUNCTION

As defined in the preceding section, the partition function Z is simply the sum over all the energy states allowed (or the sum over all energy levels accounting for degeneracy). It is a particularly useful function because all the thermodynamic properties of a system may be calculated once the partition function is known. To illustrate, consider a system consisting of one particle. The entropy of this particle as it exists among the various energy states (Eq. 10.34) is

$$S = k \ln Z + \frac{U}{T} \tag{10.41}$$

The internal energy of this particle is

$$U = \sum_i E_i P_i = \sum_i E_i \frac{\exp(-E_i/kT)}{Z} = \frac{1}{Z} \sum E_i \exp\left(-\frac{E_i}{kT}\right)$$

Note that

$$Z = \sum_i \exp\left(-\frac{E_i}{kT}\right)$$

$$\left(\frac{\partial Z}{\partial T}\right)_V = \sum_i \exp\left(-\frac{E_i}{kT}\right) \frac{E_i}{kT^2}$$

$$\left(\frac{\partial Z}{\partial T}\right)_V = \frac{1}{kT^2} \sum_i E_i \exp\left(-\frac{E_i}{kT}\right)$$

Therefore,

$$U = \frac{1}{Z} kT^2 \left(\frac{\partial Z}{\partial T}\right)_V = kT^2 \left[\frac{\partial(\ln Z)}{\partial T}\right]_V \tag{10.42a}$$

Then

$$S = k \ln Z + kT \left(\frac{\partial \ln Z}{\partial T}\right)_V \tag{10.42b}$$

Other functions such as the Helmholtz free energy are (from Eq. 10.41):

$$F = U - TS = -kT \ln Z \quad \textbf{(10.43)}$$

10.10 DISTINGUISHABILITY OF PARTICLES

When considering many particles, the question of distinguishability among them arises. In a solid material where atoms are localized—that is, they vibrate about some fixed location—the particles (the atoms) are distinguishable. If the atoms are noninteracting—that is, if we do not have to account for the energies of combinations of atoms—then the grand partition function for all the atoms, Φ, is the product of the partition functions of each of the individual atoms (Eq. 10.44). To understand this relationship, consider the grand partition function for two atoms. The grand partition function must take into account all combinations of the energy states of both atoms. For each state of the first, we must account for Z states of the second. Thus for the two, we have Z times Z, or Z^2 states. Taking N particles into account yields Eq. 10.44.

$$\Phi = Z^N \quad \textbf{(10.44)}$$

In the case of atoms or molecules in a gas, the particles are not localized and are indistinguishable. The grand partition function for this collection of N *identical,*

Table 10.2 Partition Function for Many (N) Noninteracting Particles (Φ) based on Partition Function for One Particle (Z)

Indistinguishable Particles (Gas)	Distinguishable Particles (Solid)
$\Phi = \dfrac{Z^N}{N!}$	$\Phi = Z^N$
$\ln \Phi = N \ln \dfrac{Z}{N} + N$	$\ln \Phi = N \ln Z$
$\left(\dfrac{\partial \ln \Phi}{\partial T}\right)_V = N\left(\dfrac{\partial (\ln Z)}{\partial T}\right)_V$	$\left(\dfrac{\partial \ln \Phi}{\partial T}\right)_V = N\left(\dfrac{\partial (\ln Z)}{\partial T}\right)_V$
$S = kN\left[\ln\left(\dfrac{Z}{N}\right) + 1\right] + NkT\left(\dfrac{\partial (\ln Z)}{\partial T}\right)_V$	$S = kN \ln Z + NkT\left(\dfrac{\partial (\ln Z)}{\partial T}\right)_V$
$F = -NkT\left(\ln \dfrac{Z}{N} + 1\right)$	$F = -NkT \ln Z$
$P = -\left(\dfrac{\partial A}{\partial V}\right)_T = NkT\left(\dfrac{\partial (\ln Z)}{\partial V}\right)_T$	$P = -\left(\dfrac{\partial A}{\partial V}\right)_T = NkT\left(\dfrac{\partial (\ln Z)}{\partial T}\right)_V$

noninteracting particles is the product of the partition function of each of the particles divided by the term $N!$ to take account of the number of ways in which the particles may be rearranged indistinguishably.

$$\Phi = \frac{Z^N}{N!} \qquad \text{for gases} \tag{10.45}$$

Table 10.2 is a tabulation of some thermodynamic functions in terms of the partition function for localized and nonlocalized particles.

10.11 IDEAL GAS

The equations and methods illustrated in this chapter can now be applied to calculating the thermodynamic properties of a monatomic ideal gas. In this case we will consider the energy levels involved in the translational motion of the ideal gas. Because there is only one atom in the molecule, we need not be concerned with rotation or vibration. Rotational and vibrational states must be considered in more complex molecules, but the monatomic gas can be analyzed without them. The energies involved in electronic excitations are usually much larger than the kinetic energies and can be neglected in a first approximation.

To calculate the properties of the gas, consider the motion of a particle of mass m in a cubical box of side L and volume L^3. If we consider the velocity of the particle to be quantized, λ, its wavelength in the x direction, is defined by the DeBroglie relationship (Figure 10.7):

$$mv = \frac{h}{\lambda}$$

Because the box is of size L, the geometrical constraint on the values of λ are:

$$\lambda = \frac{2L}{i} \qquad \text{where} \quad i = 1, 2, 3, 4, 5, \ldots$$

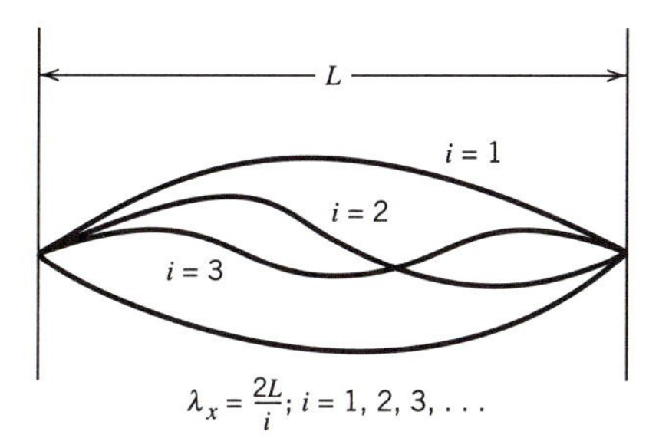

Figure 10.7 De Broglie wavelengths.

The velocity in the x direction and the various energy levels are:

$$v_x = \frac{h_i}{2mL} \quad \text{where} \quad i = 1, 2, 3, \ldots$$
$$E_x = \frac{1}{2} m v_x^2 = \frac{h^2 i^2}{8mL^2} \quad \text{where} \quad i = 1, 2, 3, \ldots \tag{10.46}$$

The partition function is:

$$Z_x = \sum_{i=1}^{i=\infty} \exp\left(-\frac{E_i}{kT}\right) = \sum_{i=1}^{i=\infty} \exp\left[-\left(\frac{h^2}{8mL^2kT}\right) i^2\right] \tag{10.47}$$

Recognizing that the energy levels are very close together, the summation may be evaluated using an integral as follows:

$$Z_x = \int_0^\infty \exp\left[\left(-\frac{h^2}{8mL^2kT}\right) i^2\right] di \tag{10.48}$$

Noting that the value of the definite integral:

$$\int_0^\infty e^{-ax^2}\, dx = \frac{1}{2}\left(\frac{\pi}{a}\right)^{1/2} \tag{10.49}$$

Thus the partition function is equal to:

$$Z_x = \left(\frac{2\pi mkT}{h^2}\right)^{1/2} L \tag{10.50}$$

The partition function in all three directions, x, y, and z, is the product of the partition function of the individual partition functions in each direction:

$$Z_{xyz} = Z_x Z_y Z_z = \left(\frac{2\pi mkT}{h^2}\right)^{3/2} L^3 = \left(\frac{2\pi mkT}{h^2}\right)^{3/2} V \tag{10.51}$$

To explore the results of our study of statistical thermodynamics so far, let us make use of the partition function of an ideal gas derived above to derive an expression for the pressure of the gas in terms of other variables. We know that the pressure of the ideal gas can be calculated from an expression for the Helmholtz free energy, F, because:

$$P = -\left(\frac{\partial F}{\partial \underline{V}}\right)_T$$

The Helmholtz free energy in terms of the partition function for one mole of gas is (see Table 10.2):

$$\underline{F} = -N_A kT \left(\ln \frac{Z}{N_A} + 1 \right) \tag{10.52}$$

Hence:

$$P = \left(\frac{\partial \left[N_A kT \left(\ln \frac{Z}{N_A} + 1 \right) \right]}{\partial \underline{V}} \right)_T = N_A kT \left(\frac{\partial \ln Z}{\partial \underline{V}} \right)_T$$

Combining with Eq. 10.51:

$$P = \frac{N_A kT}{\underline{V}} = \frac{RT}{\underline{V}}$$

This derivation of the ideal gas equation using statistical methods should give us additional confidence in the conclusions drawn from statistical thermodynamics.

10.12 MAXWELL–BOLTZMANN DISTRIBUTION: IDEAL GAS

Earlier in this chapter (Section 10.1), an equation was derived for the average energy of an atom in a monatomic gas. The equation as derived gave no information concerning the distribution of energies in the gas, that is, the distribution of velocities. Now we use the equations derived describing the energy of an individual particle and the principles laid down in Section 10.10 to derive this energy and velocity distribution.

The energy of a monatomic gas molecule imparted by its motion (translation only) is:

$$E_{i,j,k} = \frac{h^2}{8mL^2}(i^2 + j^2 + k^2) \tag{10.53}$$

For each value of i, j, and k, there is a quantum energy level. Notice that there may be some degeneracy because different values of i, j, and k can yield the same value of energy. For example, the energy levels are the same for i, j, and k values of (1, 2, 1) and (1, 1, 2). Similarly for levels (5, 1, 1) and (3, 3, 3), the energy values are the same.

One way to handle this degeneracy is to assign a unit volume to each point in

i-j-k space. In two dimensions this can be represented as in Figure 10.8. Imagine a sphere with its center at the origin of the i-j-k axes. The radius of such a sphere is

$$R^2 = i^2 + j^2 + k^2 = \frac{8mL^2E_{i,j,k}}{h^2} \tag{10.54}$$

where i, j, and k are numbers 1, 2, 3, 4, ...

The volume of that portion of the sphere that lies along the positive i-j-k axes is one-eighth the volume of the sphere in i-j-k space

$$V = \frac{1}{8}\left(\frac{4}{3}\pi R^3\right) = \frac{\pi}{6}\left[\frac{8mL^2E_{i,j,k}}{h^2}\right]^{3/2} \tag{10.55}$$

Because in i-j-k space each state has a volume of 1 associated with it, the number of states, N_s, with energy less than E is:

$$N_s = \frac{\pi}{6}\frac{V}{h^3}(8m)^{3/2}E^{3/2} \tag{10.56}$$

The density of states $g(E_i)$ is the rate of change of N_s with E, that is,

$$g(E_i) = \frac{\pi}{6}\frac{V}{h^3}(8m)^{3/2}\left(\frac{3}{2}\right)E_i^{1/2} = 4\sqrt{2}\pi\frac{V}{h^3}m^{3/2}E_i^{1/2} \tag{10.57}$$

where $g(E_i)$ is the number of states between E and dE.

The probability of occupancy of the state E_i, $P(E_i)$ is

$$P(E_i) = g(E_i)\exp\frac{(-E_i/kT)}{Z} = \frac{N(E_i)}{N} \tag{10.58}$$

And the partition function, Z, of an ideal gas is

$$Z = \frac{V}{h^3}(2\pi mkT)^{3/2}$$

Hence, the number of atoms in each energy state $[N(E_i)]$ in a mole (N) of gas is the product of the density of states and the probability of occupancy:

$$N(E_i) = \frac{N\,4\sqrt{2}\pi E_i^{1/2}\exp(-E_i/kT)dE}{(2\pi kT)^{3/2}} \tag{10.59}$$

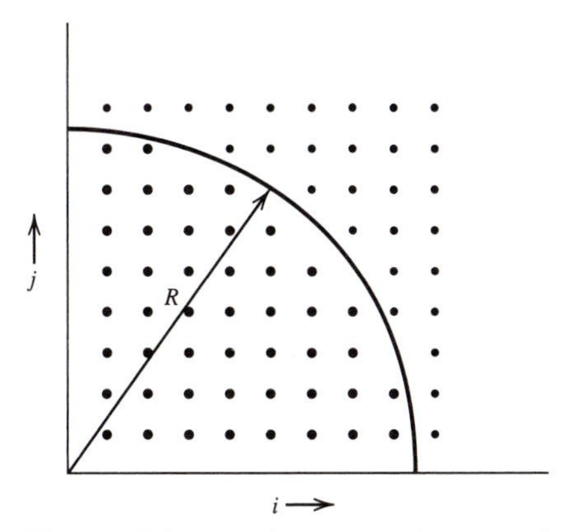

Figure 10.8 Two dimensions of $i - j - k$ space.

Equation 10.59 is the Maxwell–Boltzmann distribution for an ideal gas in terms of energy. To convert it to a distribution of velocities, we note that the energy of a particle is $\frac{1}{2}mv^2$.

$$E^{1/2} = \frac{1}{\sqrt{2}}\, m^{1/2} v \qquad \text{and} \qquad dE = mv\, dv$$

Thus:

$$N(v) = 4\pi N \left(\frac{m}{2\pi kT}\right)^{3/2} \exp\left(-\frac{mv^2}{2kT}\right) v^2\, dv \tag{10.60}$$

A graphical representation of this distribution is shown in Figure 10.9. The most probable velocity (v_{mp}) is at the point at which the derivative of the distribution with

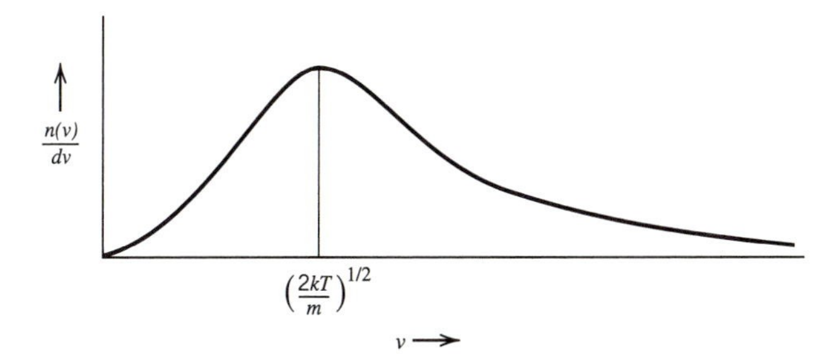

Figure 10.9 Maxwell–Boltzmann distribution of velocity.

respect to the velocity is zero

$$\left(\frac{\partial(N(v))}{\partial v}\right)_T = 0$$
$$v_{\text{mp}} = \left(\frac{2kT}{m}\right)^{1/2} \tag{10.61}$$

To calculate the average root-mean-square velocity:

$$\bar{v}^2 = \frac{\int_0^\infty v^2[N(v)]}{N}$$
$$\bar{v}^2 = 4\pi\left(\frac{m}{2\pi kT}\right)^{3/2}\int_0^\infty \exp\left(-\frac{mv^2}{2kT}\right) v^4\,dv \tag{10.62}$$

The definite integral is

$$\int_0^\infty x^4 e^{-\beta x^2}\,dx = \frac{1}{2}\sqrt{\pi}\,\frac{3}{4}\,\beta^{-5/2}$$
$$\beta = \frac{m}{2kT}$$
$$\bar{v}^2 = \frac{3kT}{m} \quad \text{or} \quad \frac{1}{2}mv^2 = \frac{3}{2}kT$$

It is comforting to note that the expression derived from this more extensive treatment is the same as the one derived using the simpler approach of Section 10.1. The additional effort expended in the calculation provides some extra information, which is useful in the study of the dynamics of atom movement.

10.13 FERMI–DIRAC DISTRIBUTION

The Maxwell–Boltzmann distribution (Section 10.12) was derived based on the assumption that there is no limit to the number of particles that may occupy a specific state defined by the values of i, j, and k; that is, they are not subject to the Pauli exclusion principle. In the case of "free" (conduction) electrons in a metal, this is an unjustified assumption. The Pauli exclusion principle, as it applies to an electron gas, states that no two electrons may have the same set of quantum numbers.[8] Thus, we must modify our approach to statistics.

Assume that an energy level E_i has a degeneracy g_i; that is, g_i states may exist at

[8]This assumes that we take into account the electron spin quantum numbers.

the energy level E_i. At this energy level, we can think of the states being occupied as in Figure 10.10.

To arrive at a value for the function Ω in Eq. 10.14, we need an expression for the number of microstates per macrostate. In this case, the macrostate is defined by g_i, the number of states at E_i, and the number of them that are occupied, N_i. A microstate consists of one way of arranging the N_i particles among the g_i places available. The total number of ways of arranging them is equivalent in statistical terms to the number of combinations of N objects taken M at a time, which, from probability considerations, is

$$C_N^M = \frac{N!}{(N - M)!M!}$$

Thus for the ith level with g_i states, N_i of which are occupied,

$$C_{g_i} = \frac{g_i!}{N_i!(g_i - N_i)!} \qquad \textbf{(10.63)}$$

The function Ω is the product of these terms for each energy level:

$$\Omega = \Pi_i \frac{g_i!}{N_i!(g_i - N_i)!} \qquad \textbf{(10.64)}$$

The entropy, $k \ln \Omega$, is

$$S = k \sum_i [g_i \ln g_i - N_i \ln N_i - (g_i - N_i) \ln(g_i - N_i)] \qquad \textbf{(10.65)}$$

To derive an expression for the population of the various states N_i, we maximize S (or $\ln \Omega$) subject to two constraints:

$$\sum_i E_i N_i = U$$
$$\sum_i N_i = N \qquad \textbf{(10.66)}$$

Applying the Lagrange method (Section A.19)

	Multiplier	
$\sum_i [-\ln N_i + \ln (g_i - N_i)]\, \delta N_i = 0$	1	**(10.67)**
$\sum_i E_i\, \delta N_i = 0$	$-\beta$	**(10.68)**
$\sum_i \delta N_i = 0$	$-\alpha$	**(10.69)**

Adding Eqs. 10.67, 10.68, and 10.69, we have

$$\sum_i [-\ln N_i + \ln(g_i - N_i) - \alpha - \beta E_i]\, \delta N_i = 0 \qquad \textbf{(10.70)}$$

For Eq. 10.70 to be true for all δN_i, we must have

$$-\ln N_i + \ln(g_i - N_i) - \alpha - \beta E_i = 0$$

$$\ln\left(\frac{g_i - N_i}{N_i}\right) = \alpha + \beta E_i \qquad \textbf{(10.71)}$$

This yields

$$N_i = \frac{g_i}{1 + \exp(\alpha + \beta E_i)} \qquad \textbf{(10.72)}$$

Noting the expression for entropy, $S = k \ln \Omega$, substituting the expression for N_i in Eq. 10.65, and differentiating, we have:

$$dS = k \sum_i \ln\left(\frac{g_i - N_i}{N_i}\right) dN_i$$

Combining with Eq. 10.71 yields

$$dS = k \sum_i (\alpha + \beta E_i)\, dN_i$$

$$dS = k\alpha \sum_i dN_i + k\beta \sum_i E_i\, dN_i \qquad \textbf{(10.73)}$$

$$dS = k\alpha\, dN + k\beta\, dU$$

Note the difference between the changes in N_i in Eqs. 10.73 and 10.68. In Eq. 10.68 we were dealing with perturbations in N_i, δN_i, the sum of which must be zero. In Eq. 10.73 we are asking how the function, S in this case, changes as the number of particles, N_i, changes. The sum of the changes in N_i, $\Sigma\, dN_i$, is equal to the total change in the number of particles, dN.

To evaluate the α and β multipliers, recall the definition of the Helmholtz free energy function $F = U - TS$:

$$S = -\frac{F}{T} + \frac{U}{T}$$

Differentiating S at constant T:

$$dS = -\frac{dF}{T} + \frac{dU}{T} \tag{10.74}$$

Comparing Eqs. 10.73 and 10.74, we conclude that

$$\begin{aligned} k\alpha\, dN &= -\frac{dF}{T}; \qquad k\beta = \frac{1}{T} \\ \alpha &= \frac{1}{kT}\frac{dF}{dN} = -\frac{\mu}{kT}; \qquad \beta = \frac{1}{kT} \end{aligned} \tag{10.75}$$

Thus,

$$N_i = \frac{g_i}{1 + \exp\left(\dfrac{E_i - \mu}{kT}\right)} \tag{10.76}$$

This is the Fermi–Dirac distribution.

It is interesting to note the similarity between Eq. 10.76 and the equation derived from Boltzmann statistics, namely:

$$N_i = \frac{g_i \exp(-E_i/kT)}{Z}$$

Note that

$$\mu = \left(\frac{\partial F}{\partial N}\right)_{V,T} \qquad \text{and} \qquad F = -NkT \ln Z$$

Hence:

$$Z = \exp\left(-\frac{\mu}{kT}\right)$$

$$N_i = \frac{g_i \exp(-E_i/kT)}{\exp(-\mu/kT)}$$

The Maxwell–Boltzmann distribution then can be expressed as

$$N_i = \frac{g_i}{\exp\left(\dfrac{E_i - \mu}{kT}\right)} \tag{10.77}$$

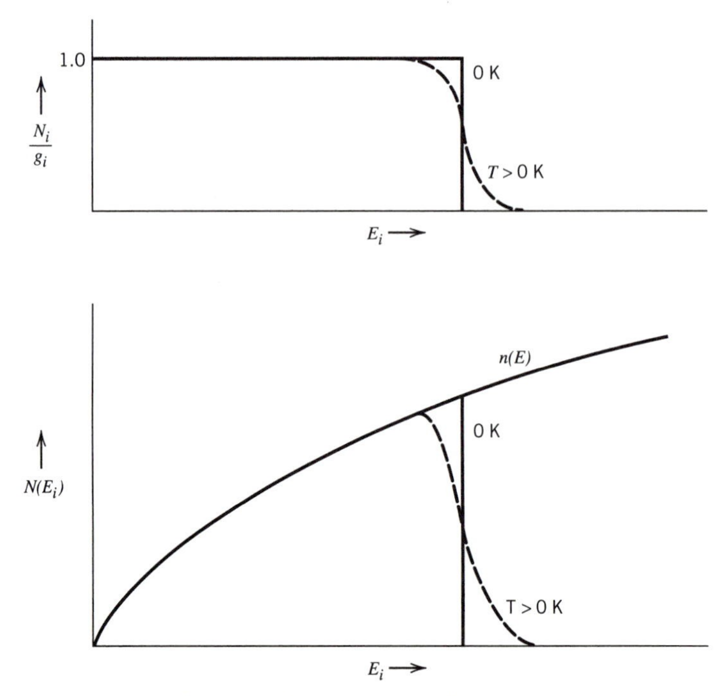

Figure 10.10 Fermi–Dirac distribution.

Comparing Eqs. 10.76 and 10.77, we can see that the Boltzmann distribution and the Fermi–Dirac distribution are equivalent when

$$\exp\left(\frac{E_i - \mu}{kT}\right) \gg 1$$

Figure 10.10 shows the Fermi–Dirac distribution expressed as fractional occupancy N_i/g_i, at 0° K and at moderate temperatures. In contrast with the Maxwell–Boltzmann distribution, the population of the states in the Fermi–Dirac distribution changes very little with temperature (i.e., only near the high energy end of the distribution). This means that the ''heat capacity'' of a collection of Fermi–Dirac particles will be low compared to the corresponding value for distinguishable classical particles. To change the temperature, it is necessary for only a few particles to change energy levels. This contrasts with a collection of Maxwell–Boltzmann particles, in which many particles change energy levels upon changing temperature.

10.14 HEAT CAPACITY OF SOLIDS

Based on Sections 10.9 and 10.10, it can be shown that the heat capacity of a material (gas, liquid, or solid) can be derived if the partition function of the material is known.

The internal energy of the material can be expressed in terms of the partition function using Eq. 10.42a:

$$U = kT^2 \left(\frac{\partial (\ln Z)}{\partial T} \right)_V \quad \textbf{(10.42a)}$$

The heat capacity is the derivative of the internal energy with respect to temperature at constant volume.

$$C_v = \left(\frac{\partial \underline{U}}{\partial T} \right)_V$$

For an ideal gas, the partition function for one molecule is:

$$Z = \left(\frac{2\pi mkT}{h^2} \right)^{3/2} V$$

For an assembly of N molecules, the partition function is:

$$\Phi = \frac{Z^N}{N!}$$

The natural logarithm of the partition function is:

$$\ln \Phi = \ln \left(\frac{2\pi m}{h^2} \right)^{3/2} + \ln \left(\frac{V^N}{N!} \right) + \frac{3}{2} N \ln(kT)$$

Thus the internal energy of an ideal gas is

$$U = kT^2 \left(\frac{\partial (\ln \Phi)}{\partial T} \right)_V = kT^2 \frac{3}{2} \frac{N}{T} = \frac{3}{2} RT$$

The heat capacity is the derivative of the internal energy with respect to temperature, which is simply $\frac{3}{2}R$. The heat capacity of an ideal gas is thus independent of temperature and equal to $\frac{3}{2}R$ throughout the temperature range.

It has been observed that the heat capacity of most solids (either C_P or C_V) at high temperatures is approximately three times the gas constant R. This relationship, $C = 3R$, is known as the law of Dulong and Petit. While most solids have a value of heat capacity of about $3R$ at high temperatures, it is also known that the heat capacity of solids is much lower than that at low temperatures. In particular, at temperatures approaching absolute zero, the heat capacity of the solid materials approaches zero. This phenomenon cannot be understood in terms of partition functions of the kind used to describe ideal gases. In addition to the vanishingly small heat capacities near absolute zero, another observation was made. Lewis and Gibson in 1917 observed that if one assigned to each material a characteristic temperature value Θ, the constant volume heat capacities of a number of materials (aluminum, copper, lead, and diamond), when plotted as the logarithm of T/Θ, all fell on the same curve.

In an effort to explain this phenomenon, Einstein, in the early 1920s, proposed that solids be analyzed as sets of identical harmonic oscillators. Each atom was considered to be an independent oscillator in three dimensions. Einstein relied on quantum mechanical concepts for permitted energies of one-dimensional oscillators, which are of the form:

$$E = (i + \tfrac{1}{2})h\nu \qquad i = 0, 1, 2, \ldots \tag{10.78}$$

The partition function for the one-dimensional oscillator is

$$Z = \sum_i \exp\left[-\left(i + \frac{1}{2}\right)\frac{h\nu}{kT}\right]$$

$$Z = \exp\left[-\frac{h\nu}{2kT}\right] + \exp\left[-\frac{3}{2}\frac{h\nu}{kT}\right] + \exp\left[-\frac{5}{2}\frac{h\nu}{kT}\right] + \cdots$$

Substituting

$$y = \exp\left[-\frac{h\nu}{kT}\right]$$

Thus

$$Z = y^{1/2}(1 + y + y^2 + \cdots) = y^{1/2}\sum_1^{\infty} y^i = y^{1/2}\left(\frac{1}{1 - y}\right)$$

The partition function is then

$$Z = \frac{\exp\left[-\dfrac{h\nu}{2kT}\right]}{1 - \exp\left[-\dfrac{h\nu}{kT}\right]} \tag{10.79}$$

If one defines a characteristic temperature Θ as $h\nu/k$, then the partition function is:

$$Z = \frac{\exp[-\Theta/2T]}{1 - \exp[-\Theta/T]} \tag{10.80}$$

Applying Eq. 10.42a, the internal energy of N oscillators is

$$U = NkT^2 \left[\frac{\partial(\ln Z)}{\partial T}\right]_V = \frac{1}{2} Nk\Theta \left[\frac{1 + \exp(-\Theta/T)}{1 - \exp(-\Theta/T)}\right] \quad \textbf{(10.81)}$$

If we assume that the oscillations occur independently in three dimensions, then N becomes $3N$. Thus the internal energy is

$$U = \frac{3}{2} Nk\Theta \left[\frac{1 + \exp(-\Theta/T)}{1 - \exp(-\Theta/T)}\right] \quad \textbf{(10.82)}$$

The constant volume heat capacity C_V is the temperature derivative of the internal energy at constant volume:

$$C_V = 3R \left(\frac{\Theta}{T}\right)^2 \frac{\exp(-\Theta/T)}{(1 - \exp(\Theta/T))^2} \quad \textbf{(10.83)}$$

Equation 10.83 has the two characteristics of the observed heat capacities: that is, they approach zero near absolute zero of temperature, and they approach $3R$ at higher temperatures. At very low temperatures the value of Θ/T becomes very large (approaches infinity). The exponential terms in Eq. 10.83 will dominate the value of C_v, and the value of C_v approaches zero. At higher temperatures, where T is much larger than Θ, the exponential terms can be approximated as $\exp(-\Theta/T) = 1 - \Theta/T$, and the value of heat capacity can be shown to approach $3R$.

Although this approach successfully explained the law of Dulong and Petit and was consistent with the experimental observation that the heat capacity approached zero at absolute zero of temperature, the form of the equation at very low temperatures did not match exactly the measured heat capacity curves. It remained for Debye to suggest another form of the vibration equations for the individual atoms. Debye realized that the individual vibrating atoms are not really independent, but are strongly coupled. He analyzed the situation by treating an entire crystal as an elastic medium with a range of frequencies, varying from zero to some maximum value ν_D which is characteristic of the material and the crystal. He defined the ν_D such that the total number of vibrational modes is equal to $3N$. Debye then proceeded in a manner similar to Einstein's approach and defined a Θ_D as the Debye characteristic temperature. The resulting equation is:

$$C_V = 9R \left(\frac{T}{\Theta_D}\right)^3 \int_0^{\Theta_D/T} \frac{x^4 e^{-x}}{(1 - x)^2}\, dx \quad \textbf{(10.84)}$$

At very low temperatures, this equation can be approximated by

$$C_V = 1943 \left(\frac{T}{\Theta_D}\right)^3 \quad \textbf{(10.85)}$$

This Debye T^3 law is useful for fitting experimental data for a particular material and provides a rational basis for extrapolating data to zero kelvin. This approach also yields a heat capacity that is $3R$ at high temperatures.

10.15 MOLECULAR WEIGHT DISTRIBUTION IN POLYMERS

Polymers are large molecules composed of many repetitive units (mers). The polymers are formed by joining the mers into molecules of varying length (or molecular weight). A particular polymer may be characterized by an *average* length or molecular weight, just as the velocity of gas molecules at a specific pressure and temperature may be characterized by an average speed or absolute velocity. But, as we demonstrated in Sections 10.11 and 10.12, all the gaseous atoms or molecules do not move at the average absolute velocity. Each has its own velocity; that is, there is a distribution of velocities. Similarly a polymeric material with an average molecular weight is composed of molecules varying in length or molecular weight. The molecular weight *distribution* may be calculated using the Lagrange method.

For this calculation, let us use a polymeric material consisting of Avogadro's number of mers, N_A. Other definitions needed in the derivation are as follows:

X = segments (mers) in a molecule
N_X = number of molecules of size X
N = total number of molecules
n_X = N_x/N = mole fraction of molecules with x mers
N_A = Avogadro's number of mers
N_b = number of bonds
P = degree of polymerization (fraction polymerized), where P is defined as the number of bonds between segments (mers) divided by the total number of segments

From Figure 10.11 it should be apparent that a molecule of length X has $X - 1$ bonds in it. Thus the total number of bonds is

$$N_b = \sum_X (X - 1)N_X \tag{10.86}$$

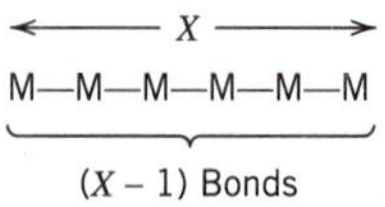

$$\overset{\longleftarrow\; X \;\longrightarrow}{\text{M—M—M—M—M—M}}$$
$$(X-1)\ \text{Bonds}$$

Figure 10.11 Demonstration that a polymer with X segments has $(X - 1)$ bonds.

The degree of polymerization defined as the number of bonds divided by segments is:

$$P = \frac{\Sigma_X (X - 1)N_X}{N_A} = \frac{\Sigma_X XN_X - \Sigma_X N_X}{N_A} = 1 - \frac{N}{N_A}$$

$$1 - P = \frac{N}{N_A} \qquad \textbf{(10.87)}$$

At this point if we were to assume that all of the molecules were of one size, the average size $\overline{X}$, we could derive an equation for the degree of polymerization as a function of this average molecular size. The conclusions would be:

$$\overline{X}N = N_A$$

$$P = 1 - \frac{1}{\overline{X}}$$

If $\overline{X} = 10$, $\quad P = 0.90$
If $\overline{X} = 100$, $\quad P = 0.99$

Of course, we know that all the molecules are not of the average size. To calculate the molecular size distribution, we use the Lagrange method. One constraint is that the sum of the mole fractions of various sizes must be 1:

$$\sum_X n_X = 1 \qquad \textbf{(10.88)}$$

A second constraint is that the sum of the number of molecules of size X multiplied by X equal Avogadro's number, the total number of mers.

$$\sum_X XN_X = N_A \quad \text{or} \quad \sum X\frac{N_X}{N} = \frac{N_A}{N} \quad \text{or} \quad \sum_X Xn_X = \frac{N_A}{N} \qquad \textbf{(10.89)}$$

The third equation is based on the principle enunciated in Section 10.5: the state of knowledge we must assume is the one that maximizes S relative to the information given. In this case we are trying to maximize the function:

$$-\sum_X n_X \ln n_X \qquad \textbf{(10.90)}$$

Using the Lagrange method on these three relationships, we have:

			Multiplier
Maximize:	$\sum_x n_X \ln n_i$	$\Rightarrow \sum_X \ln n_X \, \delta n_X = 0$	1
Constraint:	$\sum_X n_X = 1$	$\Rightarrow \sum_X \delta n_X = 0$	λ_0
Constraint:	$\sum_X X n_x = \frac{N_A}{N}$	$\Rightarrow \sum_X X \, \delta n_X = 0$	λ

$$\sum_X (\ln n_X + \lambda X + \lambda_0) \, \delta n_X = 0$$

Based on the Lagrange method, the coefficient of each δn term must be zero. Thus:

$$\ln n_X = -\lambda_0 - \lambda X \qquad \textbf{(10.91)}$$

$$n_X = e^{-\lambda_0} e^{-\lambda X}$$

To solve for the various undetermined multipliers, we apply Eqs. 10.88 and 10.89.

$$\sum_X n_X = 1$$

$$e^{-\lambda_0} \sum_1^\infty e^{-\lambda X} = 1 \Rightarrow e^{\lambda_0} = \sum_1^\infty e^{-\lambda X}$$

$$n_X = \frac{e^{-\lambda X}}{\sum_1^\infty e^{-\lambda X}}$$

and from Eq. 10.89,

$$\sum_1^\infty X n_X = \frac{N_A}{N} = \frac{\sum_1^\infty X e^{-\lambda X}}{\sum_1^\infty e^{-\lambda X}}$$

To evaluate the expressions for n_X and $X n_X$, we must deal with some infinite series:

$$\sum_1^\infty e^{-\lambda X} = e^{-\lambda} + e^{-2\lambda} + e^{-3\lambda} + \cdots$$

The infinite series:

$$S = a + ar + ar^2 + ar^3 = \frac{a}{1 - r} \qquad \text{for} \quad |r| < 1$$

If we set $a = 1$ and $r = e^{-\lambda}$, we conclude that

$$\sum_1^\infty e^{-\lambda X} = \frac{1}{1 - r} - 1 = \frac{r}{1 - r}$$

$$\sum_1^\infty e^{-\lambda X} = \frac{e^{-\lambda}}{1 - e^{-\lambda}}$$

Similarly, the infinite series is written:

$$\sum_1^\infty Xe^{-\lambda X} = e^{-\lambda} + 2e^{-2\lambda} + 3e^{-3\lambda} + \cdots$$

Note that:

$$\frac{1}{(1 - r)^2} = 1 + 2r + 3r^2 + 4r^3 + \cdots$$

If we set $r = e^{-\lambda}$, then

$$\sum_1^\infty Xe^{-\lambda X} = r + 2r^2 + 3r^3 + 4r^4 + \cdots$$

$$= r(1 + 2r + 3r^2 + 4r^3 + \cdots)$$

$$\sum_1^\infty Xe^{-\lambda X} = r\left[\frac{1}{(1 - r)^2}\right] = e^{-\lambda}\left[\frac{1}{1 - e^{-\lambda}}\right]^2$$

Therefore:

$$\sum XN_X = \frac{N_A}{N} = \frac{e^{-\lambda}\left[\dfrac{1}{1 - e^{-\lambda}}\right]^2}{e^{-\lambda}\left[\dfrac{1}{1 - e^{-\lambda}}\right]} = \frac{1}{1 - e^{-\lambda}} \qquad \textbf{(10.92)}$$

or

$$e^{-\lambda} = 1 - \frac{N}{N_A}$$

But

$$n_X = \frac{N_X}{N} = e^{-\lambda_0}e^{-\lambda X} = \frac{(e^{-\lambda})^X}{e^{\lambda_0}}$$

Note, also, that

$$e^{\lambda_0} = \sum_{1}^{\infty} e^{-\lambda X} = \frac{e^{-\lambda}}{1 - e^{-\lambda}}$$
$$\frac{N_X}{N_0} = \frac{(e^{-\lambda})^X}{e^{-\lambda}/(1 - e^{-\lambda})} = (e^{-\lambda})^{(X-1)}(1 - e^{-\lambda}) = n_X \tag{10.93}$$

Combining Eqs. 10.92 and 10.93, we have

$$n_X = \left(1 - \frac{N}{N_A}\right)^{(X-1)} \left(\frac{N}{N_A}\right)$$

Considering Eq. 10.87, and noting that $1 - P = N/N_A$, we have:

$$n_X = P^{(X-1)}(1 - P) = \frac{N_X}{N}$$
$$N_X = NP^{(X-1)}(1 - P) \tag{10.94}$$

As a result

$$N_X = N_A P^{(X-1)}(1 - P)^2 \tag{10.95}$$

The equation thus derived gives the number of molecules of length X as a function of the degree of polymerization. Note that the maximum in this equation occurs at $X = 1$. A distribution plot of the number versus molecular size (X) is shown as Figure 10.12.

To display the relationship above as a *weight* distribution, rather than a *number* distribution, we calculate the function

$$W_X = \frac{XN_X}{N_0} = XP^{(X-1)}(1 - P)^2 \tag{10.96}$$

where W_X is the total weight of molecules of dimension X in the polymer. This function has a different shape. In fact, there is a maximum in the function as follows:

$$\ln W_X = \ln X + (X - 1) \ln P + 2 \ln(1 - P)$$

$$\frac{dW_X}{W_X} = \frac{dX}{X} + \ln P \, dX$$

$$\frac{dW_X}{dX} = W_X \left[\frac{1}{X} + \ln P \right] = 0 \qquad \text{at maximum } W_X$$

$$X = \frac{1}{\ln P} \qquad \text{at maximum } W_X$$

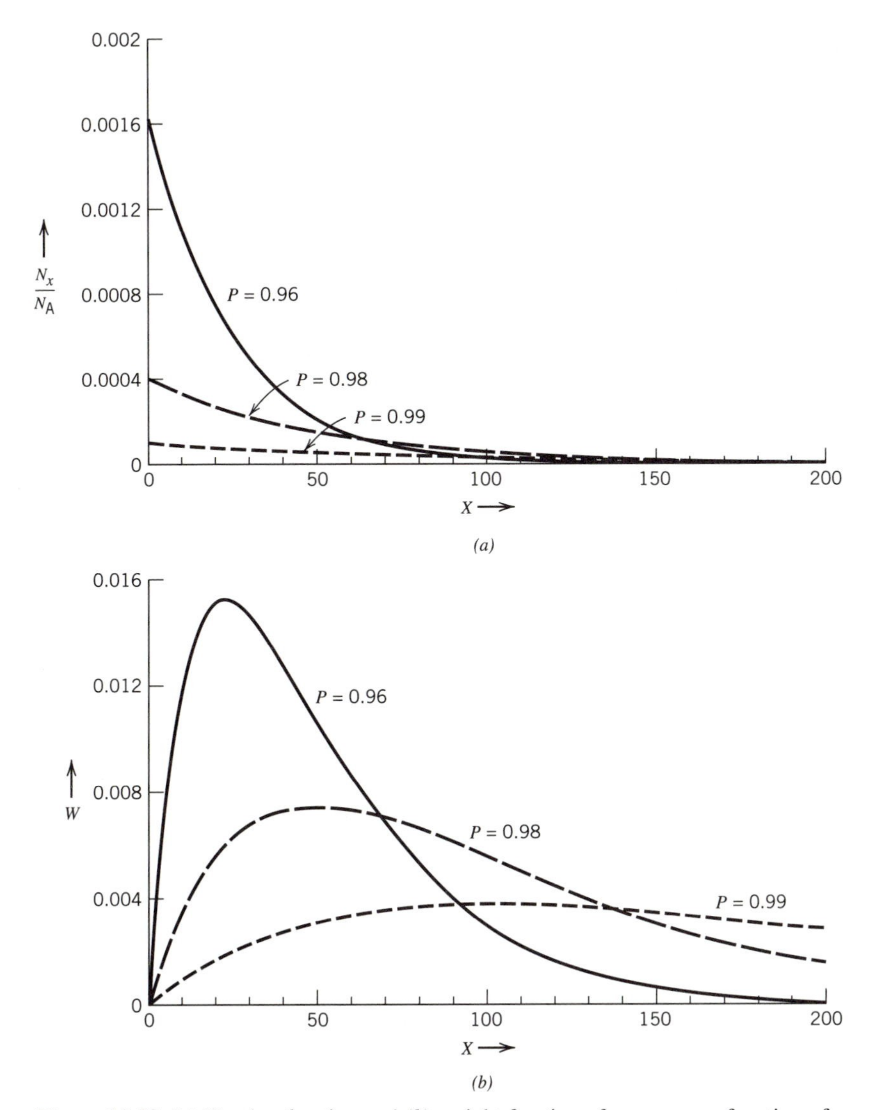

Figure 10.12 (*a*) Number fraction and (*b*) weight fraction of x-mers as a function of x for several values of P.

10.16 ENTROPY OF MIXING OF POLYMER SOLUTIONS

In Section 10.3 the Boltzmann equation (Eq. 10.14) was used to derive an expression for the entropy of mixing of two (or more) components using a lattice model. The Ω function was evaluated by considering the number of ways in which the species could be arranged randomly on a regular lattice. It was implicitly assumed that the species involved would fit on the lattice: that is, that they all had about the same atomic (or molecular) volume. This condition does not hold true for solutions of polymers (very large molecules) in solvents (small molecules by comparison). To calculate the entropy of mixing of such solutions a modified technique must be used, which will lead to an expression different from Eq. 10.17 ($\underline{S}_M = -R\,\Sigma_i x_i \ln x_i$).

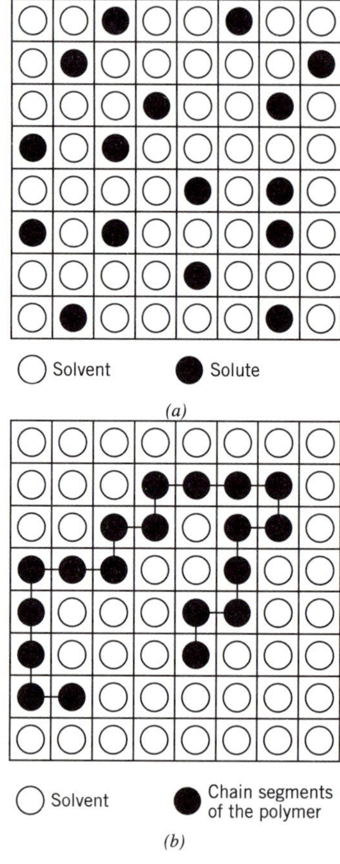

Figure 10.13 Two-dimensional lattice representations: (*a*) an ideal solution of components with equal atomic or molecular volumes and (*b*) a polymer molecule in solution.

The lattice model is also used to calculate the entropy of mixing of polymer solutions, but with a variation illustrated in Figures 10.13*a* and 10.13*b*. The method of calculation, known as the Flory–Huggins theory, is based on the assumption that the volume of each segment of the polymer is equal to the volume of the solvent molecule, $\underline{V}$. The volume of one polymer molecule is $n\underline{V}$. The difference between this calculation and the one that resulted in Eq. 10.17 is that the positions of the polymer segments on the lattice are not random (Figure 10.13*b*). Once the position of the first segment of a chain has been chosen, the next segment must lie on one of the neighboring sites, Z in number, where Z is the coordination number. The total number of sites in the lattice is

$$N = N_1 + nN_2 \tag{10.97}$$

where N_1 is the number of solvent molecules, N_2 is the number of polymer molecules, and n is the number of segments in the polymer.

To calculate the Ω function to be used in the Boltzmann equation, we must write an expression for the number of ways of placing a polymer molecule on the lattice. Begin with the placement of polymer molecule number $i + 1$, with i molecules already in place. The first segment in the chain may be placed in any one of $N - n_i$ positions. The second segment must reside on one of the Z neighboring positions, some of which may already be occupied by the segments of the i molecules already in place. The fraction of the Z sites unoccupied is $(N - n_i)/N$. The third and subsequent segments may be placed on any of $(Z - 1)(N - n_i)/N$ sites. The $Z - 1$ factor arises because one of the sites is occupied by the polymer molecule itself. Thus the number of ways of placing a polymer molecule with n segments on the lattice is expressed as follows:

$$\omega_{i+1} = (N - ni)\, Z \left(\frac{N - ni}{N}\right)\left[(Z - 1)\left(\frac{N - ni}{N}\right)\right]^{n-2}$$

or

$$\omega_{i+1} = Z(Z - 1)^{n-2} N \left(\frac{N - ni}{N}\right)^{n}$$

For the molecule number i,

$$\omega_i = Z(Z - 1)^{n-2} N \left[\frac{N - n(i - 1)}{N}\right]^{n} \tag{10.98}$$

The function Ω in the Boltzmann equation ($S = k \ln \Omega$) is the product of ω_i functions for each of the molecules divided by $N_2!$, because the N_2 polymer molecules are interchangeable:

$$\Omega = \frac{1}{N_2!} \prod_{i=1}^{N_2} \omega_i \quad \text{and} \quad S_c = k \ln\left[\left(\frac{1}{N_2!}\right) \prod_{i=1}^{N_2} \omega_i\right] \tag{10.99}$$

where S_c is the configurational entropy of the polymer solution.

A considerable amount of algebraic manipulation is required at this point (Refs. 5 and 6). The result is:

$$\frac{S_c}{k} = -N_1 \ln\left(\frac{N_1}{N}\right) - N_2 \ln\left(\frac{nN_2}{N}\right) + N_2[\ln Z + (n-2)\ln(Z-1) + (1-n) + \ln n] \tag{10.100}$$

The entropy of mixing is the difference between the expression for configurational entropy of the polymer solution, above, and the entropy of the two constituents before mixing. The entropy of the pure solvent ($N_2 = 0$) is $S_1 = 0$. The entropy of the pure polymer ($N_1 = 0$, and $nN_2 = N$) is

$$S_2 = kN_2[\ln Z + (n-2)\ln(Z-1) + (1-n) + \ln n] \tag{10.101}$$

The entropy of mixing, S_M, is the difference between Eqs. 10.100 and 10.101:

$$S_M = -k\left[N_1 \ln\left(\frac{N_1}{N}\right) + N_2 \ln\left(\frac{nN_2}{N}\right)\right]$$

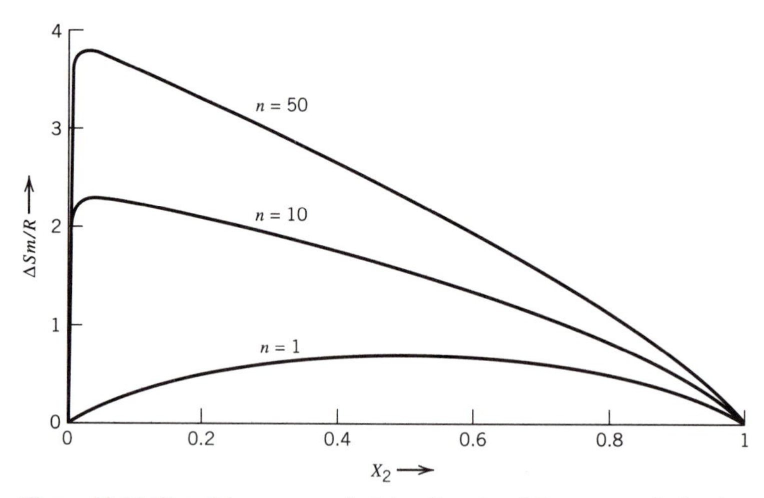

Figure 10.14 Plot of the entropy of mixing (in units of R) versus mole fraction for $X = 1$, 10, and 50 (the Flory–Huggins model).

On a molar basis, $N_1 + N_2 = N$:

$$\underline{S}_M = -R\left[x_1 \ln\left(\frac{N_1}{N}\right) + x_2 \ln\left(\frac{nN_2}{N}\right)\right]$$

Because the volumes of the solvent molecule and the polymer segment are equal, the term N_1/N is equal to the volume fraction of the solvent, ϕ_1, the term nN_2/N is equal to the volume fraction of the polymer, ϕ_2, and the entropy of mixing is

$$\underline{S}_M = -R[x_1 \ln \phi_1 + x_2 \ln \phi_2] \qquad \textbf{(10.102)}$$

The entropy of mixing for a polymer solution depends on the volume fractions of polymer and solvent as well as their fractions. This entropy of mixing expression is not symmetrical about $x = 0.5$ because the volume fractions and mole fractions are not equivalent, as illustrated in Figure 10.14.

In the limit, when there is only one segment in the polymer ($n = 1$), the entropy of mixing of the polymer–solvent solution becomes the same as the equation derived for mixing of an ideal solution derived earlier as Eq. 10.17 ($\underline{S}_M = -R\,\Sigma_i x_i \ln x_i$).

REFERENCES

1. Chandler, David, *Introduction to Modern Statistical Mechanics,* Oxford University Press, London, 1987.
2. Hill, Terrell L., *An Introduction to Statistical Thermodynamics,* Dover, New York, 1986.
3. Mayer, J. E., and Mayer, M. G., *Statistical Thermodynamics,* Wiley, New York, 1959.
4. Dole, Malcolm, *An Introduction to Statistical Thermodynamics,* Prentice-Hall, Englewood Cliffs, NJ, 1954.
5. Flory, P. J., *Principles of Polymer Chemistry,* Cornell University Press, Ithaca, NY, 1960.
6. Hiemenz, P. C., *Polymer Chemistry,* Dekker, New York, 1984.

APPENDIX 10.A

Proof That $\beta = \dfrac{1}{kT}$

The process analyzed was one of constant volume, when the energy states E_i were constant.

$$dU = T\,dS - P\,dV \Rightarrow \left(\frac{\partial U}{\partial S}\right)_V = T \qquad \text{or} \qquad \left(\frac{\partial S}{\partial U}\right)_V = \frac{1}{T}$$

From Eq. 10.40,

$$S = k \ln Z + k\beta\langle U\rangle \qquad \text{or} \qquad k \ln Z + k\beta U$$

$$\left(\frac{\partial S}{\partial U}\right)_V = \frac{k}{Z}\left(\frac{\partial Z}{\partial U}\right)_V + k\beta + kU\left(\frac{\partial \beta}{\partial U}\right)_V$$

chain rule

$$\left(\frac{\partial S}{\partial U}\right)_V = \frac{k}{Z}\left(\frac{\partial Z}{\partial \beta}\right)_V\left(\frac{\partial \beta}{\partial U}\right)_V + k\beta + kU\left(\frac{\partial \beta}{\partial U}\right)_V$$

$$Z = \sum e^{-\beta E_i}$$

$$\left(\frac{\partial Z}{\partial \beta}\right)_V = -\sum E_i e^{-\beta E_i}$$

But

$$e^{-\beta E_i} = ZP_i$$

$$\left(\frac{\partial Z}{\partial \beta}\right)_V = -\sum E_i ZP_i = -Z\sum E_i P_i$$

$$\left(\frac{\partial Z}{\partial \beta}\right)_V = -Z\langle E_i\rangle = -ZU$$

$$\left(\frac{\partial S}{\partial U}\right)_V = \frac{k}{Z}(-ZU)\left(\frac{\partial \beta}{\partial U}\right)_V + k\beta + kU\left(\frac{\partial \beta}{\partial U}\right)_V$$

$$\left(\frac{\partial S}{\partial U}\right)_V = k\beta = \frac{1}{T}$$

Therefore,

$$\beta = \frac{1}{kT}$$

PROBLEMS

10.1 The isotopic composition of lead, in atoms percent, is ^{204}Pb, 1.5; ^{206}Pb, 23.6; ^{207}Pb, 22.6; ^{208}Pb, 52.3. Calculate the entropy of mixing per mole of Pb at a temperature of 0 K.

Note: The Third Law states that $S° = 0$ at 0 K. How do you rationalize the difference between your calculation and the Third Law?

10.2 Atoms of an element E are being adsorbed on the surface of a material M. Material M has N adsorption sites avaialble per square centimeter of surface.

(a) Derive an expression for the entropy of N_E atoms of the element E adsorbed on M (per square centimeter of M), assuming that the adsorbing atoms have zero interaction energy with one another. They do, of course, have an interaction energy with the surface M atoms.

(b) At what ratio of E atoms to adsorptions sites (N_E/N) will this entropy be a maximum?

(c) Can the number of atoms adsorbed rise above this level? If so, why?

10.3 In a particular system, atoms can exist in two positions on a surface. The energy in state 1 is 0.1 eV. The energy in state 2 is 0.2 eV. The energy of the atoms in the vapor surrounding the surface is large by comparison, that is, 10 eV.

(a) At absolute zero, what is the fraction of atoms found in each of the three states?

(b) At 300 K, what is the fraction of atoms in each state?

(c) Why are some atoms not in the lowest energy state at 300 K?

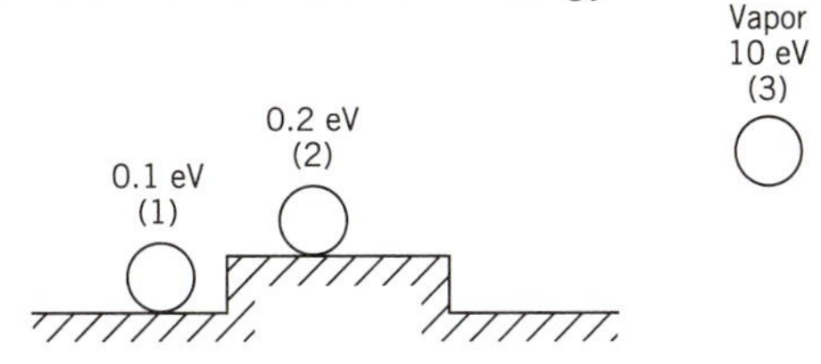

10.4 Compound AB exhibits substitutional disorder; that is, some A atoms can occupy sites on the B lattice and vice versa.

Derive an equation for the contribution to the entropy of the solid (the configurational entropy) if n atoms of A exist on the B lattice sites and n atoms of B exist on the A lattice sites per mole (N_A) of the compound AB.

What will the equation become if $n \ll N$?

10.5 When 10^{-10} g · mol of an ideal gas occupies a 10 m^3 vessel, what is the probability that all the molecules will collect momentarily in a particular 9 m^3 of this space?

Repeat the calculation for 10^{-22} g · mol.

10.6 Using the partition function for a monatomic gas (translation mode), derive an equation for the internal energy and the constant volume heat capacity (C_v) of the gas. Show that the equation of state is $P = NkT/V$, where N is the number of atoms.

10.7 A system consists of N noninteracting atoms. Each atom may be in two states: a low energy state with energy = 0 and an "excited" state with energy E.

(a) How many atoms are in the excited state?

(b) What is the total internal energy (U) of this system as a function of N, E, k (the Boltzmann constant), and T?

(c) What value will the internal energy attain at very high temperatures (at T approaches infinity)?

(d) Derive an equation for the heat capacity (C_v) of this system as a function of temperature (in terms of N, E, k and T).

10.8 *Helmholtz free energy.* Starting from the Gibbs definition of entropy and the definition of the partition function, prove that the Helmholtz free energy is given by

$$F = -kT \ln(Z)$$

10.9 *Volume dependence of the characteristic frequency.* Assume that the characteristic Einstein frequency for a particular solid depends on the molar volume as

$$\nu = \nu_0 - A \ln\left[\frac{V}{V_0}\right]$$

where ν_0 is the characteristic frequency when the volume equals V_0 and A is a constant.

Compute the isothermal compressibility of the crystal.

10.10 *The one-dimensional Ising model.* Consider a one-dimensional chain of atoms. Each atom (labeled with i) carries a magnetic moment (m_i) that can point either up ($m_i = +\mu$) or down ($m_i = -\mu$).

(a) If there are N atoms in the chain and every moment can be either up or down, how many microstates are possible for this system?

(b) The interaction energy between two neighboring moments is given by

$$-Km_i m_{i+1} = -K\mu^2 s_i s_{i+1}$$

where s_i is a variable that has the value $+1$ when the magnetic moment at atom i is pointing up and -1 when the magnetic moment at atom i is pointing down.

The total energy for a given state of all the magnetic moments in the chain is thus:

$$E = -\sum_{i=1}^{N} K\mu^2 s_i s_{i+1}$$

Compute the partition function for this system. Assume that N is very large so that you can neglect what happens at the edges of the chain.

Although it may seem complicated, this problem is clearly feasible for the one-dimensional chain. For magnetic moments arranged on a two-dimensional square lattice, the problem is much harder. In one of the major scientific achievements of the twentieth century, Onsager, in the 1940s, computed the partition function for the two-dimensional problem. The math for the two-dimensional problem takes a whole chapter in a standard textbook on statistical mechanics! Onsager was rewarded for this work with a Nobel prize. Now, 50 years later, nobody has yet been able to compute the partition function for the equivalent three-dimensional problem (spins arranged on a three-dimensional cubic lattice). Surely a Nobel prize awaits the one who accomplishes this feat.

10.11 The energy levels of a harmonic oscillator are given by

$$E_n = (n + 1/2)h\nu = (n + 1/2)k\Theta$$

where $\Theta = h\nu/k$ is the Einstein temperature of the oscillator.

For a one-dimensional oscillator with $\Theta = 500$ K at $T = 1000$ K, compute:

(a) The *value* of the partition function.
(b) The *probability* that a given oscillator is in the state with $n = 1$.
(c) The average value of n, $\langle n \rangle$.

10.12 The semiconducting material GaAs forms an ordered compound. The arrangement of Ga and As in the crystal structure is perfectly ordered. What is the entropy of mixing when the GaAs compound forms from pure Ga and pure As at constant temperature. Explain your answer briefly!

10.13 The ideal solution formula for the entropy of mixing is an approximation. For a real system, is the ideal solution model better at high temperature or at low temperature? Explain your answer *briefly, but clearly.*

10.14 A system consists of N noninteracting atoms. Each atom may be in two states: a low energy state with energy $= 0$, and an ''excited'' state with energy $= \varepsilon$.

(a) How many atoms are in the excited state?
(b) What is the total internal energy (U) of this system as a function of N, ε, k (the Boltzmann constant), and T?
(c) What value will the internal energy attain at very high temperatures (at T approaches infinity)?
(d) Derive an equation for the heat capacity (C_v) of this system as a function of temperature (in terms of N, ε, k and T).

Appendix A

Background

The presentation in this text assumes that the reader has completed courses in college-level chemistry, physics, and calculus, and is thus familiar with the basic ideas on which thermodynamics is built, and the units used in describing various physical quantities. This appendix reviews the various units of measurement and some of the mathematical concepts and techniques used in the text.

UNITS

It would be convenient indeed to have the entire world using one uniform set of units, just as it would be convenient to have one currency worldwide. We do not have the luxury of one currency, nor do we have the luxury of a uniform set of units in all fields of technology and in the technological literature.

Basically, this text uses the international system of units (SI). This is the modern version of the metric system. Nevertheless, students of materials science and engineering should recognize that in the practice of their profession, they will be required to understand and apply the knowledge generated by physicists, chemists, and other scientists. They will also be called on to communicate with, and serve the needs of, engineers from diverse fields. This means that they will be exposed to a great variety of measurement units. For example, in most engineering work in the United States today, the unit for pressure is pounds per square inch. Most scientific texts and other publications up through the middle of the twentieth century expressed pressure in atmospheres. The SI system of units, however, expresses pressure in pascals (newtons per square meter). Practicing engineers and scientists should be able to use data in all forms of units with equal facility. A brief discussion of the units used to measure various physical parameters follows.

The SI system recognizes seven fundamental, or base, units, those for length, mass, time, electric current, thermodynamic temperature, amount of material, and luminous intensity. The units from this list that are of use to us are discussed in turn. Then we present other units that are derived from them, such as force, pressure, and energy.

A.1 LENGTH

The unit of length in the SI system is the meter. Other units in use are centimeters (10^{-2} meter), feet (0.3048 meter), and inches (0.02540005 meter). In the scientific literature, especially in materials science, one often finds the micrometer (μm) which is 10^{-6} meter, and the angstrom unit (Å), which is 10^{-8} centimeter or 10^{-10} meter.

A.2 MASS AND AMOUNT OF MATERIAL

The standard SI unit for mass is the kilogram. Some engineering publications express mass in terms of pounds mass. A pound mass is 0.4536 kilogram.

The unit of mass generally adopted in the study of chemical reactions is the mole. The mole, strictly speaking, is a number of particles and is defined as "the amount of substance of a system which contains as many elementary entities as there are atoms in 0.012 kilogram of carbon-12." In practical terms this means that the mole consists of 6.022045×10^{23} particles (Avogadro's number). One gram-mole of a chemical substance contains 6.022045×10^{23} molecules of that substance. The weight of the mole is called its molecular weight, or sometimes its gram formula weight. For example, atomic mass of iron, scaled to carbon-12, is 55.847. This means that there are 55.847 grams of iron per mole of iron.

Some confusion may arise when reading engineering publications in which other types of "mole" are used, such as the kilogram-mole, or the pound-mole. A *kilogram*-mole is 1000 gram-moles, 6.022045×10^{26} particles. A kilogram-mole of iron weighs 55.847 kilograms. Similarly a *pound*-mole of iron weighs 55.847 pounds.

We also note the value of the unit used to measure the mass of precious stones, since significant quantities of diamond are now manufactured from graphite using a high pressure process, and diamond films are deposited at lower pressures. Thus it is useful to know that a "carat" of diamond is 0.20 gram.

A.3 TIME

The SI unit of time is the second. Fortunately, publications in almost all fields measure time in seconds, or units easily converted, such as hours, days, or years.

A.4 ELECTRIC CURRENT

The unit of electric current is the ampere, which is defined in terms of the force between two parallel conductors of infinite length placed one meter apart in a vacuum.

A.5 TEMPERATURE

The most fundamental measurement in thermodynamics is that of temperature. In the SI system, the unit of temperature is the kelvin. It is defined as the fraction 1/273.16 of the absolute temperature of the triple point of water.

Much of the older literature in thermodynamics specifies temperature in degrees Celsius (or centigrade). The temperature span covered by one Celsius degree is the same as that covered by one kelvin degree. For example, a temperature difference of 10 K is the same as a difference of 10°C. The two scales differ only in their zero points. On the kelvin scale, zero degrees is the absolute zero of temperature. In the Celsius scale zero degrees is the melting point of ice (under one atmosphere pressure in air), and 100°C at the boiling point of water at one atmosphere total pressure (101.325 kPa, or kilopascals: see Section A.7). Absolute zero on the Celsius scale is −273.15°C. The so-called triple point of water, at which ice, water, and water vapor exist in equilibrium, occurs at a temperature of 0.01°C at a pressure of 6.04×10^{-3} atmosphere.

The relationship between the kelvin and Celsius scales is

$$T\ (\mathrm{K}) = t + 273.15,$$

where t is the temperature in degrees Celsius.

The temperature scale used in English-speaking countries for many years, and still used in much of engineering in the United States, is the Fahrenheit scale. In this

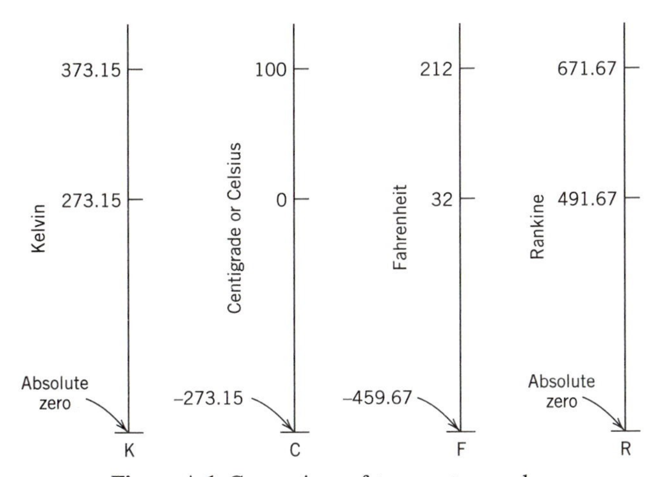

Figure A.1 Comparison of temperature scales.

scale the melting point of ice is 32°F, and the boiling point of water is 212°F. Fahrenheit, a naturalist, set the zero on his scale at the coldest temperature he could reach with a salt–water solution. The absolute zero on this scale is −459.67°F. The absolute Fahrenheit scale is the Rankine scale. On that scale the melting point of ice is 491.67°R. Figure A.1 illustrates the relation among these scales.

A.6 FORCE

The SI unit of force is the newton, the force required to accelerate a one-kilogram mass at the rate of 1 meter per second per second.

Another unit used in some calculations is the dyne, the force required to accelerate a one-gram mass at the rate of 1 centimeter per second per second.

In the English system, a force of one pound imparts an acceleration of 1 foot per second per second to a mass of one slug. A slug is 14.6 kilograms.

A newton is 10^5 dynes, and is also equal to 0.224 pound force.

A.7 PRESSURE

In the SI system, pressure is measured in newtons per square meter (pascals). The unit used in many scientific publications is the "atmosphere." The standard atmosphere is 1.01325×10^5 pascals.

In engineering units, pressure is often measured in pounds per square inch. When using the engineering units, one must be careful to differentiate between *absolute* pressure and *gauge* pressure. Absolute pressure, as the words imply, measures the absolute force divided by area. Gauge pressure is the pressure measured by a device that measures pressure *differences*, one end of which is open to the atmosphere. The standard atmosphere is about 14.7 pounds per square inch absolute pressure. A pressure of 24.7 psia (pounds per square inch absolute pressure) can also be called 10.0 psig (pounds per square inch gauge pressure).

A.8 VOLUME

The SI unit for volume is the cubic meter. Other units sometimes found in the scientific literature are liters (10^{-3} cubic meter), cubic centimeters or milliliters (10^{-6} cubic meter). Engineers often measure volume in cubic feet or in gallons. Another measure of volume is the barrel, which in the case of oil, is 42 U.S. gallons (0.1590 cubic meter).[1]

A.9 ENERGY

The unit of energy in the SI system is the joule. A joule is, in mechanical terms, a newton-meter, the energy expended by a force of one newton moving through a

[1]The definition of the barrel varies with the commodity being measured.

distance of one meter. In electrical terms, a joule is a volt-coulomb, the energy required to move a charge of one coulomb through a potential difference of one volt. The dyne-centimeter (erg), is 10^{-7} joule.

In many scientific publications, especially those in the chemical sciences, energy is expressed in calories.[2] The calorie is defined as 4.1868 joules.

The calorie, by its original definition, is the amount of energy required to heat one gram of water one degree Celsius (between 14.5 and 15.5°C at one atmosphere pressure). The ''heat capacity'' of water, between the specified temperatures is one calorie per gram degree Celsius. The engineering equivalent of this unit is the Btu (British thermal unit), which, by its original definition, is the quantity of energy required to heat one pound mass of water one degree Fahrenheit (between 59.5 and 60.5°F).

A.10 POWER

Power is an energy rate. Its units are energy units divided by time units. In the SI system, the unit of power is the watt. It is equivalent to one joule per second, or one newton-meter per second. A common engineering unit for power is the horsepower, which is 746 watts.

A.11 COMPOSITION

In thermodynamic calculations involving solutions or mixtures, it is most convenient to express composition in terms of mole fraction or mole percent of the components involved. The mole fraction (x_i) of a component i is simply:

$$x_i = \frac{n_i}{\sum_i n_i}$$

where n_i is the number of moles of component i. Of course $\Sigma\, x_i = 1$.

The composition of gases is usually stated in terms of volume fraction (or volume percent) of the components. If the individual gases behave ideally and the components may be assumed not to interact, the volume fraction in the gaseous mixture is the same as the mole fraction because volume is linearly proportional to the number of moles (at the same temperature and pressure).

The composition of solid mixtures or solutions is often stated in weight percent. The weight percent of a component i is

$$\text{wt }\%_i = 100\left(\frac{w_i}{\sum_i w_i}\right)$$

[2]The ''calories'' listed in conjunction with foods are really *kilo*calories.

where w_i is the weight of component i in the mixture or solution, and, of course, Σ wt $\%_i = 100$.

To convert from weight percent to mole fraction (x), use

$$x_i = \frac{\dfrac{(\text{wt }\%)_i}{(\text{MW})_i}}{\displaystyle\sum_i \frac{(\text{wt }\%)_i}{(\text{MW})_i}}$$

where $(\text{MW})_i$ is the molecular weight of component i.

As an example, let us determine the mole percent of gold in a gold–copper alloy containing 75% by weight of gold. This is, incidentally 18 karat gold. The karat rating of gold is the weight fraction of gold in an alloy given in twenty-fourths. An 18 karat gold is 18/24 gold by weight, that is, 75%.

$$\text{mol \% gold} = \frac{\dfrac{75}{197.0}}{\dfrac{75}{197.0} + \dfrac{25}{63.54}} = 0.4918 \text{ or } 49.18\%$$

A.12 CHECKING UNITS

A great number of mathematical manipulations are used in thermodynamics. At times they can become intricate and confusing. It is useful during the course of a derivation to stop periodically to check the units on both sides of an equation to make sure they are the same. This sort of ''sanity check'' is a very useful procedure.

The ideal gas law provides a simple example. This law, relating the pressure, volume, temperature and the mass of a gas is:

$$PV = nRT$$

where P = pressure (N/m^2)
V = volume (m^3)
n = number of moles
R = universal gas constant [$\text{J/(mol} \cdot \text{K)}$]
T = temperature (K)

Substituting the units of measurement in the ideal gas equation:

$$\left(\frac{\text{N}}{\text{m}^2}\right)(\text{m}^3) = (\text{moles})\left(\frac{\text{joules}}{\text{mole} \cdot \text{K}}\right)(\text{K})$$

$$\text{N} \cdot \text{m} = \text{joules}$$

The equation is valid dimensionally.

If pressure had been stated in pounds per square inch, and volume in cubic feet, appropriate conversion factors would have to be provided as follows:

$$P = \left(\frac{\text{lb}}{\text{in.}^2}\right) \times \frac{(\text{N/m}^2)}{(\text{lb/in.}^2)} = \frac{\text{N}}{\text{m}^2}$$

$$V = (\text{ft}^3) \times \underbrace{\left(\frac{\text{m}^3}{\text{ft}^3}\right)}_{\frac{1}{35.15}} = \text{m}^3$$

$$PV = \left(\frac{\text{lb}}{\text{in.}^2}\right)(\text{ft}^3) \times \frac{6897}{35.15} = (\text{N} \cdot \text{m})$$

MATHEMATICAL MANIPULATIONS

The study of thermodynamics relies heavily on mathematical concepts and procedures. A review of some of the important mathematical relationships used in the text follows.

A.13 LOGARITHMS

$$\text{If} \quad \ln x = y, \quad \text{then } x = \exp(y) \quad \text{or} \quad e^y \tag{A.1}$$

$$\ln(xy) = \ln x + \ln y \tag{A.2}$$

$$a \ln x = \ln x^a \tag{A.3}$$

$$\ln\left(\frac{x}{y}\right) = \ln x - \ln y \tag{A.4}$$

In terms of a series:

$$\ln x = (x - 1) - \tfrac{1}{2}(x - 1)^2 + \tfrac{1}{3}(x - 1)^3 \pm \cdots \tag{A.5}$$

Also

$$\ln(1 + x) = x - \frac{x^2}{2} + \frac{x^3}{3} - \frac{x^4}{4} + \cdots \tag{A.6a}$$

An approximation often used is

$$\ln(1 + x) = x \qquad \text{when} \quad x \text{ is small} \tag{A.6b}$$

A.14 DIFFERENTIALS

The differential of a function, $f(x)$, is written as: $d\,f(x) = f'(x)\,dx$

$$d(x^n) = nx^{n-1}dx \tag{A.7}$$

$$d(\ln x) = \frac{dx}{x} \tag{A.8}$$

$$d(e^{f(x)}) = e^{f(x)}f'(x)\,dx \tag{A.9}$$

A technique often used in thermodynamics involves the determination of the *slope* of a curve.

If $d[f(x)] = A\,d[f(y)]$, then A is the slope of the curve of a graph of $f(x)$ plotted as the ordinate and $f(y)$ plotted as the abscissa. For example, if $d[\ln x] = A\,d[1/T]$, then A is the slope of the graph of $\ln x$ versus $1/T$.

The fractional change of related variables can often be determined using the derivatives of logarithmic functions. For example, the volume of a cube (V) is the side of the cube (l) raised to the third power.

$$V = l^3$$

Taking the logarithm of both sides:

$$\ln V = \ln l^3 = 3 \ln l$$

Differentiating:

$$\frac{dV}{V} = 3\,\frac{dl}{l} \tag{A.10}$$

The fractional change of the volume of the cube is, thus, three times the fractional change of the length of a side, for small fractional changes.

An interesting interpretation of Eq. A.10 is in terms of measurement error. If Δl is the error in the measurement of l, the fractional error is $\Delta l/l$. The fractional error in volume would be $\Delta V/V$, or $3\,\Delta l/l$.

A.15 PARTIAL DERIVATIVES

Consider a function $z = f(x,y)$. The variation of z with x at constant y is the partial derivative of z with respect to x and is written as

$$\left(\frac{\partial z}{\partial x}\right)_y$$

The variation of z with y at constant x is the partial derivative of z with respect to y:

$$\left(\frac{\partial z}{\partial y}\right)_x$$

In a rather simple manipulation of a partial derivative, called the "chain rule," another variable may be introduced into a partial differential as follows:

$$\left(\frac{\partial U}{\partial P}\right)_T = \left(\frac{\partial U}{\partial V}\right)_T \left(\frac{\partial V}{\partial P}\right)_T \qquad \textbf{(A.11)}$$

A.16 TOTAL DIFFERENTIAL

If x and y are independent variables, and $z = f(x,y)$ is a single-valued function, then:

$$dz = \left(\frac{\partial z}{\partial x}\right)_y dx + \left(\frac{\partial z}{\partial y}\right)_x dy \qquad \textbf{(A.12)}$$

This can be written as $dz = M\,dx + N\,dy$. Because the order of differentiation does not influence the value of the differential,

$$\left(\frac{\partial M}{\partial y}\right)_x = \left(\frac{\partial N}{x}\right)_y = \left(\frac{\partial^2 z}{\partial x\, \partial y}\right) \qquad \textbf{(A.13)}$$

This technique is used extensively in the study of thermodynamic properties. Another expression that results from the total differential is:

$$\left(\frac{\partial x}{\partial y}\right)_z \left(\frac{\partial z}{\partial x}\right)_y \left(\frac{\partial y}{\partial z}\right)_x = -1 \quad \text{or} \quad \left(\frac{\partial x}{\partial y}\right)_z = -\frac{\left(\frac{\partial z}{\partial y}\right)_x}{\left(\frac{\partial z}{\partial x}\right)_y} \qquad \textbf{(A.14)}$$

A.17 FACTORIALS

A convenient way to denote the product of the form

$5 \times 4 \times 3 \times 2 \times 1$ is $5!$ or 5 factorial

In more general form, N factorial is

$$N! = N(N-1)(N-2)(N-3)\cdots(2)(1)$$

In the study of statistical mechanics, one encounters the factorial of very large numbers. In this case the Stirling approximation for factorial, and the natural logarithm of the factorial is useful:

$$N! \cong N^N e^{-N}(2\pi N)^{1/2}$$

or

$$\ln N! = N \ln N - N \qquad \text{(approximately)} \tag{A.15}$$

A.18 HOMOGENEOUS FUNCTIONS

A function is homogeneous of the nth degree in x and y if

$$f(kx,ky) = k^n[f(x,y)] \tag{A.16}$$

The Euler theorem of homogeneous functions is

$$x\left(\frac{\partial f}{\partial x}\right)_y + y\left(\frac{\partial f}{\partial y}\right)_x = n\,f \tag{A.17}$$

If a function is homogeneous of the first order, then:

$$x\left(\frac{\partial f}{\partial x}\right)_y + y\left(\frac{\partial f}{\partial y}\right)_x = f \tag{A.18}$$

This last expression is especially useful because many of the thermodynamic functions, such as internal energy and Gibbs free energy, are homogeneous functions of the first order.

A.19 LAGRANGE METHOD OF UNDETERMINED MULTIPLIERS

The Lagrange method can be used to solve problems of the following type. Given an objective function with n variables,

$$\Phi(x_1, x_2, x_3, \ldots, x_n)$$

and a set of m constraints on the variables,

$$\psi_1(x_1, x_2, x_3, \ldots, x_n) = 0$$

$$\psi_2(x_1, x_2, x_3, \ldots, x_n) = 0$$

$$\vdots$$

$$\psi_m(x_1, x_2, x_3, \ldots, x_n) = 0$$

Find the values of $x_1, x_2, x_3, \ldots, x_n$ that maximize (or minimize) the function Φ. Because there are n variables and m constraints, there are $n - m$ degrees of freedom in the system, that is, $n - m$ ways to satisfy the objective.

To find the maximum (or minimum) of Φ, perturb each of the values of x_i by δx_i. At the maximum,

$$d\Phi = 0 = \sum_{i=1}^{n} \frac{\partial \Phi}{\partial x_i} \delta x_i \tag{A.19}$$

Each of the constraints becomes:

$$d\psi_j = 0 = \sum_{i=1}^{n} \frac{\partial \psi_j}{\partial x_i} \delta x_i \tag{A.20a}$$

where $j = 1, 2, 3, \ldots, m$.
Each constraint equation is multiplied by an arbitrary function, λ:

$$\lambda_1\, d\psi_1 = 0 = \lambda_1 \sum_{i=1}^{n} \frac{\partial \psi_1}{\partial x_i} \delta x_i \tag{A.20b}$$

$$\lambda_2\, d\psi_2 = 0 = \lambda_2 \sum_{i=1}^{n} \frac{\partial \psi_2}{\partial x_i} \delta x_i \tag{A.20c}$$

$$\vdots \tag{A.20m}$$

$$\lambda_m\, d\psi_m = 0 = \lambda_m \sum_{i=1}^{n} \frac{\partial \psi_m}{\partial x_i} \delta x_i$$

Adding Eqs. A.19 and A.20b through A.20m:

$$\sum_{i=1}^{n} \left(\frac{\partial \Phi}{\partial x_i} + \lambda_1 \frac{\partial \psi_1}{\partial x_i} + \lambda_2 \frac{\partial \psi_2}{\partial x_i} + \cdots + \lambda_m \frac{\partial \psi_m}{\partial x_i} \right) \delta x_i = 0$$

For this to be true for all δx_i, the term in parentheses must be zero.

$$\frac{\partial \Phi}{\partial x_i} + \lambda_1 \frac{\partial \psi_1}{\partial x_i} + \lambda_2 \frac{\partial \psi_2}{\partial x_i} + \cdots + \lambda_m \frac{\partial \Phi_m}{\partial x_i} = 0 \tag{A.21}$$

These n equations can then be solved for the functions $\lambda_1, \lambda_2, \lambda_3, \ldots, \lambda_m$. Because there are n equations involving the m unknown multipliers, λ, we can solve for the $m\lambda$ functions and substitute them in the $(n - m)$ remaining equations. The sum of the two, we write $m + (n - m) = n$. Thus the values of the n variables in the objective function can be determined.[3]

Let us try a simple example to illustrate the use of the Lagrange method. For a rectangle of sides x and y, maximize the area, $A = xy$, subject to the constraint that the perimeter $P = 2x + 2y$.

Differentiate the objective function A:

$$dA = x\,dy + y\,dx$$

Differentiate the constraint function $P - 2x - 2y = 0$:

$$-2\,dx - 2\,dy = 0 \qquad \text{multiply by } \lambda \text{ and add to } dA$$

$$(x - 2\lambda)\,dy + (y - 2\lambda)\,dx = 0$$

From which $x = y = 2\lambda$, and $P = 4\lambda + 4\lambda = 8\lambda$, or $\lambda = P/8$. Then:

$$x = \frac{P}{4} \qquad \text{and} \qquad y = \frac{P}{4}$$

and we conclude that the rectangle is a square.

Let us try another illustration of the Lagrange method on a simple problem, this time using two constraints. Now we want to maximize the function $-\Sigma\, x_i \ln x_i$, the principle stated in Section 10.5. A store sells three items labeled 1, 2, and 3, priced at \$1, \$2, and \$3, respectively. We have one piece of information, namely, that the average sale is \$2.20. What is the most probable distribution of sales among these items?

The principle to be applied is, "The state of knowledge that we must assume is the one that maximizes the function S relative to the information given," and S is written as follows:

$$S = -k \sum P_i \ln P_i \qquad \textbf{(A.23)}$$

where P_i is the probability of occurrence i.

There are two constraints on this function. First, the sum of the probabilities must be unity.

$$\sum P_i = 1 \qquad \textbf{(A.24)}$$

---

[3]The method outlined above follows the presentation of the subject in the text by M. Tribus, *Thermostatics and Thermodynamics*, Van Nostrand, New York, 1961, pages 69–72.

Second, the average value of the sales, $\langle V \rangle$ is \$2.20.

$$\sum P_i V_i = \langle V \rangle = \$2.20 \tag{A.25}$$

Each of the equations (A.23, A.24, and A.25) is perturbed by δP_i as follows:

$$-k \sum (\ln P_i \, \delta P_i + \delta P_i) = 0 \Rightarrow \sum \ln P_i \, \delta P_i = 0 \tag{A.23a}$$

$$\sum \delta P_i = 0 \tag{A.24a}$$

$$\sum V_i \, \delta P_i = 0 \tag{A.25a}$$

For the equations, the following undetermined multipliers, λ, are assigned

$$\sum \ln P_i \, \delta P_i = 0 \qquad < 1$$

$$\sum dP_i = 0 \qquad < \lambda_0$$

$$\sum V_i \, \delta P_i = 0 \qquad < \lambda_1$$

Adding the three:

$$\sum_i (\ln P_i + \lambda_0 + \lambda_1 V_i) = 0$$

Then:

$$\ln P_i + \lambda_0 + \lambda_1 V_i = 0$$

$$P_i = e^{-\lambda_0} e^{-\lambda_1 V_i}$$

Because

$$\sum_i P_i = 1,$$

$$\sum_i P_i = e^{-\lambda_0} \sum e^{-\lambda_1 V_i} = 1$$

$$e^{\lambda_0} = \sum e^{-\lambda_1 V_i}$$

Therefore:

$$P_i = \frac{e^{-\lambda_1 V_i}}{\sum_i e^{-\lambda_1 V_i}}$$

We know that the average value $\langle V \rangle = \$2.20$.

$$\langle V \rangle = \sum_i P_i V_i = \frac{\sum V_i e^{-\lambda_1 V_i}}{\sum e^{-\lambda_1 V_i}}$$

In our case,

$$\langle V \rangle = \frac{e^{-\lambda_1} + 2e^{-2\lambda_1} + 3e^{-3\lambda_1}}{e^{-\lambda_1} + e^{-2\lambda_1} + e^{-3\lambda_1}} = \$2.20$$

Solving for $e^{-\lambda_1}$ (by letting $x = e^{-\lambda_1}$), we have

$$e^{-\lambda_1} = 1.38$$

Then

$$P_1 = \frac{e^{-\lambda_1(1.00)}}{e^{-1.00\lambda_1} + e^{-2.00\lambda_1} + e^{-3.00\lambda_1}} = \frac{1.38}{5.913} = 0.2334$$

$$P_2 = \frac{e^{-2.00\lambda_1}}{5.913} = 0.3221$$

$$P_3 = \frac{e^{-3.00\lambda_1}}{5.913} = \underline{0.4445}$$

$$= 1.0000$$

EXERCISES IN UNIT CONVERSIONS

A.1. The linear thermal expansion coefficient (α_l) of iron is 11.8 microinches per inch per degree centigrade (μ in. in.$^{-1}$ °C^{-1}).

(a) What is α_l in microinches per inch per degree Fahrenheit?
(b) What is α_l in microcentimeters per centimeter per kelvin?
(c) What is the volumetric thermal expansion coefficient in each of these units?

cm^3 cm^{-3} °C^{-1}
in.^{3} in.$^{-3}$ °C^{-1}
cm^3 cm^{-3} K^{-1}

A.2. The density of water is 1 gram per cubic centimeter.

(a) What is the specific volume of water in cubic centimeters per gram?
(b) What is the specific volume of water in cubic centimeters per mole?

A.3. The density of iron is 7.87 g/cm^3 at 298 K.

- **(a)** What is the density in pounds per cubic inch at 298 K?
- **(b)** If the thermal expansion coefficient given in Problem A.1 is valid from 298 to 1000 K, what is the length of a bar at 1000 K if the bar is 1.00 cm at 298 K?
- **(c)** What is the density of iron at 1000 K (in grams per cubic centimeter)?

A.4. An iron bar 1 cm long is restrained so that it cannot elongate. What stress is generated in the bar if it is heated to 600 K?

The Young's modulus for iron is 210,000 MN/m^2 or 30×10^6 psi.

Young's modulus is the slope of the stress versus strain line. The units of stress are force per unit area. Strain is dimensionless (i.e., in./in. or cm/cm).

Index